Advanced Hybrid Powertrains for Commercial Vehicles

Advanced Hybrid Powertrains for Commercial Vehicles

By

Haoran Hu

Rudy Smaling

Simon J. Baseley

Warrendale, Pennsylvania
USA 400 Commonwealth Drive

400 Commonwealth Drive
Warrendale, PA 15096-0001 USA

E-mail: CustomerService@sae.org
Phone: 877-606-7323 (inside USA and Canada)
724-776-4970 (outside USA)
Fax: 724-776-0790

ISBN 978-0-7680-3359-5
SAE Order Number R-396
DOI 10.4271/R-396

Library of Congress Cataloging-in-Publication Data
Hu, Haoran.
Advanced hybrid powertrains for commercial vehicles / by Haoran Hu, Rudy Smaling, Simon J. Baseley.
p. cm.
"SAE order number R-396."
Includes bibliographical references.
ISBN 978-0-7680-3359-5
1. Hybrid electric vehicles—Power trains—Design and construction. 2. Commercial vehicles—Design and construction. I. Smaling, Rudy. II. Baseley, Simon J. III. Title.
TL260.H825 2012
629.25'04--dc23
2012012047

To purchase bulk quantities, please contact:
SAE Customer Service
Email: CustomerService@sae.org
Phone: 877-606-7323 (*inside USA and Canada*)
724-776-4970 (*outside USA*)
Fax: 724-776-0790

Visit the SAE International Bookstore at http://books.sae.org

Contents

Preface

Commercial vehicles cover a wide range of applications from light delivery vehicles and vocational trucks to Class 8 long-haul trucks, city buses, and intercity coaches. Commercial vehicles are used for a wide range of duties, including transporting goods or people and infrastructure service. In 2010, the worldwide production of commercial vehicles exceeded 19 million units. Commercial vehicles transported over half of the value traded by the United States with Canada and Mexico.

Over the last 100 years, the powertrain of commercial vehicles has seen continued development since the late 19th Century invention of the gasoline- or diesel-fueled internal combustion engine. In 1904, Ferdinand Porsche built the first hybrid vehicle that housed electric motors in the front wheel hubs while a gasoline-engine-powered generator provided power. During the first decade of the 1900s, both General Electric in the United States and Siemens in Germany produced hybrids as commercial vehicles. In the 1920s, heavy-duty commercial vehicles were firmly established as a major means of freight conveyance. Diesel engines, invented and first built by Dr. Rudolf Diesel in 1897, have been used in large trucks since the 1930s in the United States due to their larger torque and higher thermal efficiency compared to gasoline-powered engines.

In the early 1970s, concerns emerged over growing dependence on imported petroleum and deteriorating air quality due to emissions from fossil fuel combustion. In response, the U.S. Congress passed the Electric and Hybrid Vehicle Research, Development & Demonstration Act of 1976. This legislation renewed the public interest in the development of electric and hybrid vehicles, and led to the advancement of hybrid technologies.

Hybrid vehicles are vehicles with two or more power sources in the drivetrain. Current hybrids use both an internal combustion (IC) engine and a secondary drive system to improve fuel economy and performance, and to reduce emissions. The secondary drive systems of commercial vehicles are electric motor/generator drive systems or hydraulic pump/motor drive systems. Other combinations of energy storage and conversion are possible although not yet in commercial production for commercial vehicles.

The recent development of hybrid powertrain components and system integration technologies can reduce fuel consumption by up to 50% compared to conventional IC-engine-powered vehicles.

However, growing demand of energy consumption outpaces the energy production; worsening pollution and traffic congestion in major cities demand that sustainable transportation solutions be explored.

In addition to the IC engine hybrid powertrain, further powertrain improvement technologies may include the utilization of various fuel sources with optimized energy conversion efficiency. This can include plug-in hybrid, range extender electric, and pure electric vehicles. Other options are multimodal transportation with intelligent transportation technologies, smart grid, advanced energy storage technologies, and intelligent electric charging infrastructure. Sustainable transportation may also include renewable resources such as biofuels, or energy derived from solar, wind, or geothermal sources. Reduction of energy demand through government policy and incentives, advanced transportation planning, and public awareness and participation is another option.

This book, *Advanced Hybrid Powertrains for Commercial Vehicles*, provides a broad and comprehensive insight into hybrid powertrain technologies for commercial vehicles. Based on the authors' extensive experiences in developing advanced powertrain technologies, this book explains the fundamentals of hybrid powertrain systems, government regulations, and driving cycles. It provides design guidelines and describes key components to prospective engineers and developers, powertrain researchers, engineering students, policymakers, and business executives in the commercial vehicle and transportation industries.

Chapter 1 introduces the basics of commercial vehicles, including the development history of commercial vehicles, classification, driving cycles, regulations, the fundamentals of powertrain hybridization, and various hybrid system technologies, including the hybrid-electric powertrain, hybrid hydraulic powertrain, and hybrid pneumatic powertrain. This chapter also analyzes the advantages and disadvantages of various hybrid technologies for different driving cycles and applications.

Chapter 2 describes basic concepts and new developments of internal combustion engines for the commercial vehicle market, such as HCCI and low-temperature combustion, emission reduction technologies, urea-based SCR and LNT NOx aftertreatment technologies, DPF systems for PM reduction, etc.

Chapter 3 introduces the basics of clutches and transmissions for commercial vehicles. It discusses the basic design principles and characteristics of clutches and manual, automated manual, and automatic transmissions; double clutch transmissions (DCTs); and continuously variable transmissions (CVTs).

Chapter 4 discusses the fundamentals of energy storage systems, which include electric chemical batteries (i.e., lead-acid battery, NiMH batteries, Li-ion battery, ultracapacitors, hydraulic accumulators) and flywheel energy storage devices. Various battery management systems for Li-ion batteries are also reviewed.

Chapters 5 and 6 focus on the hybrid-electric powertrain and study various configurations of the powertrain. Chapter 5 first discusses the design characteristics of series hybrid, parallel hybrid, and dual-mode hybrid, and the basics of the system integration of motors, batteries, and the IC engine. The key components of the hybrid-electric powertrain, including motors and drivers, are then discussed in Chapter 6.

Chapters 7 and 8 discuss the fundamentals of the hydraulic hybrid powertrain, introduce the basic design guidelines, and discuss the advantages and disadvantages of various hydraulic hybrid architectures, such as parallel hydraulic hybrid, series hydraulic hybrid, and dual-mode hydraulic hybrid. The key components of the hybrid hydraulic powertrain are covered in Chapter 8, which includes the hydraulic accumulator specification and design, hydraulic pumps, motor design, bent-axle pump design, and the control systems.

Chapter 9 introduces the fuel cell hybrid hydraulic powertrain. It discusses the fundamentals of PEM fuel cells and solid oxide fuel cells and their applications.

Vehicle electrification and key components of electrification of commercial vehicles are discussed in Chapter 10. This chapter also discusses the design and performance characteristics of truck APUs, electric boosting devices, plug-in hybrids, and charging systems.

Chapter 11 discusses modeling and simulation, testing and certification of electric and hydraulic hybrid powertrains.

The future of advanced hybrid powertrains and sustainable transportation solutions are discussed in Chapter 12. Future transportation technologies, such as plug-in hybrid, range extender electric, and pure electric vehicles; and multimodal transportation with intelligent transportation technologies are discussed in this chapter. Future market needs of hybrid powertrains for commercial vehicles, emissions regulations, and government incentives that may have significant impact on the development of future hybrid technologies are also discussed in this chapter.

Acknowledgments

I would like to express sincere appreciation to my co-authors of this book, Dr. Rudy Smaling and Simon Baseley, for their commitments and contributions to this book; to Dr. Dave Turner for providing an early draft of Chapter 6; to the manuscript reviewers, Dave Merrion, Darren Gosbee, and Dennis Assanis for their valuable comments and suggestions; and to Martha Swiss of SAE for her support during the entire process of writing this book.

I would like to thank Lennart Jonsson, Thomas Stover, Yannis Tsavalas, Ramanath Ramakrishnan, Daryll Fogal, Dr. Joe Lin, Gerard Devito, Dr. Zhanjiang Zou, Hanyun Yang, Yolanda Washington, and many colleagues at Eaton Corporation for their support in preparing the manuscript of this book.

I also would like to thank Dimitri Kazarinoff, Seth Deutsch, and Helena Fu of Aecom, Kin-Peng Sin of IBM, Robin Wu of Foton Auto, Yahe Wang of MGL Battery, and Charles Lu of Broad-Ocean Motor for the exciting collaborative experience of establishing the New Energy Sustainable Transportation International Alliance (NESTIA).

Last but not least, I would like to thank my beloved wife, Ping, for her help with editing the manuscripts and drawings, and my wonderful children, Caroline (a doctoral student in Biology at Stanford University) and Kevin (a sophomore in Physics at Massachusetts Institute of Technology) for their cheering and support.

Haoran Hu, Sc.D.,
May 12, 2012

Chapter 1

Introduction of Hybrid Powertrains for Commercial Vehicles

1.1 Introduction

Commercial vehicles include light commercial vehicles, heavy trucks, coaches, and buses that are used for transporting goods or passengers. The powertrain of a commercial vehicle usually refers a group of components, such as the engine, transmission, and driveshafts, which transforms stored energy (chemical, kinetic, potential, etc.) into kinetic energy for propulsion purposes.
The production of worldwide commercial vehicles, including light-, medium-, and heavy-duty trucks, coaches, and buses, exceeded 19 million units in 2010 [1.1]. The global annual production of commercial vehicles is expected to reach 23 million units by 2015 due to the rapid growth in China and other emerging markets, as shown in Fig. 1.1.

The commercial transportation system is essential to the health of the U.S. economy. Currently, there are approximately 30 million commercial vehicles registered for use on North America highways. The annual production of medium- and heavy-duty commercial vehicles for 2011 will be over 400,000

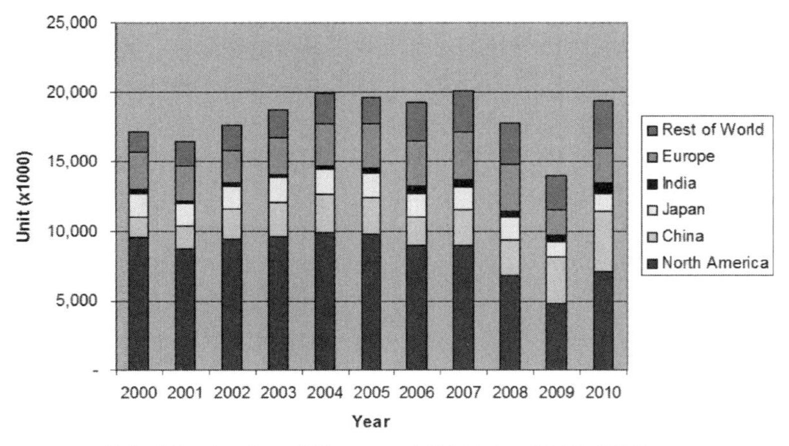

FIG. 1.1 Global Production of Commercial Vehicles (2000–2010).

units. The trucking industry is a major employer in the United States. In 2010, there were 7 million people employed in trucking-related jobs, nearly 3 million of which were truck drivers. According to the Bureau of Transportation Statistics, commercial vehicles transport 63.7% of the value of all commercial freight activities in the United States. In 2010, commercial vehicles transported over 63% of the value of trade between the United States and Canada and transported over 66% of the value of trade between the United States and Mexico [1.2]. Commercial vehicles in 2009 logged 274 billion miles and consumed 38 billion gallons of fuel for business purposes. They account for 12% of all oil consumption and nearly 6% of all greenhouse gas emissions in the United States [1.3].

The commercial hybrid vehicle market is still in its early stages, with production growing only marginally over the last few years. The two largest hybrid bus fleets are in the City of New York and the City of Beijing. NYC Transit has a hybrid-electric bus fleet of 1171 units [1.4], and the City of Beijing has 960 hybrid-electric buses as of 2010 [1.5]. The global commercial hybrid vehicle market is predicted to grow at a Compound Annual Growth Rate (CAGR) of 26% in the next few years to reach approximately 27,000 units by 2013 [1.6, 1.7].

1.2 History of Commercial Vehicles

In 1769, France's Nicholas Cugnot built a steam-powered motor carriage capable of traveling six mph (Fig. 1.2). The Gurney steam car, built by Sir Goldsworthy Gurney of Great Britain in 1825, completed an 85-mile round-trip journey in ten hours. In 1830, Gurney then designed a large stagecoach driven by a steam engine that may have been the first motor-driven bus (Fig. 1.3).

The first electric "railway" was built by Dr. Werner von Siemens in 1879 (Fig. 1.4). The wheels of the car were driven by an electric motor that drew its electricity from the rails, which were connected to a generator [1.10].

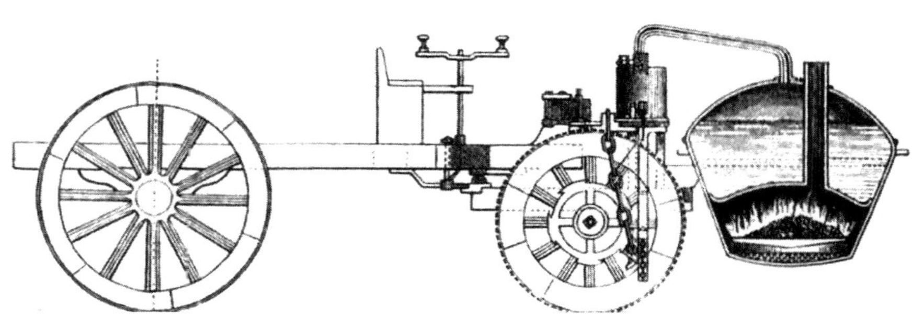

FIG. 1.2 Cut-away drawing of the first steam-powered vehicle [1.8].

FIG. 1.3 Gurney Steam Car [1.9].

FIG. 1.4 Siemens electric rail car [1.10].

In 1896, Gottlieb Daimler built the first motor truck. It had a four-horsepower engine and a belt drive with two forward speeds and one reverse. The first American company to build a truck was the Winton Company in 1898. The Winton truck was, in effect, a delivery wagon with a single-cylinder, six-horsepower engine. In the early 1900s there were more electric-powered vehicles on the road than the gas-powered ones, which encouraged the development of the hybrid-electric vehicle (HEV). In 1904, Ferdinand Porsche built a vehicle that housed electric motors in the front wheel hubs while the gasoline engine provided the power [1.10]. The Lohner-Porsche gasoline-electric 'Mixte' used a gasoline engine rotating at a constant speed to drive a generator to produce the electricity for powering the electric motors, which were contained within the hubs of the front wheels (Fig. 1.5).

During the first decade of the 1900s, General Electric in America and Siemens in Germany both produced electric cars and hybrids as commercial vehicles. In 1905, H. Piper, an American engineer, filed a patent for a gas-electric hybrid vehicle in which an electric motor would assist an internal combustion engine to improve the vehicle acceleration [1.10]. Electric car makers Baker of Cleveland and Woods of Chicago produced hybrid cars in 1916. Woods claimed that their hybrids could reach speeds of 35 mph with a fuel efficiency of 48 miles per gallon [1.11]. In the meantime, significant progress was made in engine development for gasoline-powered vehicles. When Piper's

FIG. 1.5 Lohner-Porsche gasoline-electric "Mixte" hybrid.

patent was issued in 1908, gasoline engines had become powerful enough to achieve the patent-claimed standalone performance of 25 mph in 10 seconds. The premier of more-powerful gasoline engines, along with equipment that allowed them to be started without cranks, contributed to the decline of the electric vehicle (EV) and of the nascent HEV between 1910 and 1920.

During the 1920s, heavy-duty commercial vehicles were firmly established as a major means of freight conveyance. Diesel engines, invented and developed by Dr. Rudolf Diesel in 1897, were used in large trucks since the 1930s in the United States because they possessed larger torque and higher thermal efficiency than gasoline-powered engines.

In the early 1970s, concerns over growing dependence on imported petroleum and deteriorating air quality due to fossil fuel combustion spurred the U.S. Congress to pass the Electric and Hybrid Vehicle Research, Development & Demonstration Act of 1976. The bill established a five-year, $160 million research, development, and demonstration project within the Energy Research and Development Administration (ERDA) to promote the development of an electric vehicle that could function as a practical alternative to the gasoline-powered automobile [1.12]. This legislation renewed the public interest in the development of electric and hybrid vehicles, and led to the advancement of hybrid technologies.

1.3 Commercial Vehicle Classification

There is a wide range in design and application of commercial vehicles: pickup trucks, delivery trucks, transit buses, school buses, mixers, dump trucks, tankers, bulk haulers, interstate line haul, and heavy-haul trucks. Industry classifies trucks and buses by weight using the vehicle's gross vehicle weight rating (GVWR), the maximum in-service weight set by the manufacturers, or the gross vehicle weight (GVW) plus the average cargo weight in the trucking industry. The use categories of vehicles are not as well-defined as weight classes, and they depend on widely varying industry usage. For example, the same vehicle may be called heavy-duty by one segment of the industry and medium-duty by another.

1.3.1 Commercial Vehicle Classification in the United States

Table 1.1 shows the widely used truck classes and categories in the United States. Some truck classifications used by the U.S. EPA and the California Air Resources Board (CARB) for emissions regulations differ from Table 1.1. In other cases, the Vehicle Inventory and Use Survey (VIUS) categories are used, in which "heavy truck" is the term used for vehicles over 10,000 lb GVW.

TABLE 1.1 Widely Used Truck Weight Classes and Categories [1.13]

Weight Class	Minimum GVWR (lb)	Maximum GVWR (lb)	VIUS Category	Common Category
Class 1	NA	6,000	Light-duty	Light-duty
Class 2	6,001	10,000	Light-duty	Light-duty
Class 3	10,001	14,000	Medium-duty	Light-duty
Class 4	14,001	16,000	Medium-duty	Medium-duty
Class 5	16,001	19,500	Medium-duty	Medium-duty
Class 6	19,501	26,000	Light-heavy	Medium-duty
Class 7	26,001	33,000	Heavy-duty	Heavy-duty
Class 8	33,001	NA	Heavy-duty	Heavy-duty

NOTE: GVWR, Gross Vehicle Weight Rating; VIUS, Vehicle Inventory and Use Survey; NA, not available to the committee.

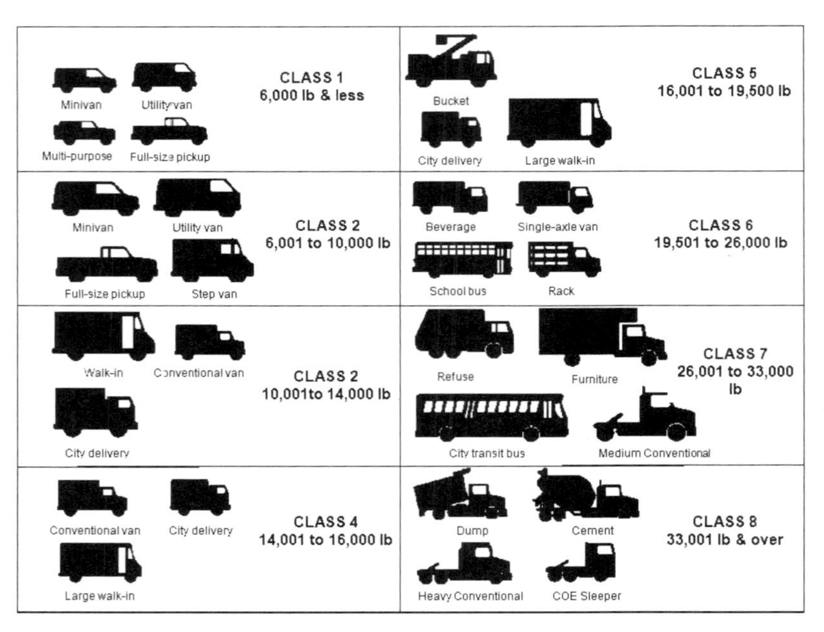

FIG. 1.6 Commercial vehicle classification by the Federal Highway Administration (FHWA).

Figure 1.6 shows the classification schemes used for commercial vehicles by the Federal Highway Administration (FHWA) [1.14].

1.3.2 Commercial Vehicle Classification in Europe

In Europe, according to the Economic Commission of Europe (ECE), the classifications of commercial vehicles are from the three major categories, M, N, and O, as shown in Table 1.2, Table 1.3, and Table 1.4 [1.15].

TABLE 1.2 Category M: Motor Vehicles with at least 4 Wheels and Intended for Passenger Transport

Grading	Total Seats	Total Weight (Tons)
M1	≤9	≤3.5
M2	>9	≤5
M3	>9	>5

TABLE 1.3 Category N: Motor Vehicles with at least 4 Wheels and Intended for Freight Transport

Grading	Total Weight (tons)	
N1		≤3.5
N2	>3.5	≤12
N3	>12	

TABLE 1.4 Category O: Trailers and Semitrailers

Grading	Total Weight (tons)	
O1		≤0.75
O2	>0.75	≤3.5
O3	>3.5	≤10
O4	>10	

Vehicles in categories M, N, and O may be equipped for special purposes (e.g., caravans, ambulances). Categories M2 and M3 can be further divided with subcategories such as vehicles with seats only, and vehicles with seats and standing rooms, etc.

1.3.3 Commercial Vehicle Classification in China

The commercial vehicle classification standards in China, which were introduced in 2005, are weight-based classifications for commercial vehicles (Table 1.5) and length-based classifications for city buses (Table 1.6).

The vehicles (buses) for carrying passengers are classified as:

TABLE 1.5 Weight-based Classification of Commercial Vehicles in China [1.16]

	Total Weight (tons)	
Mini		≤1.8
Light	>1.8	≤6
Medium	>6	≤14
Heavy	>14	

TABLE 1.6 Length-based Classification of Buses in China

	Total Length (meters)	
Mini		≤3.5
Small	>3.5	≤7
Medium	>7	≤10
Large	>10	

1.4 Energy Consumption of Commercial Vehicles

Commercial vehicle manufacturing and operation is a major source of energy consumption globally. In 2009, the United States consumed 23% of the global petroleum production [1.3]. According to the U.S. Department of Energy, 72%

of the U.S. petroleum consumption is for transportation. Commercial vehicles consumed up to 18.7% of the total energy consumption in transportation in the United States. In other words, commercial vehicles in the United States alone consumed over 3% of the global petroleum production in 2009 [1.3, 1.34].

1.4.1 Energy Demand of Commercial Vehicles

In addition to the fundamental energy needed to move a certain mass of payload or to perform a certain chore (i.e., tilling a field), there are a number of factors that determine the total energy demand of a vehicle. Specifically, for a vehicle traveling at constant speed on level ground, the force required to propel the vehicle is the sum of aerodynamic drag and rolling resistance. This force is often referred to as road load.

In addition to the road-load work, fuel energy is used in overcoming a number of powertrain-related losses. Specifically, these losses can be categorized in Table 1.7 as:

1. Engine losses, which in turn can be broken down into the following:
   a. Thermal losses in the engine (no heat engine can operate at 100% efficiency)
   b. Parasitic losses (fuel consumed to overcome the engine's internal friction, pumping losses, etc.)
2. Auxiliary loads (air-conditioning, power steering and brakes, etc.)
3. Driveline losses (transmission, differential, axle, and wheel bearings)

In addition to the powertrain-related energy losses, significant kinetic energy can be dissipated into heat when drivers brake. Therefore, braking is an energy loss, the magnitude of which is given by the vehicle inertia (or weight) and the specific driving pattern.

1.4.2 On-road Heavy-duty Commercial Vehicles

On-road commercial vehicle fuel efficiency is influenced by several factors including basic vehicle design, zone of operation, driver technique, and weather factors.

Engine losses, aerodynamic losses, and tire-rolling resistance account for approximately 94% of the energy used to sustain vehicle speed at 65 mph (Table 1.7). Because these factors are all dependent on vehicle speed, terrain, traffic conditions, etc., the expected benefits to fuel economy will be highly dependent on zone of operation. Driveline friction and engine-based accessories such as compressors and alternators account for the remaining 6%. Therefore, improvements in engine efficiency, aerodynamic drag, and tire-rolling resistance will have a significant impact on fuel efficiency, while improvements in driveline and accessory efficiency will have a small influence [1.13].

TABLE 1.7 Energy Audit of On-highway Commercial Vehicles [1.17]

Energy Balance (kWh)		Line Haul Truck (full-loaded, level road @ 65 mph)		Med Duty Truck (full-loaded, 26,000 lb GVW, level@ 40 mph)		Transit Bus (14,515 kg, over CBD-14 @ 1/2 seated load with A/C on)	
Base		% Total Energy	Base	% Total Energy	Base	% Total Energy	
Energy loss per hour		240	60	71.7	67	14.16	60
Auxiliary Loads	Energy Losses (sinks)	15	4	7	7	6.06	25
Drivetrain Losses		9	2	3	3	0.87	3
Braking Losses		0	0	0	0	1.37	6
Aerodynamic Drag	Road Load	85	21	15.7	15	0.22	1
Rolling Resistance		51	13	9.7	9	1.2	6
Total Energy used per hour		400 kWh	100	107.7 kWh	100	24.2 kWh	100

1.4.3 Medium-duty Trucks

Medium truck fuel efficiency is also influenced by basic vehicle design, mode of operation, driver technique, and weather factors, as for the on-highway long-haul trucks. The energy audit shown in Table 1.7 is a breakdown of energy requirements of a typical Class 6 delivery truck. The vehicle is operating at a steady speed of 40 mph, typical of the average speed in a suburban environment, with a GVW of 26,000 lb (11,791 kg).

Engine losses, inertial resistance, and tire-rolling resistance have significant effects on vehicle efficiency, whereas drivetrain friction is less significant. Aerodynamics, a major factor for line-haul trucks, plays only a minor role for delivery trucks because they have much slower operating speeds. However, relative energy usage for engine-based accessories such as compressors and alternators can be a major factor for medium truck operations. For low-power operations, such as city missions, the auxiliary loads (ranging up to 15 hp) can represent a fairly large percentage of overall missions.

1.4.4 Transit Buses

Transit bus power requirements can be broken into five categories: accelerating the vehicle to speed (a function of weight), operating auxiliary systems ("hotel load"), overcoming aerodynamic drag, and overcoming rolling resistance (a function of weight) and drivetrain losses. To increase overall propulsion efficiency and reduce emissions, engine transient operating conditions must be minimized. To accomplish this, engine speed and load must be independent of drive-wheel speed and required tractive effort. By optimizing vehicle design, together with powertrain efficiency improvements, the transit bus can be significantly improved in overall operational efficiency.

1.5 Drivers of the Efficient Powertrain for Commercial Vehicles

1.5.1 Fuel Prices and Energy Security

Crude oil prices have fluctuated dramatically during the last few years, from around US$20 to US$25 per barrel (in 2007 dollars) between the mid-1980s and 2002, to around US$135 per barrel in mid-2008 and back to US$36 in January 2009. It remained in the US$70 to US$85 range for most of 2010. The unrest in North Africa and Mid-Eastern countries caused renewed concern about the energy security in North America. The crude oil price surpassed $100 per barrel in March 2011. As a consequence, retail fuel prices increased from $2.80 to $3.40 in less than two weeks. As well as the fuel price issue, concerns regarding the stability and the long-term availability of oil are impetuses to improve energy security by reducing oil consumption.

1.5.2 Greenhouse Gas Emissions and Fuel Economy Regulations

1.5.2.1 Background of Greenhouse Gases

A greenhouse gas (GHG) is a gas in an atmosphere that absorbs and emits radiation within the thermal infrared range. The process of absorbing thermal radiation and re-radiating part of it back to the planetary surface is called greenhouse effect.

There are naturally occurring greenhouse gases including water vapor, carbon dioxide (CO_2), methane (CH_4), nitrous oxide (N_2O), and ozone (O_3). There are also several gases that do not have a direct global warming effect but indirectly affect terrestrial and/or solar radiation absorption by influencing the formation or destruction of greenhouse gases, including tropospheric and stratospheric ozone. These gases include carbon monoxide (CO), oxides of nitrogen (NO_x), and non-CH_4 volatile organic compounds (NMVOCs). Aerosols, which are extremely small particles or liquid droplets, such as those produced by sulfur dioxide (SO_2) or elemental carbon emissions, can also affect the absorptive characteristics of the atmosphere.

Although the direct greenhouse gases CO_2, CH_4, and N_2O occur naturally in the atmosphere, human activities have changed their atmospheric concentrations. From the pre-industrial era (ending about 1750) to 2005, concentrations of these greenhouse gases have increased globally by 36, 148, and 18%, respectively [1.18].

1.5.2.2 CO_2 Emissions by Commercial Vehicles

Transportation activities (excluding international bunker fuels) accounted for 34% of CO_2 emissions from fossil fuel combustion in 2009 [1.3]. Virtually all of the energy consumed in this end-use sector came from petroleum products. Nearly 60% of the emissions resulted from petroleum consumption for personal vehicle use and 19% from commercial vehicles. The remaining emissions came from other transportation activities, including petroleum fuel usages in rail, ship, and aircraft. In 2009, 84% of fuel use by commercial vehicles are for freight transportation, 11% for light-duty commercial vehicles, and 5% for bus transportation [1.3].

1.5.2.3 Greenhouse Gas Emission Regulations and Fuel Economy Standards for Commercial Vehicles

Regulations that set fuel economy standards for commercial vehicles exist in the European Union, Japan, China, the United States, and other countries. In the European Union, the regulations have taken the form of limits for CO_2 emissions expressed as grams of CO_2 per kilometer (g/km). For new passenger cars, including vehicle category M1 (Table 1.2), the average CO_2 emissions of those sold in Europe by each OEM in 2012 must not exceed 130 g/km. Each OEM's new sales fleet must average no more than 175 g/km by 2012, which will be reduced to 160 g/km by 2015. OEMs can trade credits to achieve the targets, but will be penalized in proportion to the number of vehicles and the average excess if the target is missed. While part of the target can be achieved by measures such as the use of biofuels, efficient tires, efficient transmissions, and so forth, the rest of the savings must be achieved through improved powertrain fuel economy, toward which EV, HEV, plug-in hybrid-electric vehicle (PHEV) and fuel cell vehicle (FCV) technologies can make a significant contribution.

The European Commission (EC) is developing CO_2 emissions targets for heavy commercial vehicles, such as trucks and buses, to account for variations in size and weight. The EC has opened a new laboratory at its Joint Research Center in northern Italy for the testing of heavy vehicle emissions, and one of its roles will be to compare CO_2 emissions from heavy-duty diesel engines running on alternative fuels and in combination with hybrid drive systems. It is expected that CO_2 emissions standards for heavy commercial vehicles will be proposed after 2014.

On August 9, 2011, the U.S. Environmental Protection Agency (EPA) and the National Highway Traffic Safety Administration (NHTSA) adopted the first U.S. GHG emission and fuel consumption standards for heavy- and medium-duty commercial vehicles under the authority of the 2007 Energy Independence and Security Act (EISA) and the Clean Air Act [1.18]. The GHG program includes CO_2 emission standards, as well as emission standards for N_2O and CH_4, and provisions to control hydrofluorocarbon leaks from air conditioning systems.

The regulation covers model years (MY) 2014–2018, with NHTSA fuel economy standards being voluntary in MY 2014–2015 to satisfy EISA lead time requirements. The affected heavy- and medium-duty fleet incorporates all on-road vehicles rated at a GVW≥8500 lb, and the engines that power them, except those covered by the GHG emissions and Corporate Average Fuel Economy (CAFE) standards for MY 2012–2016 light-duty vehicles.

Different CO_2 and fuel consumption standards are applicable to three categories of vehicles: combination tractors, heavy-duty pickups and vans, and vocational vehicles:

- For combination tractors (the semi trucks that typically pull trailers), the adopted engine and vehicle standards begin in MY 2014 and achieve from 7 to 20% reduction in CO_2 emissions and fuel consumption by MY 2017 over the 2010 baselines. The standards would phase in to the 2017 levels shown in Table 1.8.
- For heavy-duty pickup trucks and vans, the standards phase in starting in MY 2014 and achieve up to a 10% reduction in CO_2 emissions and

TABLE 1.8 Final (MY 2017) Combination Tractor Standards

	EPA CO_2 Emissions (g/ton-mile)			NHTSA Fuel Consumption (gal/1000 ton-mile)		
Category	Low Roof	Mid Roof	High Roof	Low Roof	Mid Roof	High Roof
Day Cab Class 7	104	115	120	10.2	11.3	11.8
Day Cab Class 8	80	86	89	7.8	8.4	8.7
Sleeper Cab Class 8	66	73	72	6.5	7.2	7.1

fuel consumption for gasoline vehicles and 15% reduction for diesel vehicles by MY 2018. These vehicles must meet corporate average CO_2 and fuel economy standards, in an approach similar to that taken for light-duty vehicles, but with different standards for gasoline and diesel vehicles [1.20].

- For vocational vehicles, the engine and vehicle standards start in MY 2014 and achieve up to a 10% reduction in fuel consumption and CO_2 emissions by MY 2017. The respective vehicle standards are depicted in Table 1.9.

1.5.3 Commercial Vehicle Exhaust Emissions

Emission standards have become increasingly stringent since the passage of the Clean Air Act in 1963. These increasingly stringent standards have dictated that new technologies are developed to comply with them. As an additional challenge, increasingly stringent emission standards for heavy-duty vehicles tend to adversely affect fuel economy at a time when there are challenges to improve fuel economy.

1.5.3.1 On-road Vehicle Emission Standards

The control of emissions from the engines of heavy-duty trucks with GVWR over 8500 lb began in 1973 in California, and in 1974 in the United States nationwide. Shown in Fig. 1.7 are the progressively more stringent emission standards for heavy-duty diesel engines.

TABLE 1.9 Final (MY 2017) Vocational Vehicle Standards

Category	EPA CO_2 Emissions (g/ton-mile)	NHTSA Fuel Consumption (gal/1000 ton-mile)
Light Heavy Class 2b-5	373	36.7
Medium Heavy Class 6–7	225	22.1
Heavy Heavy Class 8	222	21.8

Emission reduction in large tractor-trailer combination trucks must focus on the diesel engine. The diesel engine presently dominates this sector of commercial trucks because of its efficiency, durability, and torque/speed characteristics. In the foreseeable future, no other type of powerplant is expected to be ready for this application in spite of years of research on alternatives.

Over the past 30 years, diesel-engine manufacturers have achieved remarkable reductions in nitrogen oxide (NO_x) and particulate matter (PM) emissions in response to regulations. The first diesel emissions regulation for commercial vehicles by the U.S. EPA in the mid- to late 1970s had the typical emission values of 10 to 15 g/bhp-h of NO_x and 1 g/bhp-h of PM. Emissions reductions have been achieved by optimizing diesel electronic control, retarding fuel-injection timing, increasing injection pressures, improving air-handling systems, using oxidation catalysts, and implementing EPA's mandate for low-sulfur diesel fuel (no greater than 0.05% sulfur content) for on-highway vehicles in the early 1990s.

In 1996, the EPA, the State of California, and major engine manufacturers prepared a Statement of Principles (SOP) that required further reduction to 2.4 g/bhp-h of NO_x plus non-methane hydrocarbons (NMHC) or 2.5 g/bhp-h of NO_x plus NMHC with a maximum of 0.5 g/bhp-h of NMHC by 2004. An action by the EPA and the U.S. Department of Justice resulted in a consent decree with the diesel-engine manufacturers that moved the SOP requirements to October 2002 and placed caps on emissions at all operating conditions. In 2007, the heavy-duty diesel engines were regulated to 0.2 g/bhp-h of NO_x, 0.01 g/bhp-h of PM, and 0.14 g/bhp-h for NMHC. The PM emission standard took full effect in the 2007 heavy-duty engine model year. The NO_x and NMHC standards were to phase in for diesel engines between 2007 and 2010.

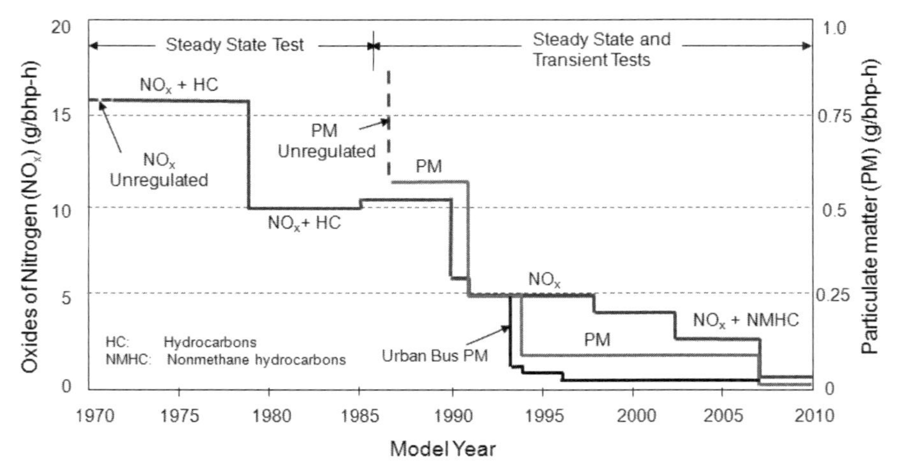

FIG. 1.7 Historical trend in exhaust emission standards for heavy-duty diesel engines.

U.S. EPA also regulated diesel fuel sulfur content in on-highway diesel fuel to 15 ppm (wt.), down from the previous 500 ppm. At the terminal level, highway diesel fuel sold as low-sulfur fuel must meet the 15 ppm sulfur standard as of July 15, 2006. For retail stations and wholesale purchasers, highway diesel fuel sold as low-sulfur fuel must meet the 15 ppm sulfur standard as of September 1, 2006.

European emission regulations for new heavy-duty diesel engines are commonly referred to as Euro I . . . VI for heavy duty vehicles and Euro 1 . . . 6 for light duty vehicles, which define the acceptable limits for exhaust emissions of new vehicles sold in EU member states. The emission standards apply to all motor vehicles with a "technically permissible maximum laden mass" over 3500 kg, equipped with compression ignition engines or positive ignition natural gas (NG) or liquefied petroleum gas (LPG) engines. The emission standards are defined in a series of European Union directives for progressive introduction of increasingly stringent standards.

Euro I standards were introduced in 1992, followed by the introduction of Euro II regulations in 1996. These standards applied to both truck engines and urban buses; the urban bus standards, however, were voluntary. In 1999, the EU introduced Euro III standards (2000), as well as Euro IV/V standards (2005/2008). This rule also set voluntary, stricter emission limits for extra-low-emission vehicles, known as "enhanced environmentally friendly vehicles" or EEVs. Euro VI emission standards were introduced in July 2009. The new emission limits, comparable in stringency to the U.S. EPA 2010 standards as shown in Fig. 1.8, become effective from 2013 (new type approvals) and 2014 (all registrations).

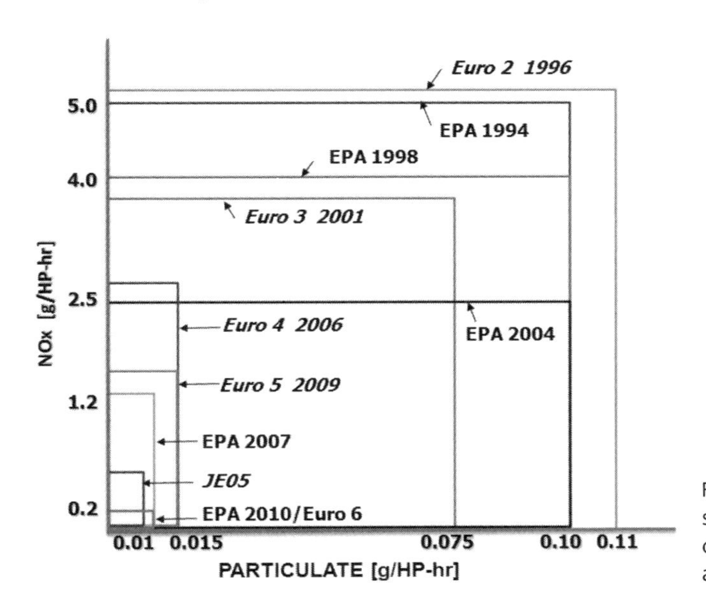

FIG. 1.8 Comparison of emission regulations for commercial vehicles in the U.S., EU, and Japan.

In Japan, emission standards for new diesel fueled commercial vehicles are also shown in Fig. 1.8. Light-duty trucks and buses are tested on the 10–15 mode cycle, which was replaced by the JC 08 mode test by 2011. The test procedure for heavy-duty engines is the JE05 mode cycle (hot start version). Before 2005, heavy-duty engines were tested over the 13-mode cycle and the 6-mode cycle. Vehicles and engines are tested using 50 ppm S fuel for the 2005 standards [1.22].

Many developing countries, such as China and India, will adopt the EU emission standards with different implementation dates as shown in Table 1.10. For example, China has a two-stage implementation process. China will implement equivalent Euro V standards in Beijing by 2012, but equivalent Euro IV standards for the rest of the country.

1.5.3.2 *Off-road Vehicles*

Off-road engines and equipment have previously not been regulated to the same level as on-road applications employing the same or similar engines. Often, cleaner burning engines from the on-road sector make their way into off-road applications, and emission reductions occur. Yet, off-road diesel engines are the largest remaining contributor to the overall mobile source emissions inventory. But recently, with the United States instituting Tier 1 requirements phased in through 2000, and the European Commission introducing a directive in February of 1998, these engines have been subjected to more restrictive regulatory mandates. In addition to regulating smaller engines through these initiatives, this is the first time off-road diesel fuel will

Global Emissions Standards - 2005-2015

	2005	2006	2007	2008	2009	2010	2011	2012	2013	2014	2015
US	EPA 2004		EPA 2007			EPA 2010					
Australia	Euro 3		Euro 4				Euro 5				
Mexico	Tier 1			Euro 4				Euro 5			
Brasil Diesel	Euro 3				Euro 4			Euro 5			
Argentina	Euro 3				Euro 4				Euro 5		
Europe	Euro 4					Euro 5			Euro 6		
China (Bejing)	Euro 2	Euro 3		Euro 4				Euro 5			
China Nationwide	Euro 2			Euro 3			Euro 4			Euro 5	
Thailand	Euro 2						Euro 3		Euro 4		
Japan	JE05										
S. Korea	Euro 3		Euro 4				Euro 5				
India-National Capital Area	Euro 3						Euro 4		Euro 5		
India-Nationwide	Euro 2						Euro 3		Euro 4		

TABLE 1.10 Worldwide Emission Regulations for Commercial Vehicles

be regulated. These measures bring the off-road applications to emission control levels similar to those of on-road engines.

Just as for on-highway engines, regulatory authorities in the EU, Japan, and United States have been working together toward worldwide harmonization of off-highway emission tests and standards, to streamline engine development and emission certification for engine manufacturers. Off-road engines are primarily tested in the different regions using the same test cycles—ISO 8178-4 cycles C1 or D2. The different cycles are used depending on whether the engine is intended for use in a constant or variable engine speed application. Some differences occur in timing implementation and in engine categorization, but the emission levels should become essentially equivalent in the next decade. To represent emissions during real-world conditions, a new transient test was established through the cooperation of the EU and the U.S. EPA. The Non-Road Transient Composite (NRTC) test cycle has been used in the U.S. since 2011, the EU since 2006, and Japan since about 2010. Tier 4 emission standards for U.S. and EU are shown in Table 1.11.

1.5.3.3 Low Emissions Zones

Some cities around the world have established low emissions zones in efforts to improve urban air quality, and although these zones have been created to reduce toxic emissions rather than CO_2, any improvement in a vehicle's fuel efficiency can help it comply with lower toxic emissions targets.

In 2003, the Tokyo Metropolitan Government launched regulations that prevent commercial vehicles that do not comply with certain emissions standards from entering the metropolitan area. In London, a zone was established in 2008 that applies to all diesel commercial vehicles of more than 3500 kg

EPA Tier 4 Interim / EU Stage   EPA Tier 4 Final / EU Stage

kW	EPA	HP	2008	2009	2010	2011	2012	2013	2014	2015	2016
0-18*		0.24		(7.5)/6.6/0.4							
19-36		25-48		(7.5)/5.5/0.3					(4.7)/5.5/0.03		
37-55		49-74		(4.7)/5.0/0.3					(4.7)/5.0/0.03		
56-129*		75-173					3.4/0.19/3.5/0.02			0.4/0.19/5.0/0.02	
130-560*		174-751					2.0/0.19/3.5/0.02			0.4/0.19/3.5/0.02	
>560		>751					3.5/0.4/3.5/0.10			3.5/0.19/3.5/0.04	

kW	EU	HP	2008	2009	2010	2011	2012	2013	2014	2015	2016
18-36		24-48		Stage IIIA (7.5)/5.5/0.6							
37-55		49-74							(4.7)/5.0/0.025		
56-129*		75-173						3.3/0.19/5.0/0.025		0.4/0.19/5.0/0.025	
130-560		174-751					2.0/0.19/3.5/0,025			0.4/0.19/3.5/0.025	

(NOx + HC) / CO / PM (Oxides of Nitrogen + Hydrocarbons) / Carbon Monoxide / Particulate Matter (g/kW-hr)
NOx / HC / CO / PM Oxides of Nitrogen / Hydrocarbons / Carbon Monoxide / Particulate Matter (g/kW-hr)
* Combines regulatory powerbands with same emission levels

TABLE 1.11 Emission Standards for Off-road Commercial Vehicles [1.21]

GVW. There was a phased introduction of the scheme from 4 February 2008 through to January 2012. Different vehicles will be affected over time, and increasingly tougher emissions standards will apply. The daily fees for heavy vehicles to enter the zone can be as much as £200 if they do not comply with Euro 3 for PM emissions or Euro 4 from 2012. Milan trialed similar systems during 2008, under which trucks and buses that failed to comply were subject to a charge of €10 per day. In Germany, Berlin started a low emission zone in the central city area on January 1, 2008, as did Cologne and Hanover.

As of June 2009, only vehicles carrying a yellow label that indicates "low emissions" will be allowed within Beijing's fifth ring road and as of October 2009, the low emissions zone was extended to the city's sixth ring road. Shanghai is planning a similar exclusion zone within its Middle Ring Road.

1.5.3.4 No-idling Laws

There are increasing efforts to reduce toxic emissions and increase fuel economy by preventing vehicles from idling the engine while stationary, particularly in urban areas. As a consequence, more and more new light vehicles are now being fitted with stop-start technology to enable the rapid restarting of the internal combustion engine (ICE) once the driver is ready to move off from rest.

There is no U.S. federal anti-idling law, but about 30 states and dozens of municipalities require vehicles to limit idling time, typically to three to five minutes. New York City, for example, has had a three-minute law since 1971, but it is not vigorously enforced, with only 526 idling violation notices issued in 2007. A study found that school buses average 20 minutes idling per day, and the city is considering a bill to limit idling outside schools to one minute.

1.6 Classification of Commercial Vehicle Hybrid Powertrains

Hybrid vehicles are vehicles with two or more power sources in the drivetrain. Current hybrids use both an IC engine or a fuel cell and a secondary drive system to improve fuel economy and performance, and to reduce emissions. The secondary drive systems of commercial vehicles are electric motor/generator drive system or hydraulic pump/motor drive system. Other combinations of energy storage and conversion are possible, although not yet in commercial production for commercial vehicles.

There are many different types of hybrid commercial vehicles. Hybrid commercial vehicles can be classified by the architecture of the hybrid system design, energy storage method, or applications. The major category and the classification are listed on Table 1.12.

TABLE 1.12 Classification of Hybrid Powertrains for Commercial Vehicles

Classification Category	Classification Type
Primary Power Source	IC Engine Hybrid
	Fuel Cell Hybrid
Architecture	Parallel
	Series
	Power split
	Plug-in Hybrid
	Range-Extender Hybrid
	In-Wheel Hybrid
Application	Micro Hybrid
	Mild Hybrid
	Full Hybrid
Energy Storage	Electric Hybrid
	Hydraulic Hybrid
	Air Hybrid
	Flywheel Hybrid

1.6.1 Types of Energy Storage

One of the primary functions of hybrid vehicles is to recover and store the kinetic engine during regenerative braking. Hybrid vehicles can be classified by the energy storage method as electric hybrid, hydraulic hybrid, pneumatic hybrid, and mechanical hybrid.

1.6.1.1 Electric Hybrid

A schematic of a hybrid-electric powertrain, shown in Fig. 1.9(a), combines an internal combustion engine propulsion system with an electric propulsion system to achieve either better fuel economy or better performance than a conventional vehicle.

One of the key features of the hybrid-electric powertrain is to convert the vehicle's kinetic energy into battery-replenishing electric energy through regenerative braking, rather than wasting it as heat energy. Some varieties of hybrid-electric powertrain use their internal combustion engine to generate electricity by spinning an electrical generator (this combination is known as a motor-generator) to either recharge their batteries or to directly power the electric drive motors.

The hybrid-electric commercial vehicle has been used in city delivery vehicles and city buses in recent years. Prototypes of hybrid-electric off-road vehicles have been reported and will be discussed in the later sections.

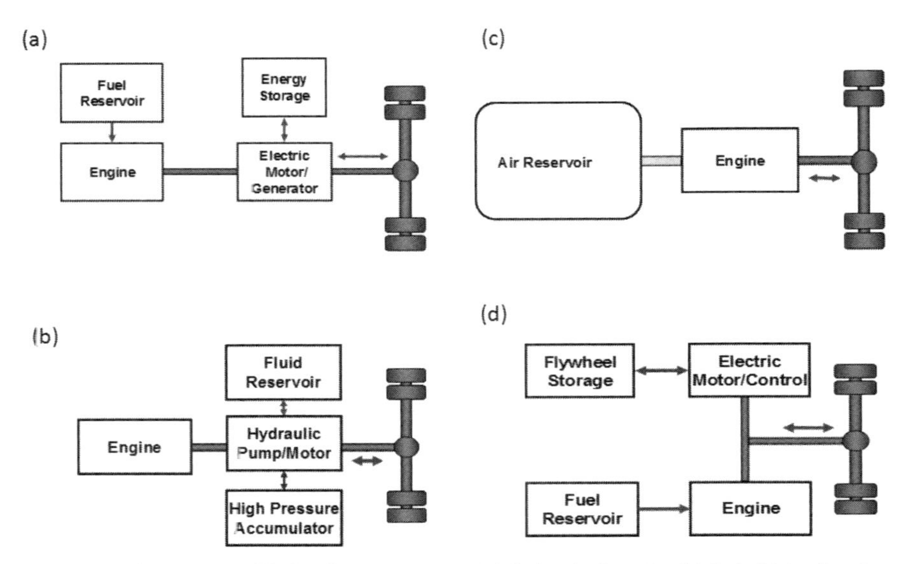

FIG. 1.9 Schematics of hybrid powertrains: (a) hybrid electric, (b) hybrid hydraulic, (c) pneumatic hybrid, and (d) flywheel hybrid.

1.6.1.2 Hydraulic Hybrid

Like the HEV, the kinetic energy can be captured through regenerative braking for the hydraulic hybrid vehicle (HHV). As the vehicle slows, a hydraulic pump is activated and moves fluid from the reservoir to a hydraulic accumulator. As hydraulic pressure in the accumulator builds, compressed gas acts to store energy in the same way as a charging battery in an HEV, ready to power a hydraulic motor to propel the vehicle.

Figure 1.9(b) shows a schematic of a hydraulic hybrid system. Hydraulic hybrids use four main components to power a vehicle; i.e., a low pressure reservoir, a high pressure accumulator, a hydraulic pump, and a hydraulic motor. A low-pressure reservoir is filled with hydraulic fluid. A pump moves the fluids from the low-pressure reservoir to a high-pressure accumulator. The accumulator holds not only the fluids brought over by the pump, but also pressurized nitrogen gas. The stored energy in the high-pressure accumulator can be released to power hydraulic motors, which in turn propel the vehicle.

1.6.1.3 Pneumatic Hybrid

The pneumatic hybrid, also called Air-Power-Assist (APA), uses the engine to convert braking energy into compressed air, which can be stored in an onboard air tank [1.23]. The stored energy in the compressed air can be used to power the vehicle. Once the vehicle has attained a cruising speed, the engine is switched to operate in a conventional diesel mode. Since the thermodynamic availability of the stored air is primarily a function of the pressure, the APA technology is expected to be robust to ambient temperature variations.

Figure 1.9(c) shows a schematic of an APA system. The energy-saving principle of this technology is similar to that of the hybrid-electric technology except that the high initial cost for a hybrid-electric drivetrain is avoided. For heavy-duty diesel engines, modeling and simulation shows that APA could provide 4 to 18% improvements in fuel economy over a wide range of driving cycles [1.23].

1.6.1.4 Mechanical (Flywheel) Hybrid

Another type of hybrid system stores the kinetic energy during the regenerative braking in a high-speed flywheel with a mechanism as shown in Figure 1.9(d). The flywheel hybrids do not require any energy conversion and yield efficiencies of about 70%. In general, flywheel systems have energy losses due mainly to bearing friction, which makes them less efficient than a battery-based system for storing energy for long periods.

In the 1950s, flywheel-powered buses, known as gyrobuses, were used in Yverdon, Switzerland [1.24], and there is ongoing research to make flywheel systems that are smaller, lighter, cheaper, and have a greater capacity. It is hoped that flywheel systems can replace conventional chemical batteries for mobile applications, such as for electric vehicles. Proposed flywheel systems would eliminate many of the disadvantages of existing battery power systems, such as low capacity, long charge times, heavy weight, and short usable lifetimes.

In 2009, Ricardo, Inc. developed a sealed high-speed flywheel to recover and store braking kinetic energy for a flywheel hybrid vehicle, called FLYBUS [1.25]. However, use of a large, high-speed flywheel for mobile applications presents major challenges due to potential imbalances and stresses introduced by mechanical vibrations.

1.6.2 Hybrid Architecture

Hybrid vehicles can be classified by their design architectures. The most common types of hybrid design architectures are parallel hybrid, series hybrid, and power-split hybrid.

1.6.2.1 Parallel Hybrid

Parallel hybrid systems, which are presently the most commonly produced, have both an internal combustion engine and a secondary power source, for example, an electrical motor connected to a mechanical transmission [1.26].

Most designs combine a large electrical generator and a motor into one unit. This unit is often located between the combustion engine and the transmission, as shown in Fig. 1.9(a). Li-Ion battery or other advanced energy storage systems are used to store the electric energy. Accessories such as power steering and air conditioning are powered by electric motors instead of being attached to the combustion engine. This allows efficiency gains, as the accessories can run at a constant speed, regardless of how fast the combustion engine is running. Parallel hybrids can be categorized by the way the two sources of power are mechanically coupled. If they are joined at some axis truly in parallel, the speeds at this axis must be identical and the supplied torques add together. When only one of the two sources is being used, the other must either also rotate in an idling manner or be connected by a one-way clutch. For commercial vehicles, the hybrid system usually provides supplement torques during acceleration, such as the hydraulic launch assist (HLA) system [1.27]. When only one of the two sources is being used, the other must still supply a large part of the torque or be fitted with a reverse one-way clutch or automatic clamp.

Parallel hybrids can be further categorized depending upon the portions of motive power provided by power sources. In some cases, the combustion engine is the dominant portion (the electric motor turns on only when a boost is needed) or vice versa. Others can run with just the electric system operating.

In a parallel electric hybrid, the electric motor and the internal combustion engine are installed so that they power the vehicle individually or together. Most commonly, the internal combustion engine, the electric motor, and gear box are coupled by automatically controlled clutches. For electric driving, the clutch between the internal combustion engine is open while the clutch to the gear box is engaged. While in combustion mode, the engine and motor run at the same speed.

In a parallel hybrid hydraulic system, the conventional vehicle powertrain is supplemented by the addition of the hybrid hydraulic system (Fig. 1.9(b)). The hybrid hydraulic system is best suited for vehicles that operate in stop-and-go duty cycles, including refuse trucks and buses, in which fuel economy improvements between 20 and 30% are typical.

1.6.2.2 *Series Hybrid*

A series electric hybrid uses electric motor(s), powered by an internal combustion engine, to propel the vehicle. While operating in its most efficient speed zone, the combustion engine drives an electric generator instead of directly driving the wheels. This engine can either charge an energy storage system such as a battery system, or directly power the electric motor or a combination of both. When large amounts of power and torque are required, the electric

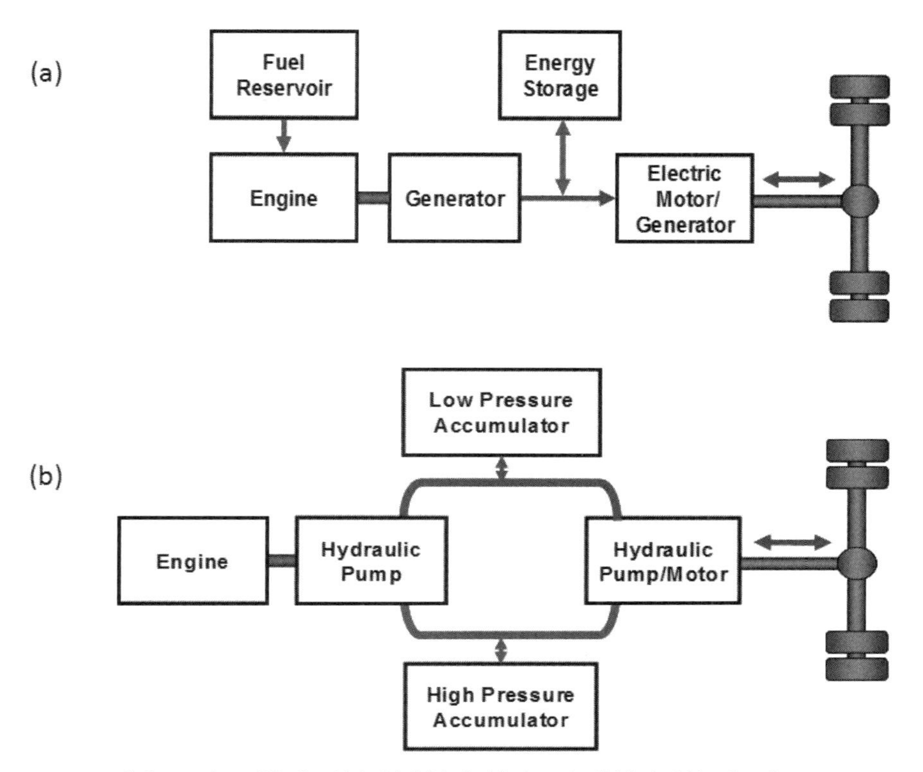

FIG. 1.10 Schematics of Series Hybrid: (a) hybrid electric, (b) hybrid hydraulic.

motor can draw electricity from a combination of batteries, capacitors, and the generator (Fig. 1.10(a)).

In a series hybrid hydraulic system, the conventional transmission and driveline are replaced by the hybrid hydraulic powertrain, as shown in Fig. 1.10(b). Energy is transferred from the engine to the drive wheels through hydraulic power.

The system is suited to a broader number of applications than parallel hydraulic hybrids, although benefits will be highest in vehicles that operate in stop-and-go duty cycles. The vehicle uses hydraulic pumps/motors and hydraulic storage tanks to recover and store energy, similar to what is done with electric motors and batteries in hybrid-electric vehicles.

A series hybrid hydraulic system developed by the U.S. EPA has demonstrated fuel economy improvements up to 70% and emission reduction up to 40% in selected driving cycles [1.28]. The fuel economy improvements can be achieved in three ways: (a) Braking energy, which otherwise is wasted, is recovered and reused; (b) The engine is operated more efficiently; and (c) The engine can be shut off when not needed, such as when stopped or decelerating.

1.6.2.3 *Power-split Hybrid*

Power-split hybrids combine the features of both series hybrid and parallel hybrid. They incorporate power-split devices, allowing for power paths from the engine to the wheels that can be either mechanical or electrical. The principle behind this system is the decoupling of the power supplied by the engine (or other primary source) from the power demanded by the driver. Figure 1.11(a) shows a schematic for a hybrid-electric power-split system, and Fig. 1.11(b) shows a hydraulic power-split system.

In a conventional vehicle, a larger engine is necessary to provide sufficient torque at lower engine speeds to accelerate the vehicle from a standstill. The larger engine, however, has more power than needed for steady speed cruising. An electric motor, on the other hand, exhibits maximum torque at standstill and is well-suited to complement the engine's torque deficiency at low engine speed.

In a power-split hybrid powertrain, two motors or pumps are needed in addition to the internal combustion engine. For a hybrid-electric powertrain, one motor mostly acts as a generator while the other one is used as a motor

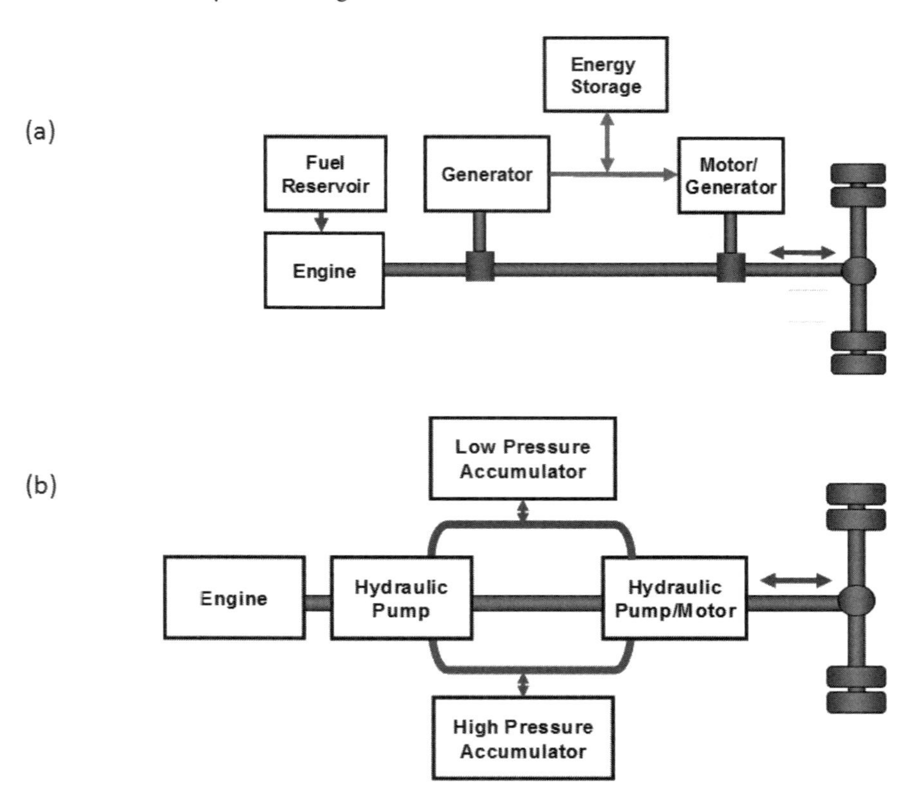

FIG. 1.11 Power-split hybrids: (a) electric and (b) hydraulic.

and generator. The two motors are connected through a planetary gear set and controlled by a powertrain controller [1.29].

1.6.2.4 Plug-in Hybrid-electric Vehicle (PHEV)

The plug-in hybrid-electric vehicle (PHEV) is another type of hybrid-electric vehicle with increased energy storage capacity. The PHEV can be either parallel or series hybrid depending on applications. The energy storage system of a PHEV can be charged by connecting to the electricity supply or using the onboard internal combustion engine. The main advantages of the PHEV include a) reducing on-road emissions by minimizing the use of the ICE during driving; b) reducing operating cost as long as the electric energy being used is cheaper than the corresponding amount of gasoline or diesel fuel; c) providing a quiet working environment for a utility service truck.

For a plug-in hybrid utility service truck, the powertrain system can be designed to provide power for use during engine-off worksite operations, such as running the bucket, power tools, lights, and accessories.

1.6.2.5 Range Extender Hybrid Vehicle

The range extender hybrid vehicle is another type of plug-in hybrid vehicle with series hybrid architecture. A typical range extender is an electric vehicle with an onboard charging unit, which includes a generator and an internal combustion engine or a fuel cell unit. In addition to charging with an external electric supply, the energy storage system can charge during operation by onboard charging. The range extender vehicle typically has a smaller energy storage capacity than comparable pure electric vehicles, and has a smaller IC engine for battery charging than a typical hybrid vehicle.

The advantages of range extender commercial vehicles include: a) cost reduction by reducing requirement of energy storage capacity; b) charging time reduction; and c) weight and size reduction of energy storage systems.

1.6.2.6 In-wheel Motors

In-wheel motor hybrid is a type of series hybrid. Instead of having one large motor to drive the wheel, the in-wheel motor hybrid has two or more motors designed inside the wheels. An in-wheel motor hybrid bus includes: a diesel generator charges a battery, which in turn supplies electricity for two motors, one in each rear wheel.

Hybrid buses with in-wheel motors are also in commercial operation [1.30]. The in-wheel motors confer additional savings by eliminating the need for a transmission, differential, and related mechanical parts. This reduces both the overall weight of the bus and energy losses due to friction. In-wheel hybrid buses typically see fuel-consumption reductions of up to 50% compared with conventional buses. In certain conditions—at low speeds in frequent stop-and-go traffic—some other hybrid buses have seen similar fuel-economy improvements. The in-wheel motors can also improve traction by allowing precise control over each wheel, and they allow for greater flexibility in vehicle design since there is no need to mechanically link the wheels to an engine.

1.6.3 Hybrid Classification by Function

A hybrid vehicle can also be classified based on its function, or the power/torque ratio to the IC engine. It can be classified as mild hybrid, power assist, or full hybrid.

1.6.3.1 Mild Hybrid

Mild hybrids are essentially conventional IC engines with oversized starter motors to provide assistance in engine starting and vehicle acceleration. They are used mostly in passenger cars and light-duty commercial vehicles.

Mild hybrids do not normally allow regenerative braking or the use of smaller, lighter, more-efficient internal combustion engines. Therefore, their fuel savings would generally be lower than expected with use of a full hybrid design. However, mild hybrids may provide some of the benefits of the application of hybrid technologies, with less of the cost-weight penalty that is incurred by installing the "full" hybrid powertrain.

1.6.3.2 Power Assist Hybrid

Power assist hybrids have an engine for primary power and an electric or a hydraulic motor that is connected to a largely conventional powertrain to provide power assist as needed. For example, the electric motor, mounted between the engine and transmission, will operate when the vehicle requires extra power to accelerate or the combustion engine needed to re-start after the shutdown at traffic stops.

The Eaton Hydraulic Launch Assist (HLA) is another example of power assist hybrid [1.27]. During braking, the vehicle's kinetic energy drives the pump/motor as a pump, transferring hydraulic fluid from the low-pressure reservoir to the high-pressure accumulator. The fluid compresses nitrogen gas in the accumulator and pressurizes the system. During acceleration, fluid in the high-pressure accumulator is metered out to drive the pump/motor as a motor. The system propels the vehicle by transmitting torque to the driveshaft.

1.6.3.3 *Full Hybrid*

A full hybrid is a vehicle that can be propelled by the power from the engine, the energy storage system, or a combination of both. The Allison Hybrid System is an example, which can move the vehicle forward on battery power alone [1.29]. A large, high-capacity battery pack is needed for battery-only operation. These vehicles have a split power path that allows more flexibility in the drivetrain by interconverting mechanical and electrical power, at some cost in complexity. To balance the forces from each portion, the vehicles use a differential-style linkage between the engine and motor connected to the head end of the transmission.

Full hybrid vehicles sometimes can use smaller IC engines than equivalent conventional vehicles, and allow the engine to operate at their highest efficiency range to improve the vehicle fuel efficiency.

1.6.4 Fuel Cell Hybrid Vehicle

Instead of the IC engine as the primary power source for the vehicles, fuel cells can also be used as primary power sources.

General Motors claims to have produced the first FCV in 1966 [1.31]. The fuel cells, powered by hydrogen and oxygen, supplied a continuous output of about 43 hp (32 kW) and peak output of 215 hp (160 kW). The fuel cell vehicle had a top speed of 70 mph (113 km/h), but it took 30 seconds to accelerate to 60 mph (97 km/h) and has a driving range about 150 miles (241 km). In Europe, Mercedes-Benz Buses has developed a fuel cell system for hybrid bus applications with lithium-ion batteries [1.32].

Although the fuel cell technology is undergoing rapid advances, it is still expensive, and the almost total lack of a refueling infrastructure means that FCVs have so far not made significant penetration into the market. However, long-term trials of fuel cell buses have been conducted for several years [1.33].

1.6.5 Hybrid Vehicles by Application

Commercial hybrid vehicles can be classified by application. Figure 1.12 shows a wide range of applications of commercial hybrid vehicles.

According to hybrid truck industry estimates, the production of medium- and heavy-duty hybrid commercial trucks is expected to reach 20,000 units in 2012, especially due to the growth in China. By 2020, if the market produces 60,000 hybrid vehicles as projected, these would translate into an estimated 52 to 130 million gallons of fuel saved and 0.6 to 1.4 million tons of CO_2 reduced per year [1.35, 1.36].

Description	Example	Status
Electric Hybrid Delivery Truck		Hybrid excavator is powered by a combination of an electric motor and diesel engine. The batteries are recharged every time the excavator swings to the side to move a scoopful of earth or debris into a waiting dump truck. The midsize excavator consumes 30% less fuel than a comparable size driven only by a diesel engine. (Photo taken at HTUF 2011 in Baltimore, MD)
Electric Hybrid Transit Bus		A hybrid diesel-electric bus, which starts and stops from block to block, can cut fuel use by up to 30%. There are over 900 hybrid buses operated in the city of Beijing and Guangzhou, China with Eaton's hybrid powertrain. The buses have dramatically lower maintenance costs. (Photo taken in Guangzhou, China, 2008)
Plug-in and Range Extender Electric Hybrid Vehicle		The Plug-in hybrid vehicle (PHEV) is usually a (parallel or serial) hybrid with increased battery capacity. Its battery may be charged by plugging-in to the outlet of electricity supply at the end of the journey or to be charged using the on-board power unit, such as an internal combustion engine. Range Extender hybrid is usually a series PHEV. (Photo taken at HTUF 2011 in Baltimore, MD)
Heavy Duty Electric Hybrid Truck		Five Peterbilt Model 386 heavy duty hybrid trucks with diesel-electric hybrid power systems developed by Eaton Corporation and PACCAR, that will be based in Dallas, Houston, Apple Valley, Calif., Atlanta and the Washington/ Baltimore regions (Photo taken at HTUF 2011 in Baltimore, MD)
Electric Hybrid Utility Trouble Truck		Utility trouble truck delivers a number of benefits, including: Up to 60 percent reduction in fuel consumption; Up to 87 percent reduction in idle times, reduced maintenance and lower life cycle costs; and quieter operations and better acceleration. (Photo taken at HTUF 2011 in Baltimore, MD)

FIG. 1.12 Example applications of hybrid commercial vehicles.

Description	Example	Status
Hydraulic Hybrid Delivery Vans		Series hydraulic hybrid delivery vehicle has demonstrated fuel economy improvements up to 50 percent. (Photo taken at HTUF 2011 in Baltimore, MD)
Hydraulic Hybrid Refuse Collector Vehicle		The parallel hydraulic launch Assist (HLA) system provides supplement torque to the conventional vehicle powertrain. The hybrid hydraulic system is best suited for vehicles that operate in stop-and-go duty cycles, including refuse and buses, with fuel economy improvements between 20 and 30 percent. (Photo taken at HTUF 2011 in Baltimore, MD)
Super-Cap Hybrid Vehicle		Super-Cap hybrid uses super-capacitor/ultra-capacitor as energy storage device. Super-capacitor allows quick change and discharge of electric energy. The Super-Cab hybrid vehicle has demonstrated over 40% fuel saving and lower cost than hybrid vehicle with Li-Ion battery. (Photo taken at HTUF 2011 in Baltimore, MD)
Fuel Cell Hybrid Vehicle		The fuel cell hybrid has the both fuel cell and battery as the power sources. When the battery is depleted, fuel cell will generate electricity to charge the battery. Proton exchange membrane (PEM) based fuel cell uses hydrogen as fuel. Solid Oxide Fuel Cell (SOFC) can use diesel reformate as fuel. (Photo taken at HTUF 2011 in Baltimore, MD)

FIG. 1. 12 *(Continued)*

1.7 References

1.1 "2010 Production Statistics," http://oica.net/category/production-statistics/, assessed, Oct. 2011.

1.2 "Freight Data and Statistics," The Research and Innovative Technology Administration, U.S. Department of Transportation, http://www.bts.gov/programs/freight_transportation/, assessed, Oct, 2011.

1.3 Annual Energy Outlook 2011 with projections to 2035, U.S. Energy Information Administration, DOE/EIA-0383(2011), April 2011.

1.4 Chandler, K. et al., "New York City Transit Diesel Hybrid-Electric Buses: Final Results," DOE/NREL Transit Bus Evaluation Project, NREL/BR-540-32427, 2002.

1.5 Xu, Changming, "China Automotive Market Analysis, 2010," China State Information Center. http://wenku.baidu.com/view/6f630642be1e650e52ea9966.html, assessed, March 7, 2011.

1.6 "Commercial Hybrid Vehicles A Global Analysis of Medium- and Heavy-Duty Hybrid Trucks and Buses," Allied Business Intelligence, Inc, 2009.

1.7 Lowe, M., Ayee, G., and Gereffi, G., "Hybrid Drivetrains for Medium- and Heavy-Duty Trucks," http://www.calstart.org, 2009.

1.8 Manwaring, L., "The Observer's Book of Automobiles," (12th ed.) 1966, Library of Congress catalog card # 62-9807, p. 7.

1.9 Porter, D. H., "The Life and Times of Sir Goldsworthy Gurney, Gentleman Scientist and Inventor," pp. 1793–1875, Lehigh University Press, 1988.

1.10 "Hybrid Vehicle History More than a Century of Evolution and Refinement," http://www.hybrid-vehicle.org/hybrid-vehicle-history.html, assessed, March 7, 2011.

1.11 "History of Hybrids An overview of the hybrid's past," http://www.pages.drexel.edu/~vld24/history.html, assessed, March 7, 2011.

1.12 "Veto of the Electric and Hybrid Vehicle Research, Development and Demonstration Bill," http://www.presidency.ucsb.edu/ws/index.php?pid=6329#axzz1UHimdtLD, assessed, August 4, 2011.

1.13 "Technology Roadmap For the 21st Century Truck Program—A Government-Industry Research Partnership," http://ntl.bts.gov/lib/14000/14800/14854/DE2001777307.pdf, Assessed, Sept. 2011.

1.14 "Truck Classification," http://changingears.com/rv-sec-tow-vehicles-classes.shtml, assessed, March 7, 2011.

1.15 Bosch, *Automotive Handbook*, Bentley Publishers, 7th edition.

1.16 China Commercial Vehicle Market, 2006, KPMG, http://www.kpmg.com.cn/en/virtual_library/Industrial_markets/China_com_vehicle_mkt.pdf, assessed, Sept. 2011.

1.17 Davis, S. et al., *Transportation Energy Data Book*, US DOE EERE, 28th Edition, 2009.

1.18 "Inventory of U.S. Greenhouse Gas Emissions and Sinks: 1990–2007," http://www.epa.gov/climatechange/emissions/downloads09/GHG2007entire_report-508.pdf, assessed Sept. 2011.

1.19 "EPA and NHTSA to Propose Greenhouse Gas and Fuel Efficiency Standards for Heavy-Duty Trucks; Begin Process for Further Light-Duty Standards: Regulatory Announcement," http://www.epa.gov/oms/climate/regulations/420f10038.htm, assessed, March 7, 2011.

1.20 "CAFE-GHG Final Rule," http://www.nhtsa.gov/staticfiles/rulemaking/pdf/cafe/CAFE-GHG_MY_2012-2016_Final_Rule_FR.pdf, assessed Oct. 2011.

1.21 "Clean Air Nonroad Diesel—Tier 4 Final Rule," US EPA, http://www.epa.gov/nonroaddiesel/2004fr.htm, assessed Oct. 2011.

1.22 "Emission Standards: http://www.dieselnet.com/standards/, assessed, Oct. 2011.

1.23 Lee, C-Y et al., "Analysis of a Cost Effective Air Hybrid Concept," SAE Paper No. 2009-01-1111, SAE International, Warrendale, PA, 2009.

1.24 "Gyrobus: a great idea takes a spin," http://photo.proaktiva.eu/digest/2008_gyrobus.html, assessed, August 5, 2011.

1.25 "Ricardo Kinergy delivers breakthrough technology for effective, ultra-efficient and low cost hybridization," http://www.ricardo.com/News--Media/Press-releases/News-releases1/2009/Ricardo-Kinergy-delivers-breakthrough-technology-for-effective-ultra-efficient-and-low-cost-hybridisation/, assessed, August 5, 2011.

1.26 Hu, H. et al., "On-board Measurements of City Buses with Hybrid Electric Powertrain, Conventional Diesel and LPG Engines," SAE Paper No. 2009-01-2719, SAE International, Warrendale, PA, 2009.

1.27 "Hybrid Power Systems, Refuse Truck," http://www.eaton.com/ecm/groups/public/@pub/@eaton/@roadranger/documents/content/ct_168261.pdf, assessed, August 11, 2011.

1.28 "Hydraulic Hybrid Research," http://www.epa.gov/oms/technology/research/research-hhvs.htm, assessed, August 5, 2011.

1.29 "Allison Hybrid Bus," http://www.allisontransmission.com/commercial/transmissions/hybrid-bus/, assessed August 6, 2011.

1.30 "Wheel Motors to Drive Dutch Buses," http://www.technologyreview.com/energy/22328/, assessed August 8, 2011.

1.31 "1966 GM Electrovan," http://www.hydrogencarsnow.com/gm-electrovan.htm, assessed August 8, 2011.

1.32 "Mercedes-Benz Citaro FuelCELL-Hybrid bus previews in Hamburg," http://www.gizmag.com/mercedes-benz-citaro-fuelcell-hybrid-bus/13390/, assessed August 8, 2011.

1.33 Eudy, L. et al., "Fuel Cell Buses in U.S. Transit Fleets: Current Status 2010," NREL/TP-5600-49379. Golden, CO: National Renewable Energy Laboratory, 2009.

1.34 "Review of the 21st Century Truck Partnership," http://www.nap.edu/catalog/12258.html, 2008. assessed, Sept. 2011.

1.35 "Commercial Hybrid Vehicles A Global Analysis of Medium- and Heavy-Duty Hybrid Trucks and Buses," Allied Business Intelligence, Inc., 2009.

1.36 Lowe, M., Ayee, G., and Gereffi, G. "Hybrid Drivetrains for Medium- and Heavy-Duty Trucks," http://www.calstart.org, 2009.

Chapter 2

Internal Combustion Engines for Commercial Vehicles

2.1 Requirements of Internal Combustion Engines for Commercial Vehicles

Internal combustion engines are the most frequently used prime movers for commercial vehicles since the early 1900s. However, internal combustion engines for the modern commercial vehicle markets must meet the following requirements: a) size and weight, b) performance, c) reliability and durability, d) government regulations [2.9, 2.10].

2.1.1 Size and Weight

Engine size is important to the commercial vehicle market as is engine configuration. The inline six-cylinder engines are the most popular configurations for on-highway long-haul trucks, and some V-type engines are used in off-road vehicles with higher power requirements. The bottom of the engine must be narrow enough to fit between the vehicle frame rails. The weight of the engine can directly impact the payload of the commercial vehicles.

2.1.2 Performance

2.1.2.1 Fuel Economy

The vehicle fuel economy requirement is one of the most important attributes of the internal combustion engine. For light-duty commercial vehicles, the fuel economy is regulated in the United States. In 2010, the U.S. Environmental Protection Agency (EPA) and the Department of Transportation's National Highway Traffic Safety Administration (NHTSA) proposed regulatory action to reduce greenhouse gas (GHG) emissions and improve fuel efficiency of medium- and heavy-duty vehicles for 2017 and later model years [2.1]. The fuel economy of internal combustion engines will also directly impact the vehicle operating economics. The engine manufacturers must contribute to the vehicle's fuel economy by providing an engine with the best brake specific fuel consumption over the entire operating range.

2.1.2.2 *Power*

The power required to operate an 80,000-lb tractor trailer on level terrain at 65 mph (105 km/h) is less than 230 hp. Power levels from 350 to 450 hp (261 to 336 kW) are required to pass and handle on rolling terrain and to provide comfort features, such as air conditioning, to the drivers. Applications with higher weight will require 500 hp (373 kW) or more. Most of the light-duty commercial vehicles use both spark ignition engines (gasoline engines) and compression ignition engines (diesel engines). For heavy-duty commercial vehicles, diesel engines are the dominant prime movers.

2.1.2.3 *Torque*

The engine and transmission must be matched to allow an adequate launch of the vehicle from a stop, allow acceleration through the gears to get the vehicle "up to speed," and allow the vehicle to be able to stay in top gear while negotiating mile grades. This last feature is enhanced by an engine that has high torque rise, i.e., increasing torque as speed decreases. High torque and high torque rise give the driver the feeling of "pulling power" and "good driveability."

2.1.3 Reliability and Durability

Reliability can be measured by the engine performance for carrying on the tasks as specified. Durability is the useful life of the internal combustion engine. For on-road commercial vehicles with internal combustion engines in the North American market, the durability requirement for the engine and the vehicle are expected to be 1 million miles.

2.1.4 Meeting Government Emissions Regulations

Since its formation in 1971, the Environmental Protection Agency (EPA) has been the primary federal agency responsible for regulating engine emissions in the United States. Engine exhaust components such as high hydrocarbon (HC), nitrogen oxides (NO_x), and particulate matter (PM) are regulated under specific test cycles. Currently, on-road commercial vehicles and off-road commercial vehicles have different emission standards. The United States Department of Transportation (U.S. DOT) also has safety regulations for on-road commercial vehicles that regulate parameters such as engine noise.

2.2 Basics of Internal Combustion Engines

The purpose of the internal combustion engine is to convert chemical energy in the fuel to mechanical power. In the internal combustion engine, the chemical energy of fuel is oxidized in a combustion chamber to generate high-temperature and high-pressure working fluid. The expansion of the working fluid directly applies force to a movable piston of the engine to generate useful mechanical energy.

There are many types of engine classifications, including reciprocating or rotary, spark ignition or compression ignition, and two-stroke or four-stroke; internal combustion and external combustion. Because the compression ignition (diesel) engines are the dominant prime movers for commercial vehicles, the diesel engines will be the subjects of discussion.

2.2.1 History of IC Engines for Commercial Vehicles

The first person to experiment with an internal combustion engine was the Dutch physicist Christian Huygens, about 1680. In 1862 Alphonse Beau de Rochas, a French scientist patented but did not build a four-stroke spark ignition engine; sixteen years later, when Nikolaus A. Otto built a successful four-stroke spark ignition engine, it became known as the "Otto cycle" [2.2, 2.3].

The reciprocating compression ignition engine was invented and patented by Dr. Rudolf Diesel in 1892. Today, the reciprocating compression ignition engine is commonly referred to as a diesel engine. The main operating difference between diesel and gasoline engines is that the diesel engine uses the heat produced by compression rather than the spark from a plug to ignite a mixture of air and fuel. Diesel engines are heavier than gasoline engines because of the extra strength required to contain the higher temperatures and compression ratios. For equal displacement volumes, diesel engines have higher torque than gasoline engines.

Commercial vehicles are major contributors of air pollution. Significant technology advancements have been made in the past few decades to limit vehicle exhaust emissions while improving engine fuel economy and driveability. Emission reduction technologies, such as aftertreatment catalysts, exhaust gas recirculation, and electronic controlled high-pressure fuel injection technologies are developed to achieve significant exhaust emission reduction and fuel economy improvement of internal combustion engines.

2.2.2 Basic Operations of Internal Combustion Engines

The internal combustion engines used in most commercial vehicles are four-stroke engines. Both gasoline engines and diesel engines are used in light and medium commercial vehicles. However, turbocharged diesel engines are used in most heavy-duty commercial vehicles.

Figure 2.1 shows the operating cycle of a four-stroke diesel engine. A single cycle of operation (intake, compression, power, and exhaust) takes place over four strokes of a piston, made in two engine revolutions.

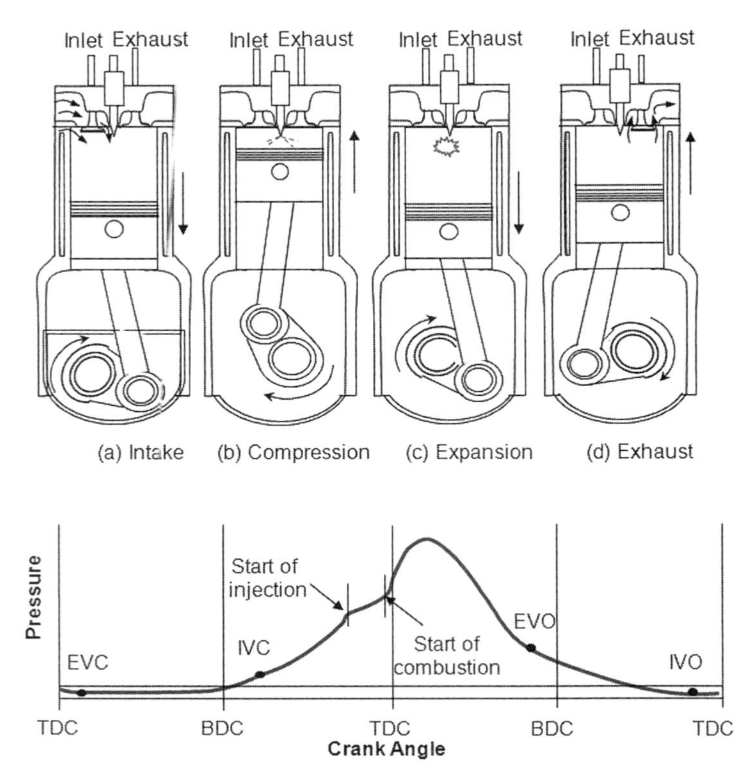

FIG. 2.1 Operating cycle of a diesel engine [2.2] and pressure-crank angle diagram for a four-stroke diesel engine [2.3].

2.2.2.1 *Intake*

The intake stroke starts when the piston is at the top of the cylinder (top dead center, TDC) and the intake valve opens. The descending piston moves downward to draw in the air and a small amount of exhaust gas (EGR) into the cylinder. When the piston reaches bottom dead center (BDC), the cylinder capacity is at its largest.

2.2.2.2 *Compression*

The piston moves upward and compresses the air while the inlet and exhaust valves are closed. For a typical on-highway heavy-duty diesel engine, the compression ratio is in the range of 14:1 to 18:1. At the end of the compression process, the air trapped in the cylinder is heated up to as high as 1652°F (900°C). Near the top dead center, the fuel injection system injects pressurized fuel into the hot, compressed air. The injected fuel will quickly mix with hot air and atomize in the cylinder.

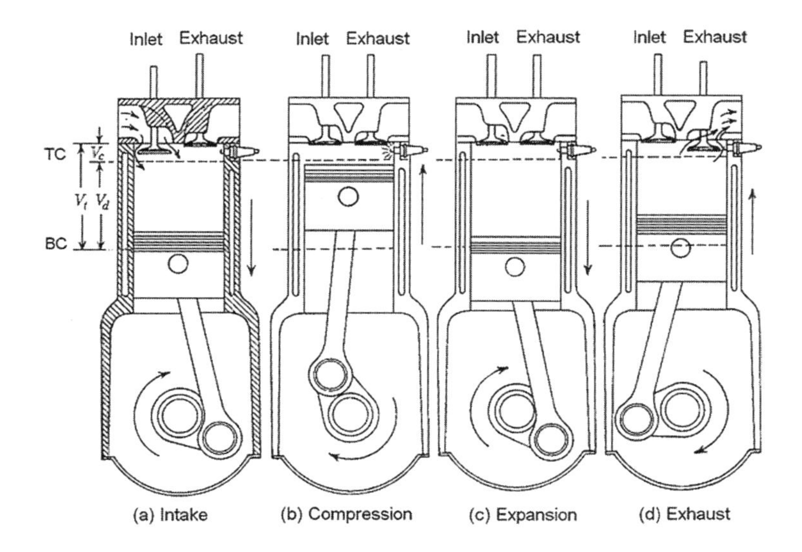

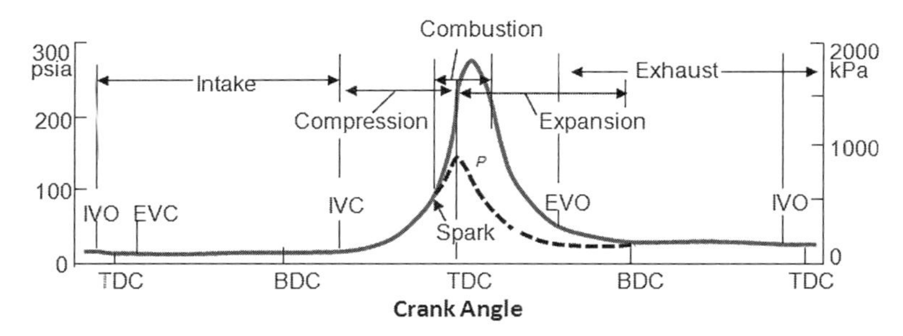

FIG. 2.2 Operating cycle of a four-stroke spark ignition engine [2.2].

2.2.2.3 *Power*

The fuel-air mixture will ignite after a short ignition delay that lapses a few degrees of crankshaft rotation, and the power stroke begins. The amount of energy released by the combustion is determined by the mass of the fuel injected and generates high pressure in the cylinder. The high pressure forces the piston to move downward. The chemical energy of diesel fuel released by combustion is thus converted into kinetic energy of the piston and crankshaft.

2.2.2.4 *Exhaust*

Before the piston reaches the BDC, the exhaust valve opens to release the pressurized gases out of the cylinder. A portion of the exhaust gas will be recirculated back to the intake manifold for emission control purposes. On completion of the exhaust stroke, the crankshaft has completed two revolutions, and the four-stroke operating cycle starts again with the intake stroke.

Figure 2.2 shows the operating cycle of a four-stroke spark ignition engine. There are major differences between spark ignition (i.e., gasoline) engines and compression ignition (i.e., diesel) engines. In the intake stoke, spark ignition engines intake premixed fuel and air into the cylinder before the compression stroke starts. Near the end of the compression stroke, the mixture is ignited by a spark plug near TDC instead of self-igniting, as with compression ignition engines.

Spark ignition (SI) engines are characterized by the spark ignition of homogeneous or a mostly well-mixed charge of fuel and air. However, diesel engines achieve their high performance and excellent fuel economy by compressing air to high pressures, then injecting a small amount of fuel into the highly compressed air. The injected fuel will evaporate and reach its auto-ignition temperature to ignite. The auto-ignition temperature of fuel depends on its chemistry. Unlike the SI system, combustion in compression-ignited engines occurs at many points where the air/fuel ratio and temperature can sustain this process. For a direct injection (DI) diesel engine, it has higher specific torque output mainly contributed by controlling the combustion rate through an advanced fuel injection system. Diesel engines also have higher thermal efficiency than gasoline engines due to their higher compression ratios.

2.2.3 Otto and Diesel Cycles

The Otto cycle is a theoretic thermodynamic cycle for spark ignition internal combustion engines. The German engineer Nicolaus Otto was the first person to develop a functioning four-stroke spark ignition engine in 1876; hence the thermodynamic cycle is known as the Otto cycle. The Otto cycle consists of two isentropic (reversible adiabatic) phases interspersed between two constant-volume phases.

The thermodynamic working fluid in the cycle is subjected to isentropic (adiabatic and reversible) compression. The ideal Otto (Fig. 2.3) includes the following processes:

1. Isentropic compression
2. Constant-volume heat addition
3. Isentropic expansion
4. Constant-volume heat rejection

The Otto cycle is represented in many millions of spark ignition engines using either the four-stroke principle or the two-stroke principle.

The thermal efficiency of the Otto cycle can be expressed as

$$\eta = \frac{work}{heat\ input} = \frac{Q_H - Q_L}{Q_H} = 1 - \frac{Q_L}{Q_H} \tag{2.1}$$

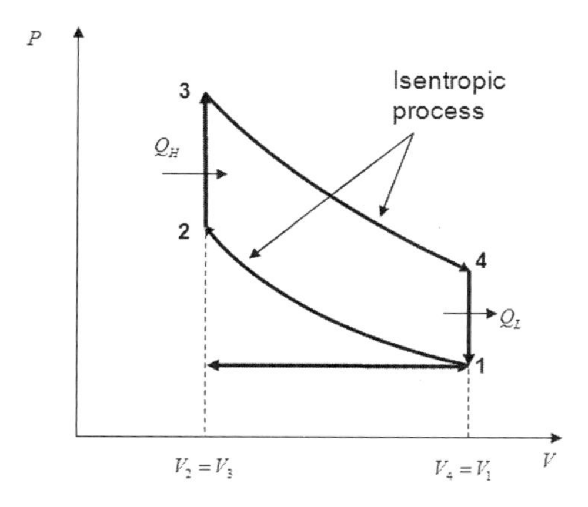

FIG. 2.3 Otto cycle.

Assuming the heat added occurs during combustion at constant volume, the heat added can be related to the temperature change from state 2 to state 3 as

$$Q_H = Q_{2\text{-}3} = \Delta U_{2\text{-}3} \quad (W_{2\text{-}3} = 0) \tag{2.2}$$

$$Q_H = \int_{T_2}^{T_3} C_v dT = C_v(T_3 - T_2) \tag{2.3}$$

The heat rejected is given by (for a perfect gas with constant specific heats)

$$Q_L = Q_{4\text{-}1} = \Delta U_{4\text{-}1} = C_v(T_4 - T_1) \tag{2.4}$$

Substituting Eq. (2.3) and Eq. (2.4) into Eq. (2.1), the thermal efficiency of the Otto cycle can be expressed as

$$\eta_{otto} = 1 - \frac{T_4 - T_1}{T_3 - T_2} \tag{2.5}$$

$$\eta_{otto} = 1 - \frac{1}{(V_1/V_2)^{k-1}} = 1 - \frac{1}{r^{k-1}} \tag{2.6}$$

where $r = V_1/V_2$ is called the compression ratio; and $k = C_p/C_v$ is the ratio of specific heats.

The diesel cycle is the thermodynamic cycle that approximates the pressure and volume of the combustion chamber of the compression ignition engine or diesel engine, invented by Rudolph Diesel in 1897. It is assumed to have constant pressure during the first part of the "combustion" phase 2–3 in the diagram shown in Fig. 2.4. The ideal diesel cycle consists of the following operations:

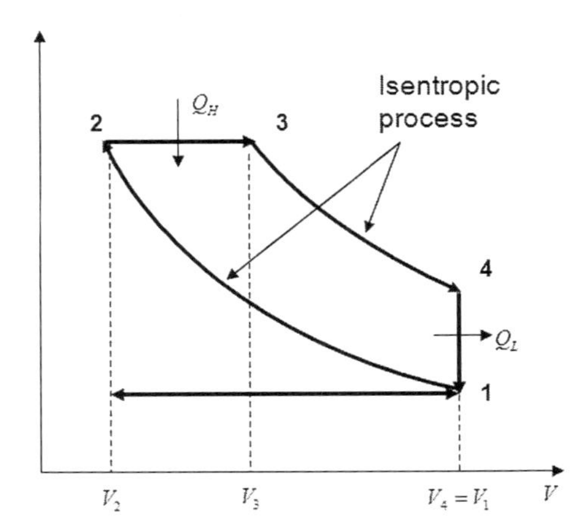

FIG. 2.4 Diesel cycle.

1. Isentropic compression
2. Reversible constant pressure heat addition
3. Isentropic expansion
4. Reversible constant volume rejection

The difference between the diesel cycle and the Otto cycle is that the diesel cycle has the heat addition at constant pressure and the heat rejection at constant volume, and the Otto cycle by comparison has both the heat addition and rejection at constant volume.

The thermal efficiency of the diesel cycle is

$$\eta_{diesel} = 1 - \frac{Q_L}{Q_H} = 1 - \frac{C_v(T_4 - T_1)}{C_p(T_3 - T_2)} \tag{2.7}$$

$$\eta_{diesel} = 1 - \frac{T_1}{kT_2}\frac{(T_4/T_1 - 1)}{(T_3/T_2 - 1)} = 1 - \frac{1}{r^{k-1}}\left(\frac{\beta^k - 1}{k(\beta - 1)}\right) \tag{2.8}$$

where $\beta = V_3/V_2$ is the cut-off ratio, the ratio between the end and start volume for the combustion phase.

2.2.4 Atkinson and Miller Cycles

The Atkinson-cycle engine is a type of internal combustion engine invented by James Atkinson in 1882. The Atkinson cycle has a larger expansion ratio than compression ratio to gain better efficiency. The goal of the modern Atkinson cycle is to allow the pressure in the combustion chamber at the end of the power stroke to be equal to atmospheric pressure to gain all the available energy from the combustion process. The operating processes of a typical Atkinson cycle (Fig. 2.5) are:

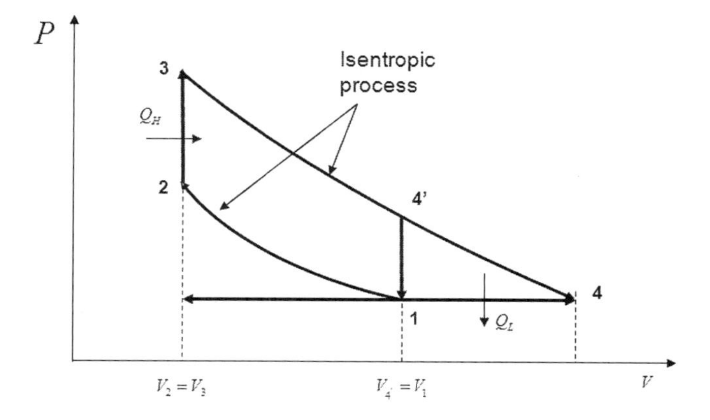

FIG. 2.5 Atkinson cycle.

1. Isentropic compression
2. Constant volume heat addition
3. Isentropic expansion
4. Constant pressure rejection

Recently, a modified Atkinson cycle has been used for both diesel- and gasoline-powered hybrid vehicles to improve vehicle efficiency: for example, a modified direct fuel injection Otto-cycle engine in which the intake valve is held open longer than normal to allow a reverse flow of intake air into the intake manifold. The effective compression ratio is reduced (for a time the air is escaping the cylinder freely rather than being compressed) but the expansion ratio is unchanged. A modified Atkinson cycle, shown in Fig. 2.6, consists of the following operations:

1. Isentropic compression
2. Constant volume heat addition
3. Constant pressure heat addition
4. Isentropic expansion
5. Constant volume heat rejection
6. Constant pressure heat rejection

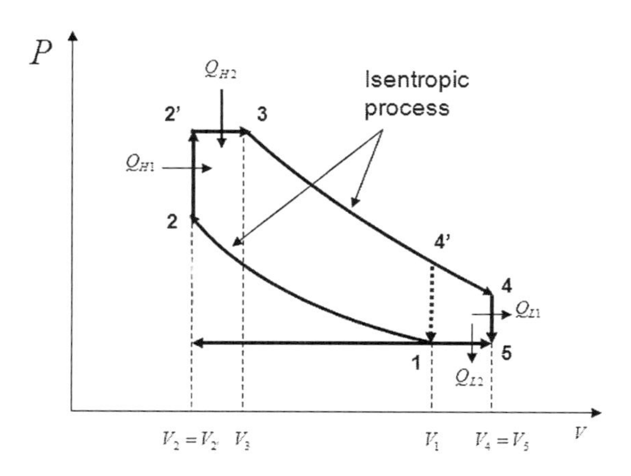

FIG. 2.6 Modified Atkinson cycle.

The thermal efficiency of the modified Atkinson cycle can be expressed as

$$\eta_{Atkinson} = 1 - \frac{Q_{L1} + Q_{L2}}{Q_{H1} + Q_{H2}} = 1 - \frac{C_v(T_4 - T_5) + C_p(T_5 - T_1)}{C_v(T_{2'} - T_2) + C_p(T_3 - T_{2'})} \tag{2.9}$$

$$\eta_{Atkinson} = 1 - \frac{\frac{1}{k} T_1 \left[\left(\frac{T_2}{T_1} \right) \gamma_p \beta \left(\frac{V_3}{V_4} \right)^{k-1} - A \right] + T_1(A - 1)}{\frac{1}{k} T_2(\gamma_p - 1) + T_2 \gamma_p (\beta - 1)} \tag{2.10}$$

$$\eta_{Atkinson} = 1 - \frac{A(\gamma_p \beta^{\gamma} A^{-\gamma} - 1) + k(A - 1)}{\gamma_c^{k-1} \left[(\gamma_p - 1) + k\gamma_p(\beta - 1) \right]} \tag{2.11}$$

where $\gamma_c = V_1/V_2$ is the compression ratio; $\gamma_p = T_{2'}/T_2 = P_{2'}/P_2$ is the pressure ratio; $\gamma_e = V_4/V_2$ is the expansion ratio; $\beta = V_3/V_2 = T_3/T_{2'}$ is the cutoff ratio; $A = V_4/V_1 = \gamma_e/\gamma_c = T_5/T_1$ is the Atkinson ratio; and $k = C_p/C_v$ is the specific heat ratio.

The disadvantage of a four-stroke Atkinson-cycle engine is the reduction of power density. Because the engine power is directly proportional to the amount of air in the engine cylinder, a smaller portion of the compression stroke will have adverse impact on the power density of an Atkinson-cycle engine. However, the loss of air mass can be compensated through supercharging to increase the air density. The process of improving the power density with a supercharger is known as the Miller cycle.

The Miller cycle was patented by Ralph Miller, an American engineer, in the 1940s. A key aspect of the Miller cycle is that the intake air is first compressed by the supercharger. In the Miller cycle, the intake valve is left open longer than it would be in an Otto-cycle engine. The compression stroke actually starts only after the piston has pushed out this "extra" charge and the intake valve closes. This happens at around 20 to 30% into the compression stroke. The loss of charge air is compensated for by the use of a supercharger. This lower intake charge temperature, combined with the lower compression of the intake stroke, yields a lower final charge temperature than would be obtained by simply increasing the compression of the piston. The advantage of the lower final charge temperature is the lower combustion adiabatic flame temperature for lower NO_x formation during the combustion process.

2.3 Major Engine Components and Subsystems

Basic components of a turbocharged diesel engine are illustrated in Fig. 2.7. The functions of these components are reviewed as follows:

The piston seals the cylinder and transmits the combustion-generated gas pressure to the crankshaft via the connecting rod, which is fastened to the piston. There are multiple cylinders for larger engines, and the engine cylinders are contained in the cylinder block. The bottom of the cylinder block acts as an oil reservoir for the lubrication system. The crankshaft is supported in main bearings and connected to the pistons by the connecting rods. A flywheel is usually attached to one end of the crankshaft to smooth the output engine torque.

The valve actuation system consists of a camshaft, cam, and valves. In four-stroke engines, camshafts turn at one-half the crank speed. Mechanical or hydraulic lifters slide in the cylinder block and ride on the cam. Advanced variable valve actuation mechanisms have been developed for large commercial engines to enable internal exhaust gas recirculation for emissions reduction, and to improve the fuel economy and retarding performance [2.21]. The valve stem moves in the valve guide, which can be an integral part of the cylinder head. Intake and exhaust manifolds, a turbocharger, which consists of a turbine and a compressor, and an exhaust gas recirculation mechanism form the air supply and exhaust systems for the engine.

Other major components such as the fuel injection system, turbocharger, and emission reduction system will be discussed more in the following sections.

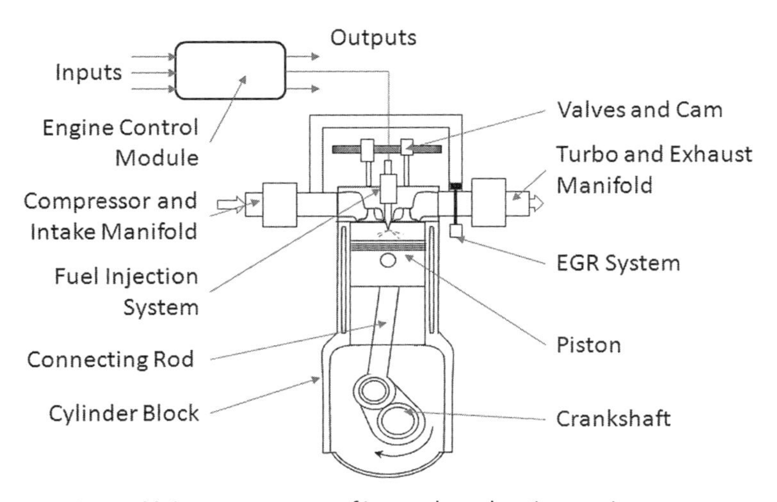

FIG. 2.7 Major components of internal combustion engines.

2.3.1 Diesel Fuel Injection System

The fuel injection system plays a key role in fuel consumption and emissions of the engine. Any fuel injection system must meet the following requirements for modern commercial vehicles [2.6].

1. Fuel injection must be precisely timed. Even small discrepancies have a substantial effect on fuel consumption, emission levels, and combustion noise.
2. It should be possible to vary the injection pressure as independently as possible to suit the demands of all engine operating conditions.
3. The injection must be reliably terminated. Uncontrolled "post injection" leads to higher emissions levels.

There are four major fuel injection systems for commercial vehicles: a) electronically controlled distributor pump system; b) electronically controlled unit pump system; c) electronically controlled unit injector; and d) high-pressure common-rail system.

The main components of an electronically controlled distributor pump consist of a pump electronic control unit (ECU), a vane-type supply pump, a high-pressure solenoid valve, a timing device, and a timing-device solenoid valve. The pump ECU controls the high-pressure solenoid valve. The start of delivery begins after the high-pressure solenoid valve has closed. High fuel pressure builds up in the high-pressure delivery lines. This nozzle-side line pressure opens the nozzle needle upon reaching the nozzle-opening pressure, and injection starts.

The unit pump system is a modular high-speed injection system. The system consists of a high-speed pump with attached solenoid valve, a high-pressure delivery line, and a nozzle-and-hold assembly. The high-speed pump is driven by an extra cam on the engine's camshaft. The solenoid valve is triggered by an ECU to enable the precise control of injection quantity and timing for each cylinder.

The unit injector combines the injector pump and the injector nozzle in a single unit, which is driven by the engine camshaft. The unit injector can be installed directly into the engine's cylinder head. Each unit injector has its own solenoid valve that controls the start and end of injection. The solenoid valve is controlled by an engine ECU.

Figure 2.8 shows a schematic of a high-pressure common-rail system, which consists of a high-pressure pump, rail, high-pressure fuel lines, injectors, and an electronic control unit (ECU). The common fuel rail supplies fuel

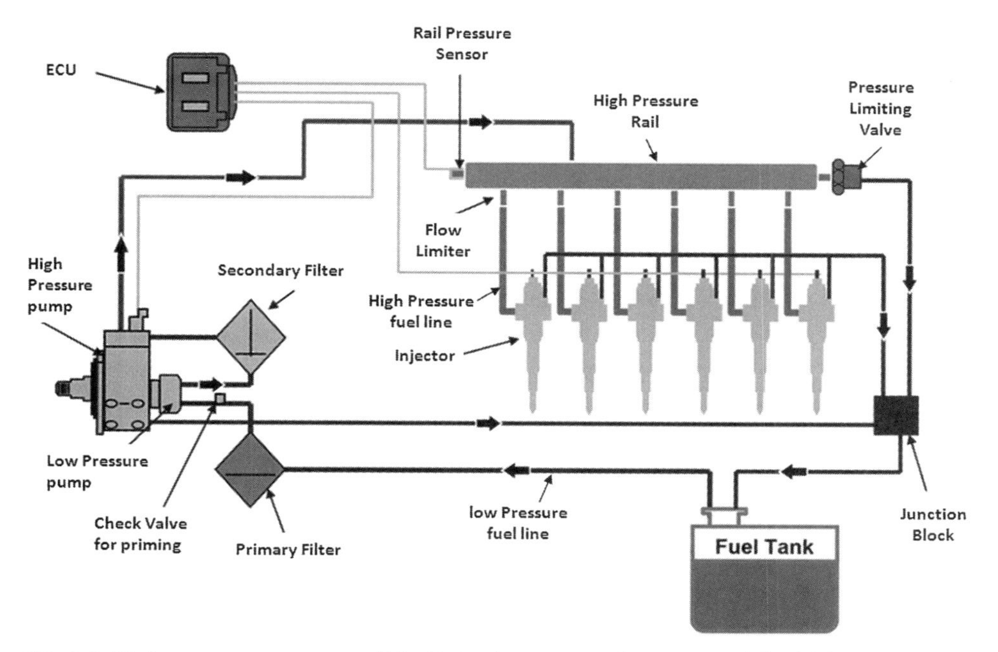

FIG. 2.8 High-pressure common-rail fuel injection systems for commercial vehicles.

to all injectors. The ECU controls the injection timing, rate shaping, and quality of the fuel injected. The common-rail system offers a significantly higher level of adaptability to engine design on the part of the fuel injection system than a cam-operated system. As shown in Fig. 2.9, the common-rail fuel injection system should have the following functions:

1. Pre-injection to reduce combustion noise, white smoke, and NO_x emissions, especially for DI engines
2. Positive pressure gradient during the main injection phase to reduce NO_x on engines without EGR
3. Two-stage pressure gradient during the main injection phase to reduce NO_x and soot on engines without EGR
4. Constant high pressure during the main injection phase to reduce soot on engines with EGR
5. Post injection immediately following the main injection phase to reduce soot emissions
6. Retarder post injection of fuel as a reducing agent for an NO_x LNT (lean NO_x trap) catalyst and/or to raise temperature for diesel particulate filter (DPF) regeneration.

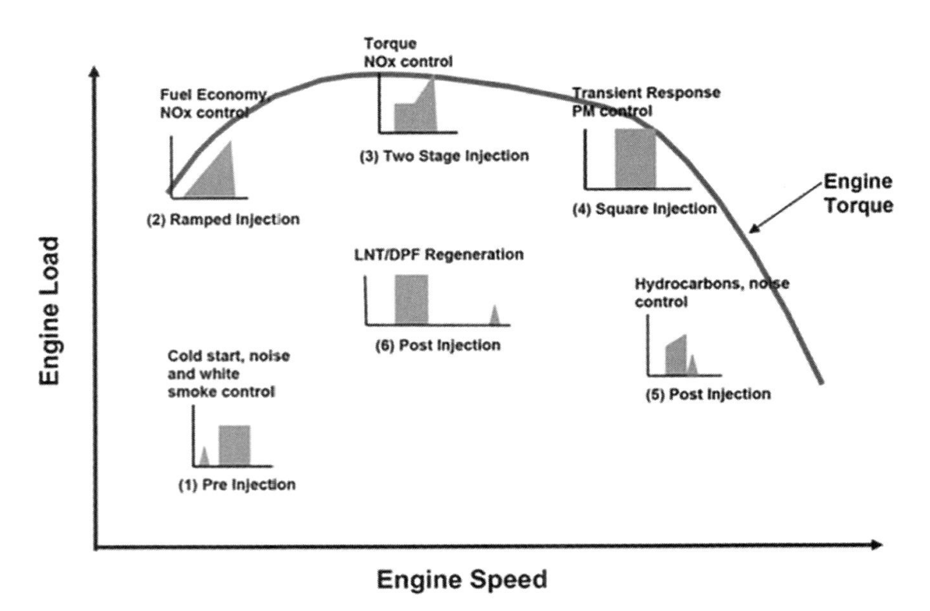

FIG. 2.9 Fuel injection profile and impact on engine performance [2.6].

Key components of the diesel fuel injection system are the electronically controlled injectors. The fuel injectors are connected with the fuel rail through the high-pressure fuel pipes. The common-rail injection is activated by a solenoid valve inside each injector through an ECU. A control signal from the ECU defines the start and end of the injection process. Injection quantity is calibrated by rail pressure, hydraulic flow of the nozzle, and the pulse length from the ECU.

The most commonly used high-pressure pumps of the common-rail fuel injection system for heavy commercial vehicles are radial piston pumps. The fuel volume between the high-pressure pump and the injector acts as an accumulator. The system's sensors are constantly monitored by the ECU. One of distinguishing features of a common-rail system is the closed-loop control for rail-pressure. The duty cycle for control of the high-pressure pump is changed by the ECU if the measured rail pressure deviates from the proper value. In the case of continual deviation, the fuel injection value for full load as well as the speed for the maximum speed governor is reduced. A relief valve mounted on the rail limits system pressure to avoid damaging the circuit.

Another important feature of the common-rail fuel injection system is that the rail pressure is always available at the nozzle seat for better engine performance and exhaust emission control. The highly flexible control of fuel injection systems results in better atomization of the fuel, with a corresponding reduction in unburnt carbon and more-efficient combustion. Currently, the high-pressure common-rail fuel systems are a key technology to achieve both emissions standards and to enhance engine performance in the commercial vehicle market.

2.3.2 Turbocharger

A turbocharger consists of a turbine and a compressor on a shared shaft (Fig. 2.10a). The compressor is powered by the turbine, which is driven by the exhaust gases of the engine, to increase the pressure of air before entering the engine combustion chamber to improve the engine's volumetric efficiency.

Turbocharger performance and operating characteristics are normally communicated through the use of the compressor and turbo maps. Figure 2.10b shows a typical compressor map, which can help to determine the required size of turbocharger based on the air-mass flow required by the engine and the necessary boost curve in an effective range. The effective range of the compressor is determined by the surge limit on the left edge and by the choke limit on the right edge of the map. The surge limit is defined as the transition from the stable to unstable range of operation, which is characterized by flow reversals or surge through the compressor. The choke limit, identifiable by the range of steeply descending speed curves, is the maximum low capability of the compressor.

A wastegate or variable nozzle turbine is usually used to control the turbo speed and the compressor mass flow rate of a turbocharger to assure the safety of the system or to better match the needs of the engine operating condition. A wastegate allows some of the exhaust to bypass the turbine when the set intake pressure is achieved. A variable geometry turbocharger (VGT) or a Variable Nozzle Turbine (VNT) is a turbocharger equipped with movable vanes for

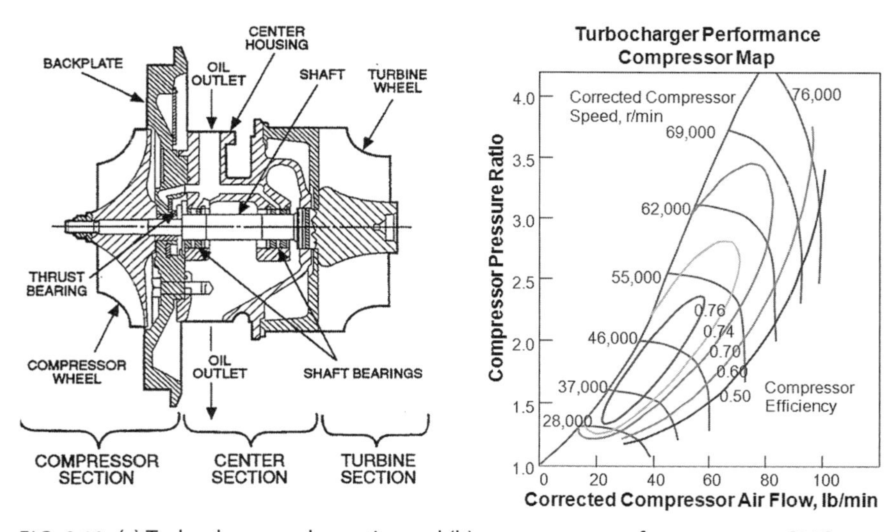

FIG. 2.10 (a) Turbocharger schematics and (b) compressor performance map [2.7].

directing exhaust flow onto the turbine blades. The vane angles are adjusted via an actuator. By varying the angle of the vanes, the turbine flow resistance is adjusted according to the corresponding required boost pressure level.

2.3.3 Exhaust Gas Recirculation (EGR) System

The Exhaust Gas Recirculation (EGR) system recirculates a portion of an engine's exhaust gas back to the engine cylinders. The purpose of EGR is to dilute intake air with combustion products, namely carbon dioxide (CO_2) and water vapor, to decrease the adiabatic flame temperature during the combustion process in order to decrease the formation rate of nitric oxide (NO). Because the specific heats of CO_2 (above 532 K) and water vapor are greater than that of air, they can absorb more heat than air, thus lowering the peak adiabatic flame temperature during the combustion process.

There are two primary types of EGR configurations on the market, a Low-Pressure-Loop (LPL) EGR system and a High-Pressure-Loop (HPL) EGR. The Low-Pressure-Loop (LPL) EGR extracts the exhaust gas downstream of the exhaust turbine and introduces it to the inlet of the intake compressor (Fig. 2.11a). LPL EGR has the advantage of having a sufficient pressure differential to drive EGR over a wide engine operating range, but LPL EGR may cause durability problems with the intake compressor and after-cooler. The High-Pressure-Loop (HPL) EGR (Fig. 2.11b) extracts the exhaust gas from the exhaust manifold and routes it to the intake manifold. The deployment of a DPF in the LPL loop minimizes the durability issues associated with the LPL method. With a VGT and a venturi in the intake manifold, the HPL is able to generate high enough pressure differential between the exhaust gas and the

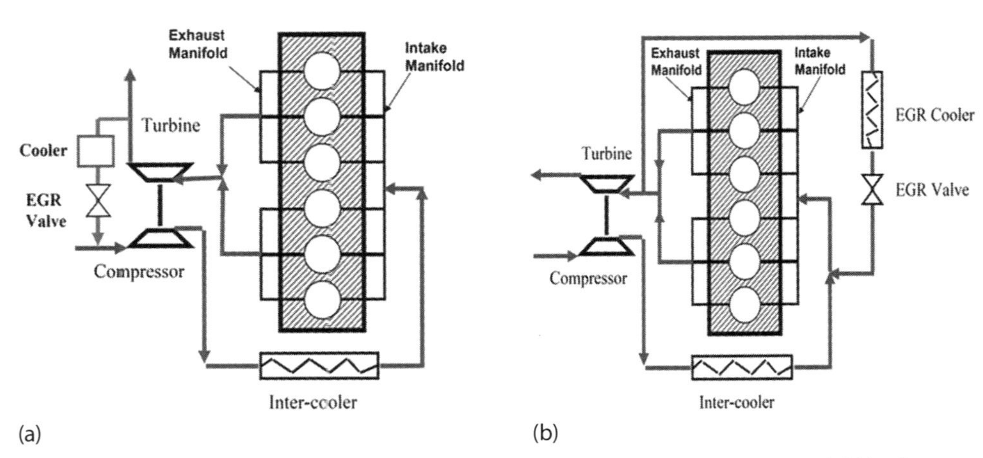

FIG. 2.11 Exhaust gas recirculation system: (a) low-pressureloop (LPL) system and (b) high-pressure-loop (HPL) system.

inlet air, but the required pressure differential is difficult to achieve at high load, and particularly in heavy-duty engine applications.

EGR is demonstrated to reduce NO_x emissions by up to 90% at light load and up to 60% at 75% load near rated speed. However, because EGR decreases the overall rate of combustion in the cylinder, EGR tends to increase *PM* emissions and brake specific fuel consumption (*BSFC*).

2.3.4 Engine Electronic Control Module

The engine electronic control module (ECM) of commercial vehicles is a vital element in optimizing engine fuel economy, improving vehicle drivability, and meeting emissions standards. For a vehicle and electronics controlled transmission, the ECM is a part of the powertrain control module that includes transmission controls. Based on the vehicle operating requirements, such as driver input, and road and traffic conditions, the ECM optimizes the fuel economy and driving performance while meeting emissions standards through precise control of fuel injection, air handling, and aftertreatment systems. At the same time, the operating environments in both on-highway and off-highway applications require a robust electronic system for reliable operation.

An electronic control module (ECM) contains the hardware and software. The hardware consists of electronic components on a printed circuit board (PCB), ceramic substrate, or a thin laminate substrate. The main component on this circuit board is a microcontroller chip (central processing unit, CPU). The software is stored in the microcontroller or other chips on the PCB, typically in electrically programmable read-only memories (EPROMs) or flash memory so the CPU can be reprogrammed by uploading updated code or replacing chips.

Modern ECMs use one or more microprocessors to process the inputs from the engine sensors and from other sources. For instance, some variable valve timing systems are electronically controlled; turbocharger waste gates can also be managed; and exhaust emission reduction devices are electrically controlled. In addition, an ECM may communicate with a transmission control unit or directly interface with electronically controlled automatic transmissions, traction control systems, and the like. The Controller Area Network, such as J1939, is often used to achieve communication between these devices. A Schematic of an ECM is shown in Fig. 2.12.

The basic operating principle of control systems can be categorized as either open loop or closed loop. The most commonly used control system in commercial vehicles is the open-loop control system with lookup tables. Measured values of speed and torque are used to construct a control table of injection timing values. At any speed and load condition, the corresponding injection timing is calculated using an interpolation algorithm and translated

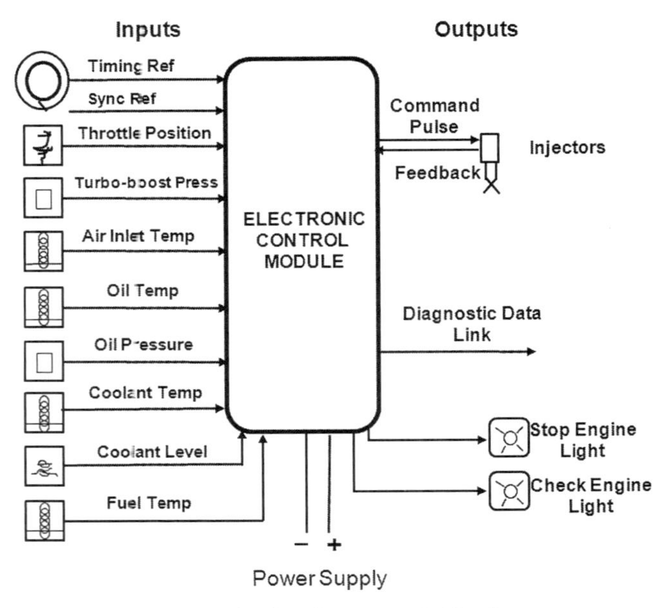

FIG. 2.12 Schematic of a diesel engine ECM [2.7].

into the output value to set the injection timing. The current engine control systems have a large number of tables and demand substantial effort to fill in the lookup tables. However, closed-loop control algorithms have also been developed and implemented for commercial vehicles in North America, such as the on-board diagnostics (OBD II) for exhaust emission controls

Model-based design (MBD) techniques are widely used for the design and development of the engine ECU. In model-based design of control systems, development usually has six steps: 1) define the requirements; 2) model a plant; 3) analyze and synthesize a controller for the plant; 4) simulate the plant and controller; 5) hardware-in-the-loop testing; and 6) integrate all these phases by deploying the controller. The model-based design paradigm is significantly different from traditional design methodology. Rather than using complex structures and extensive software code, designers can use MBD to define models with advanced functional characteristics using continuous-time and discrete-time building blocks. These built models used with simulation tools can lead to rapid prototyping, software testing, and verification. Hardware-in-the-loop simulation can test the dynamic effects on the system more quickly and much more efficiently than traditional design methodology.

As the internal combustion engines and in particular the modern DI diesel engines for on-road commercial applications move toward providing more flexibility, pressure will be applied on the control system to derive more performance, less fuel consumption, and lower emissions. These goals cannot be achieved unless sophisticated engine controls are used.

2.3.5 Exhaust Gas Aftertreatment Systems

Despite new technologies and modern electronic control devices that significantly aid in reducing engine-out exhaust emissions, NO_x and particulate emissions are still the subject of environmental concern. Additional reduction in NO_x and particulate emissions are mandatory to comply with stringent emission standards.. Various exhaust gas aftertreatment systems have been developed and deployed for commercial vehicles to meet the regulatory emission requirements.

The most commonly used forms of exhaust gas aftertreatment system are catalytic converters, which consist of an inflow and outflow funnel, and a monolith [2.8]. The monolith is made up of a large number of very fine parallel channels covered with an active catalytic coating. The monolith can be made of metal or ceramic materials.

Diesel emissions such as *CO* and *HC*, which include such *HC* materials as the organic fraction of diesel particulates (*SOF*) or polynuclear aromatic hydrocarbons (*PAH*), NO_x, and *PM*, can be controlled with high efficiency by today's catalyst technologies to meet the emissions regulations. The catalytic converters for diesel engines for commercial vehicle applications are summarized in Table 2.1.

2.3.5.1 *Diesel Oxidation Catalyst*

A diesel oxidation catalyst (DOC) is a flow-through device that consists of a canister containing a honeycomb-like structure or substrate. The substrate has a large surface area that is coated with an active catalyst layer. This layer was coated with a small, well-dispersed amount of precious metals such as platinum or palladium. As the exhaust gases flow through the catalyst, carbon monoxide, gaseous hydrocarbons, and liquid hydrocarbon particles (unburned fuel and oil) are oxidized, thereby reducing harmful emissions.

At sufficiently high exhaust temperatures, diesel oxidation catalysts can provide very effective control of *HC* and *CO* emissions, with reduction efficiencies in excess of 90%. The DOC hydrocarbon activity extends to compounds such as polynuclear aromatic hydrocarbons or the SOF fraction of diesel particulates.

These processes can be described by the following chemical reactions.

$$[Hydrocarbons] + O_2 \rightarrow CO_2 + H_2O \quad (2.12)$$

$$CO + 1/2O_2 \rightarrow CO_2 \quad (2.13)$$

TABLE 2.1 Diesel Catalyst Technologies

Catalyst Technologies	Reaction Type	Reduced Emissions	Reduction Efficiency	Issues	Commercial Status
Oxidation Catalyst	Oxidation	*CO, HCs, PM* (*SOF*), odor.	over 90%		MY 94
Selective Catalytic Reduction (SCR)	SCR by ammonia/Urea	NO_x	70 to 90%	Urea supply, Ammonia Slip	MY10 DDC, Cummins, Volvo
Lean NO_x Trap (LNT)	Adsorption of NO_x from lean exhaust, followed by release and catalytic reduction under rich condition	NO_x, *CO, HCs*	70 to 85%	Sulfur poison	MY07 Dodge Ram
Diesel Particulate Filter (DPF)	*PM* trapped by wall flow filter and oxidized	*PM* (Elemental carbon)	over 95%	Ash cleaning	MY07 engines
$DeNO_x$ (lean NO_x) Catalyst	SCR by hydrocarbon	NO_x, *CO, HCs, PM* (*SOF*)	30 to 50%	Low NO_x conversion rate	Some commercial oxidation catalysts incorporate small NO_x reduction activity
Hybrid LNT and SCR	Adsorption of NO_x from lean exhaust, followed by release and catalytic reduction under rich condition and generate NH_3. NH_3 will stored by SCR for reducing NO_x	NO_x, *CO, HCs*, NH_3	75 to 90%	Sulfur poison	Lab demonstration

Hydrocarbons are oxidized to form carbon dioxide and water vapor, and carbon monoxide is oxidized into carbon dioxide. Since carbon dioxide and water vapor are nontoxic, the above reactions bring an obvious emission benefit.

Additional benefits of the DOC include oxidation of several nonregulated, *HC*-derived emissions, such as aldehydes or *PAHs*, as well as reduction or elimination of the odor of diesel exhaust.

2.3.5.2 Selective Catalytic Reduction (SCR)

Selective Catalytic Reduction (SCR) uses ammonia as a reducing agent for NO_x over a catalyst composed of precious metals, base metals, and zeolites. A schematic of the SCR system is shown in Fig. 2.13. The urea/water fluid (prepared by dissolving solid urea to create a 32.5% solution in water), stored in a separate tank, is injected into the exhaust upstream of the catalysts. The catalysts consist of hydrolysis catalyst, selective reduction catalyst (SCR), and an oxidation catalyst in series. The hydrolysis catalyst promotes the decomposition of urea to form ammonia (NH_3) at elevated temperatures, as described in Eq. (2.14). The SCR catalyst reduces NO_x by reacting with ammonia, NH_3, to form nitrogen, N_2, and water. The oxidation catalyst converts unburned *HC* as well as unused ammonia (ammonia slip) into H_2O, N_2, and CO_2.

Typical chemical reactions that occur in the ammonia SCR system are expressed by Eqs. (2.15) through (2.17).

$$(NH_2)_2CO + H_2O \rightarrow 2NH_3 + CO_2 \tag{2.14}$$

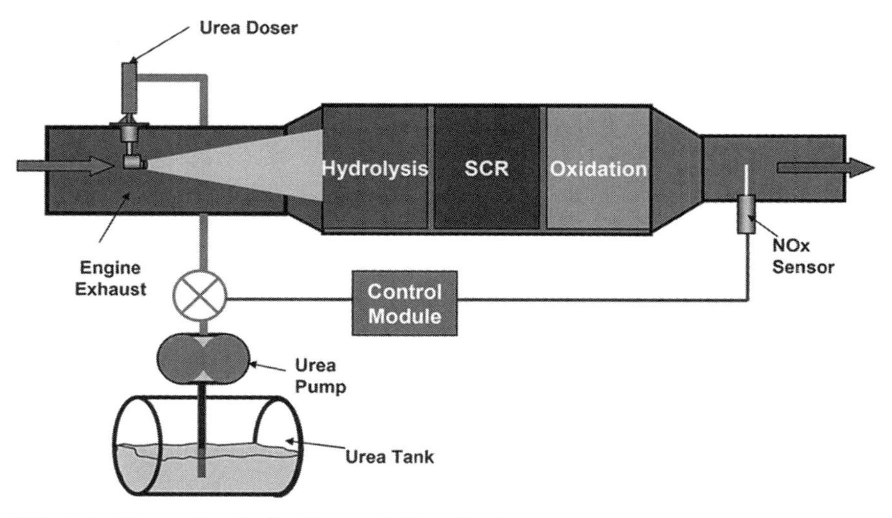

FIG. 2.13 Schematic of selective catalytic reduction.

$$4NO + 4NH_3 + O_2 \rightarrow 4N_2 + 6H_2O \quad (2.15)$$

$$NO + NO_2 + 2NH_3 \rightarrow 2N_2 + 3H_2O \quad (2.16)$$

$$6NO_2 + 8NH_3 \rightarrow 7N_2 + 12H_2O \quad (2.17)$$

All of these processes represent desirable reactions that reduce NO_x to elemental nitrogen. NO_x reductions of 70 to 90% are possible using such systems. Reactions given by Eqs. (2.16) and (2.17) involve nitrogen dioxide reactant. With the presence of NH_3, NO and NO_x are converted into N_2 and H_2O.

For each 1-g/hp-h reduction in NO_x, an SCR engine consumes urea/water fluid at a rate of approximately 2% of the amount of fuel used, depending on vehicle operation. Assuming an NO_x reduction from 1.2-g/hp-h to 0.2-g/hp-h and for an engine to consume 100 gal of diesel fuel in 600 mi (6 mpg), the urea/water fluid consumption for this period would be 2 gal.

Control of the quantity of urea fluid injection into the exhaust, particularly during transient operation, is an important issue with SCR systems. Injection of too large of a quantity of urea leads to a condition of "ammonia slip," whereby excess ammonia formation can lead to both direct ammonia emissions and oxidation of ammonia to produce (rather than reduce) NO_x. There are also a number of potential hurdles to overcome with respect to a major emission control system that requires frequent replenishing in order to function. There is currently no widespread distribution system in the United States for supplying the necessary water/urea fluids for diesel vehicles and trucks. This raises issues related to supply, quality control, tampering, and the possibility of running the urea tank dry. The on-board diagnostics (OBD) requirement to monitor the performance and function of the emission reduction system are regulated by the U.S. EPA starting in Model Year 2013.

2.3.5.3 *NO_x Adsorber Catalyst*

The NO_x adsorber catalyst, also called lean NO_x trap (LNT), is a technology developed in the late 1990s. The NO_x adsorber catalyst uses a combination of base metal oxide and precious metal coatings to effect control of NO_x. The base metal component (for example, barium oxide) reacts with NO_x to form barium nitrate—effectively storing the NO_x on the surface of the catalyst. When the available storage sites are occupied, the catalyst is operated briefly under "rich" exhaust gas conditions (i.e., the air-to-fuel ratio is adjusted to eliminate oxygen in the exhaust). This releases the NO_x from the base metal storage sites, and allows it to be converted over the precious metal components to nitrogen gas and water vapor.

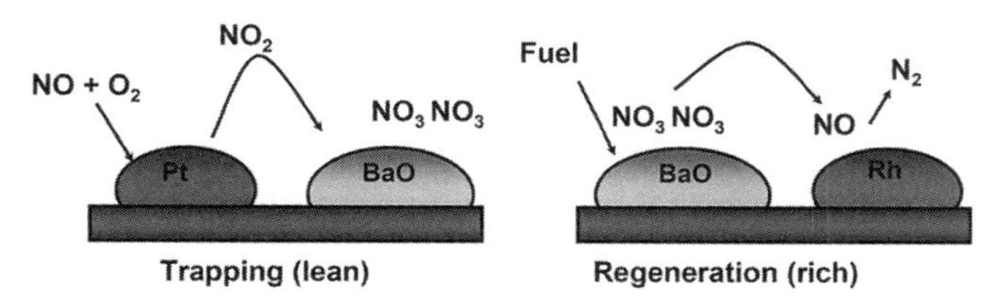

FIG. 2.14 NO_x adsorption and reduction mechanism [2.8].

The NO_x adsorption/reduction mechanism is illustrated in Fig. 2.14. The catalyst washcoat combines three active components: 1) an oxidation catalyst, for example Pt; 2) an adsorbent, for example barium oxide (BaO); and 3) a reduction catalyst, for example Rh.

Diesel engines normally operate with an excess ratio of air-to-fuel—so-called "lean" operation. Under lean operating conditions it is extremely difficult to control NO_x with a catalyst because of the excess of oxygen in the exhaust stream. Under lean operating conditions, the NO_x is simply stored in the catalyst as described in Eqs. (2.18) and (2.19). Regeneration is required to release and convert the NO_x to nitrogen gas as in Eqs. (2.20) to (2.21).

$$NO + 1/2O_2 \rightarrow NO_2 \tag{2.18}$$

$$BaCO_3 + NO_2 + 1/2O_2 \rightarrow Ba(NO_3)_2 + CO_2 \tag{2.19}$$

$$Ba(NO_3)_2 + 5CO \rightarrow N_2 + BaCO_3 + 4CO_2 \tag{2.20}$$

$$Ba(NO_3)_2 + 8H_2 + CO_2 \rightarrow 2NH_3 + 5H_2O + BaCO_3 \tag{2.21}$$

Regeneration of the LNT requires elimination of all excess oxygen in the exhaust gas for a short period of time. This can be accomplished by operating the engine in a "rich" mode, or by injecting fuel directly into the exhaust stream ahead of the adsorber to consume the remaining oxygen in the exhaust.

However, sulfur poses challenges for NO_x absorbers. In addition to storing NO_x, the LNT will also store sulfur, which reduces the capacity to store NO_x and requires the engine design to provide for a periodic desulfation process—a process to remove sulfur from the catalyst at higher temperatures. The elevated temperature during desulfation has a negative impact on durability life of the NO_x adsorber catalyst.

2.3.5.4 *Diesel Particulate Filter*

A diesel particulate filter (DPF) is a device designed to remove diesel particulate matter (*PM*) or soot from the exhaust gas of a diesel engine by a catalyst with wall-flow monoliths (Fig. 2.15). DPF captures a very high percentage of the particulate and holds it until the *PM* can be removed. Removing the *PM* from the trap, called trap regeneration, is accomplished by oxidizing (i.e., burning) the *PM*. Because diesel exhausts almost never reach the high temperatures needed to ignite the *PM*, oxidation requires either an external heat source or a catalyst material to lower the oxidation temperature of the *PM*. When needed to raise the temperature and actively manage the DPF regeneration, a small quantity of fuel is injected from a dozer or through the engine fuel injection system.

Ceramic wall-flow monoliths, which are derived from the flow-through cellular supports used for catalytic converters, became the most common type of diesel filter substrate. They are distinguished, among other diesel filter designs, by high surface area per unit volume and by high filtration efficiencies. Monolithic diesel filters consist of many small parallel channels, typically of square cross-section, running axially through the part. Diesel filter monoliths are obtained from the flow-through monoliths by plugging channels, as shown in Fig. 2.15. Adjacent channels are alternatively plugged at each end to force the diesel aerosol through the porous substrate walls, which act as a mechanical filter. To reflect this flow pattern, the substrates are referred to as wall-flow monoliths. Wall-flow monoliths are typically extrusions made from porous ceramic materials. Cordierite and silicon carbide are the most common materials used for DPFs.

Maintenance is required on diesel particulate filters. Additives in lubricating oils become ash and collect in the filter as oil is consumed and *PM* is oxidized through regeneration. Particulate trap filters need to be regenerated (cleaned) after a period of time because the filters eventually begin to fill up, creating unacceptable back pressure on the engine. The cleaning interval for a typical particulate filter for on-highway engines is 200,000 to 400,000 miles. However, the actual cleaning interval will be dependent on duty cycle and engine oil consumption.

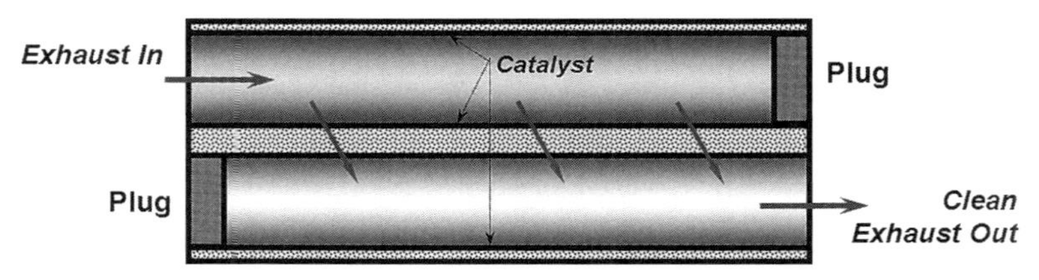

FIG. 2.15 Wall-flow monoliths.

2.4 Engine Operating Characteristics

Internal combustion engines based on the spark ignition engine and compression ignition engine will retain their dominant positions in automotive engineering for the foreseeable future. Spark ignition engines are usually used in passenger cars. Their key features are high power/weight ratio, good performance, and low combustion noise. Disadvantages are the quality of fuel required and the high part load fuel consumption.

The economy of the diesel engine is based on its low consumption, especially in the part load range, its low maintenance requirement, the low fuel quality required, and its good gaseous emissions ranges. Disadvantages are the level of particulate emissions, noise, irregular running, the lower engine speed spread (N_{max}/N_{min}), the low power output per liter, and the resultant great weight and high price. The higher capital cost means that it only becomes economical in vehicles that achieve a high mileage. Almost all heavy-duty commercial vehicles use diesel engines.

2.4.1 Indicated Power

Figure 2.16 shows a typical pressure-volume diagram of an internal combustion engine. Indicated power is work done on the surface of the piston by the gas pressure over the operating cycle.

Figure 2.16 shows that the net pressure applied to the surface of the piston is (Area A – Area B):

$$P = \frac{N}{n_R} \int P_i dV \tag{2.22}$$

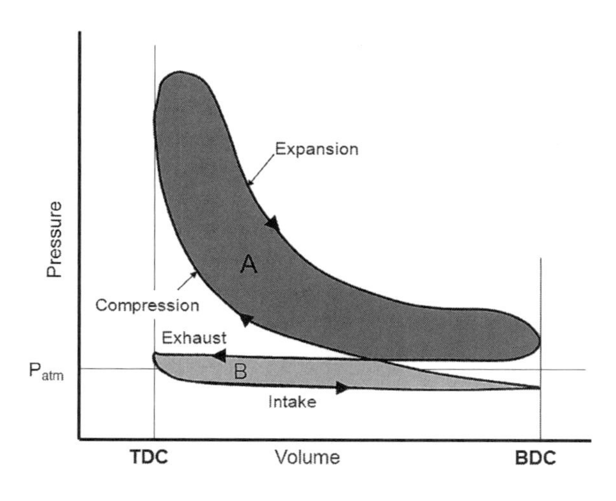

FIG. 2.16 Pressure-volume diagram.

where n_R is the number of crank revolutions for each power stroke per cylinder, two for four-stroke, one for two-stroke; N is the crankshaft rotational speed;

$$Indicated\ Power = Brake\ Power + Pumping\ Power + Friction\ Power$$

2.4.2 Mechanical Efficiency

Mechanical efficiency (η_{mech}) can be defined as the ratio of useful power (brake power) over available power (indicated power) as:

$$Mechanical\ Efficiency = (Brake\ Power)\ /\ (Indicated\ Power)$$

or

$$\eta_{Mech} = P_b / P_i = 1 - (P_f / P_i) \tag{2.23}$$

where P_i is indicated power, P_b is brake power, and P_f is friction power.

2.4.3 Indicated Mean Effective Pressure

The indicated mean effective pressure (*IMEP*) is the work exerted by the gas on the piston per unit swept volume.

$$IMEP = \int (P_i / V_d) dV \tag{2.24}$$

2.4.4 Brake Mean Effective Pressure

While *IMEP* is a form of the bulk pressure exerted on the surface of the piston, brake mean effective pressure may be described as the portion of that *IMEP* that produces useful power. It represents the network after subtracting friction, pumping, and other parasitic losses.

$$BMEP = \int (P_b / V_d) dV \tag{2.25}$$

2.4.5 Specific Fuel Consumption

Fuel consumption in mass per unit time, brake specific fuel consumption (bsfc) is a measure of how much fuel an engine consumes in the process of producing an output of one horsepower:

$$\text{bsfc} = (\text{Fuel Consumption per Unit Time})\ /\ (\text{Brake Power Output}),\ \text{lb/bhp-h}$$

2.4.6 Torque/Engine Speed Characteristics

The torque/engine speed curve at full load (100% accelerator pedal position) and the corresponding fuel load power curve are typically used to describe the engine characteristics of internal combustion engines of commercial vehicles.

Figure 2.17 shows an internal combustion engine and the characteristic points of the full load characteristic curve. The maximum braking torque (0% operator pedal position) increases almost linearly with engine speed to a maximum of approximately 30% of the normal torque T_n.

Various measures are used to facilitate comparison of different engines. The key enablers are torque increase (torque elasticity)

$$\tau = T_{\max}/T_{in} \tag{2.26}$$

and the engine speed ratio (engine speed elasticity)

$$\nu = N_n/N_{T\max} \tag{2.27}$$

An engine is considered to have greater elasticity the greater the product $\tau \cdot \nu$ is. This is apparent in the form of better engine power at low and medium engine speed, which in turn means less frequent gear changing.

Figure 2.18 gives an example of the characteristic curve of a DDC S60 14-L diesel engine. Different engine characteristics can be achieved by varying the engine design.

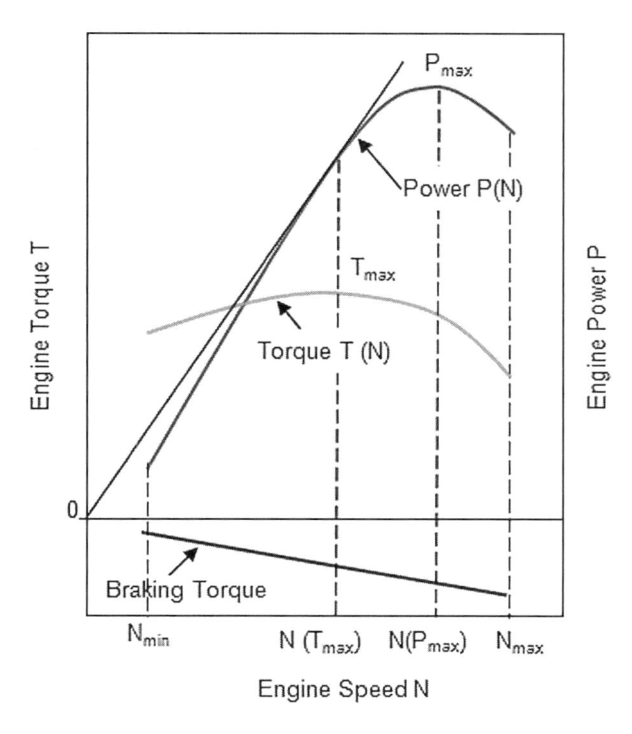

FIG. 2.17 Characteristic curves of an internal combustion engine [2.9].

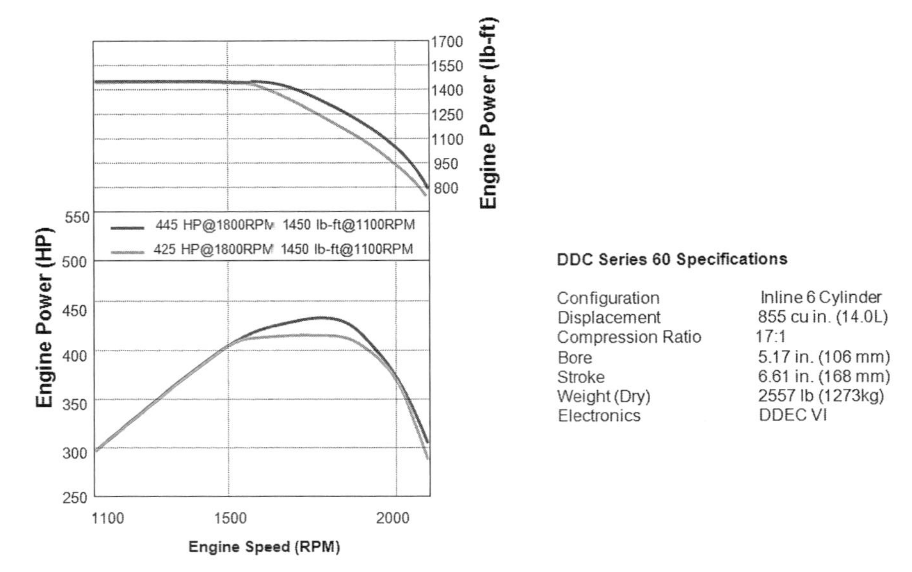

FIG. 2.18 Power and torque curve of a DDC S60 engine [2.12].

2.5 Advanced Engine Technologies

Future engines for commercial vehicles must meet:

1. Stringent government regulations in engine exhaust, such as NO_x, *PM*, *hydrocarbons*, etc.
2. Both light-duty and heavy-duty commercial vehicles must meet CO_2 emission standards
3. Must improve engine fuel economy to reduce the operating cost

Examples of advanced engine technologies are listed in Table 2.2. The functions and benefits of selected technologies are discussed in the following sections.

2.5.1 Low-temperature Combustion

Low-temperature combustion (LTC) involves developing lean or diluted mixtures in-cylinder, which after compression ignition, have peak combustion temperatures below approximately 1900 K to minimize NO_x formation in-cylinder. LTC strategies range from Homogeneous Charge Compression Ignition (HCCI) to strategies involving compression ignition of partially premixed fuel charges. LTC strategies provide the potential for engines with high, diesel-like efficiencies and dramatically lower engine-out emissions.

TABLE 2.2 Examples of Advanced Engine Technologies [2.23]

Technology	Description	Benefits
1. Advanced Combustion Technologies	• Direct injection stratified combustion and Controlled Auto Ignition (CAI) for gasoline engines • Low-temperature combustion technologies such as Homogenous Charge Compression Ignition (HCCI), Premixed Charge Compression Ignition (PCCI), and Reactivity Controlled Compression Ignition (RCCI) for diesel engines	• CAI can achieve higher efficiencies than conventional spark-ignited gasoline engines. • Low-temperature combustion significantly reduces NO_x and soot formation during the combustion process. • HCCI engines can operate on gasoline, diesel fuel, and most alternative fuels.
2. Advanced Valve Actuation Technologies	• VVA includes variable valve lift, cam phasing, cylinder deactivation, and camless technologies	• Fuel saving in part load for gasoline engines • Enabling technology for advanced combustion technologies
3. Advanced Boosting and Downsizing	• Twin turbo technologies • Super turbo (turbocharger and supercharger combination) technologies • Electric driven supercharger technologies	• Fuel economy improvement • Performance improvement in transient and part load operations
4. OBD and Energy Management Technology	• Advanced engine system control technology • Advanced sensors and diagnostic technology	• Fuel economy improvement • Emission reductions
5. Advanced Aftertreatment Technologies	• Hybrid LNT/SCR technologies	• Improved NO_x conversion efficiency without the need of urea supply infrastructure
6. Advanced Fuel Injection Technology	• Ultra-high-pressure common-rail fuel injection technology	• Fuel economy improvement • Soot reduction
7. Alternate Fuels Technologies	• Ethanol/hydrogen enhanced combustion technologies • Bio fuel technologies	• CO_2 emissions reduction
8. Thermal Energy Management and Recovery Technologies	• Turbocompound technologies • Exhaust thermal energy recovery technologies, such as bottom cycle and thermal electric technologies	• Engine fuel economy improvement
9. Friction Reduction Technologies	• Advanced lubricants • Advanced materials and designs • Accessory electrification	• Fuel economy improvement

(continued)

TABLE 2.2 (*Continued*)

Technology	Description	Benefits
10. Advanced Engine Design	• Opposite piston engines such as engine concepts developed by EcoMotors [2.24] and Scuderi Air Hybrid engines [2.25]	• Potential fuel economy improvement • Light weight • Low emissions

Figure 2.19 shows the NO_x and soot formation of diesel combustion dependent on local flame temperature and fuel air equivalence ratio as well as areas where conventional and possible alternative combustion systems are located.

HCCI is a premixed ultra-lean (equivalence ratio close to 0.2) combustion process. Fuel and air are pre-mixed together. The density and temperature of the mixture are raised by compression until the entire mixture reacts spontaneously. Because HCCI operates on lean mixtures, the peak temperatures are lower in comparison to spark ignition (SI) and diesel engines. The low peak temperatures prevent the formation of NO_x. This leads to NO_x emissions at levels far less than those found in traditional engines. However, the low peak temperatures also lead to incomplete burning of fuel, especially near the walls of the combustion chamber. This leads to high carbon monoxide and hydrocarbon emissions. An oxidizing catalyst would be effective at removing the regulated species because the exhaust is still oxygen rich.

Low-temperature combustion is a broad concept of preventing the NO_x formation during the combustion process by limiting the peak combustion temperature, which includes both HCCI and premixed controlled compression ignition (PCCI). One method of limiting the peak combustion temperature is to use a large amount of cooled EGR. Exhaust gas consists of largely carbon dioxide and water vapor. It has a higher specific heat than air and acts as a

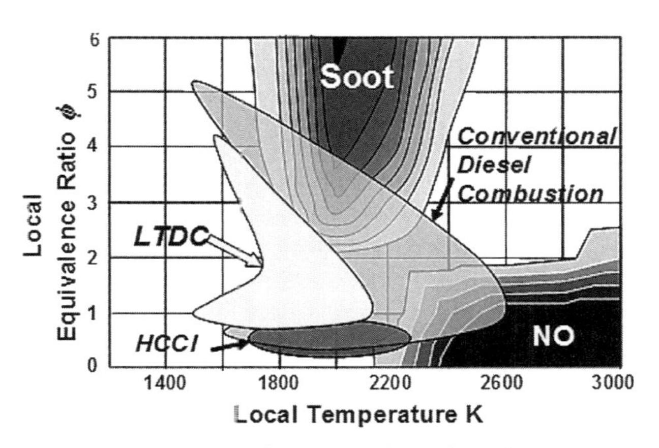

FIG. 2.19 Soot and NO_x formation of diesel combustion [2.13].

heat sink to lower peak combustion temperatures. Since diesels always operate with excess air, they benefit from EGR rates as high as 50% in controlling NO_x emissions [2.13].

2.5.2 Advanced Boost Technologies

Air boosting devices such as superchargers and turbochargers have been used to increase the density of air entering the internal combustion engine. The greater mass flow-rate provides more oxygen to support combustion and to increase the power output of the engine. Usually, a supercharger is powered mechanically by a belt, gear, shaft, or chain connected to the engine's crankshaft, and a turbocharger is powered by an exhaust gas turbine.

An advanced boost device discussed in this section is the electric-driven compressor. The compressor can be a centrifugal compressor or a roots style compressor. Because the compressor is driven by an electric motor, its speed and power are independent on turbine speed and power or the engine crankshaft speed. The application of the motor-assisted compressor provides a means of supplying the air to the engine cylinder at low idle or at low load to overcome the turbo lag of the conventional turbocharger.

The system is usually installed in series with a standard turbocharger. As shown in Fig. 2.20, the system is comprised of a standard engine with a turbocharger, a motor-driven supercharger, an air bypass to avoid air throttling

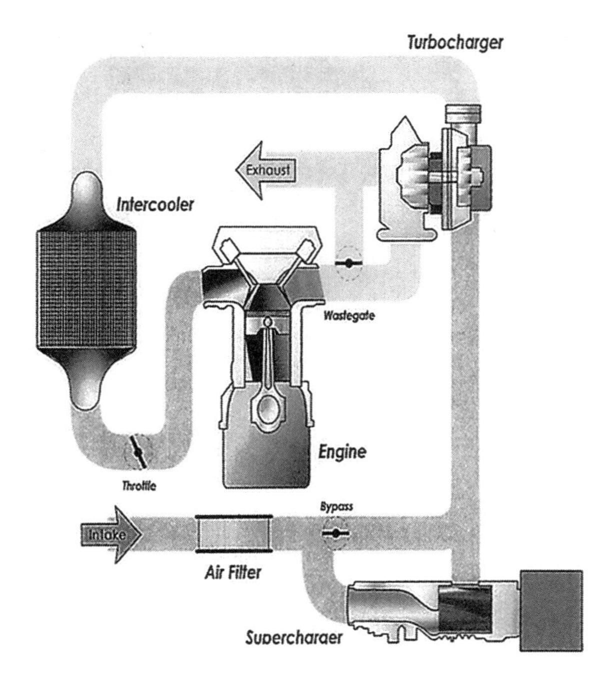

FIG. 2.20 Layout of a supercharger and turbocharger combination (Courtesy of Eaton Corporation).

at part load, an air filter, and associated control circuitry. A motor-driven supercharger can be powered by the vehicle batteries. The motor-driven supercharger supplies boost pressure to the engine until there is enough energy in the engine exhaust to accelerate the turbocharger. When the turbocharger is capable of supplying sufficient air, the unit is de-energized.

Figure 2.21 shows a road test of a supercharged heavy-duty diesel engine on a Kenworth T800 vehicle with Paccar MX-455 engine and an Eaton TVS supercharger. The supercharger bypass valve and clutch are controlled by the engine control unit (ECU).

Maintaining vehicle speed, the engine speed was reduced by 450 rpm; the acceleration was improved by 2.5%, and the fuel economy as measured by miles per gallon by 21%.

The added boost also appears to be very effective in reducing NO_x at equivalent smoke at these high EGR conditions for heavy-duty diesel engines. This occurs for two reasons. First, EGR displaces fresh air, and as the excess air ratio reaches a level of about 1.5 or lower (equivalence ratio of about 0.67 or higher), the combustion begins to degrade with excess CO emissions that imply poor combustion efficiency. The second advantage of using the supercharger is that it increases the trapped in-cylinder air density. The higher air density increases fuel-air mixing rates and allows faster mixing, reducing smoke. This allows for higher EGR rates to achieve the same smoke level, reducing NO_x at constant smoke.

2.5.3 Electric Turbocompounding

For a turbocharged diesel engine, 35% of the fuel energy is wasted in the exhaust [2.15]. Several technologies have been developed to recovery exhaust energy from the exhaust to improve the engine fuel economy. Mechanical

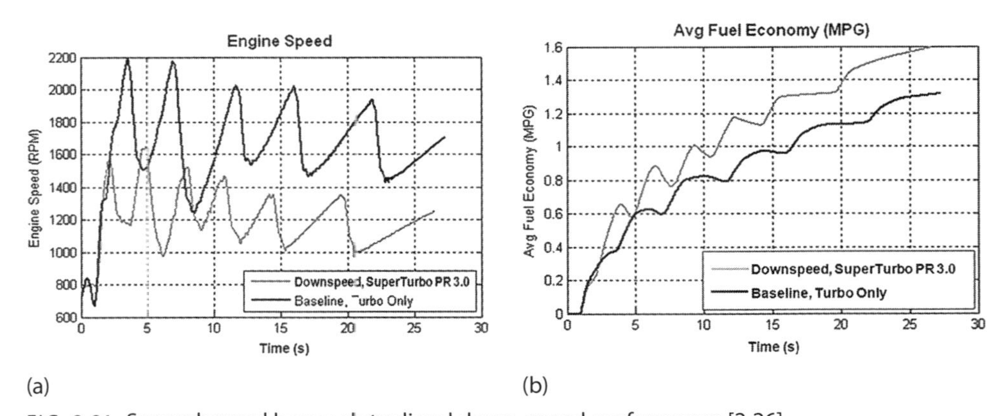

FIG. 2.21 Supercharged heavy-duty diesel down-speed performance [2.26].

turbocompounding is already established to improve fuel economy in heavy-duty diesel engines. The system includes a conventional turbocharger, which recovers exhaust energy in a turbine to boost the air coming into the engine. Downstream of the turbocharger turbine, the exhaust gas goes through a second turbine. The energy recovered here is added to the engine torque through a system of shafts, gears, and a fluid coupling. Electric turbocompounding recovers exhaust energy electrically by using high-speed generator technology. Electric turbocompounding eliminates the mechanical coupling to the engine crankshaft that is necessary in mechanical turbocompounding. This provides more flexibility in packaging. The electric turbocompound system also provides more control flexibility in that the amount of power extracted can be varied, which allows control of engine boost [2.19].

A schematic of an electric turbocompound system is shown in Fig. 2.22. The principal mode of operation is when the electric machine on the turbocharger shaft acts as a generator, and the electric machine on the engine crankshaft works as a motor to draw power from the electrical bus and assists the engine by injecting mechanical power into the crankshaft. In addition to the crankshaft-mounted motor, there could be other vehicle electrical loads that the generator could support. Furthermore, if the generator were to produce more electricity than what is being consumed, the balance would go into the energy storage block. Conversely, if the generator were unable to supply all loads fully, electrical power would be drawn from the energy storage device.

Figure 2.23 shows the surplus power of the system as a function of engine power. This surplus power can be recovered by the electric turbocompound system through an electric motor, mounted on the crank shaft, which assists

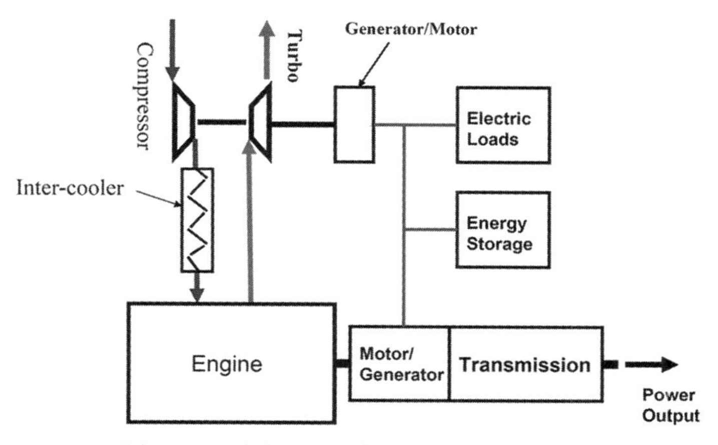

FIG. 2.22 Schematic of electric turbocompounding.

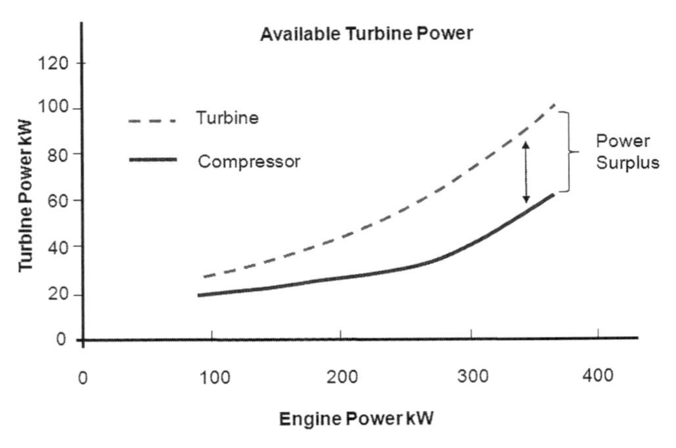

FIG. 2.23 Comparison of compressor and turbine power in an electric turbocompound engine [2.19].

the engine. The result of this process is an increase in system efficiency. Alternatively to compounding it, the surplus power can be used to drive other electrical onboard devices or truck electrical loads, or it can be stored.

The simulated benefits of high turbo efficiency and turbocompounding on a 14.7-L heavy-duty diesel engine shows that fuel consumption can be reduced by 10% at the rated power. Considering a typical road load for an on-highway truck with different weighting factors for the prevailing operating speeds and loads, fuel consumption is reduced by approximately 5%.

The challenges of commercializing the electric compound technology include the system cost and durability of turboshaft-mounted motor generator. However, the electric turbocompound system can be integrated with a hybrid powertrain for lower system cost.

2.5.4 Hybrid LNT/SCR Aftertreatment Technologies

LNT and SCR aftertreatment systems are used for reducing engine exhaust emissions in the commercial vehicle industry. However, sulfur, primarily as sulfur dioxide, in the diesel exhaust causes the LNT performance gradually to decline. Periodic desulfation is required to maintain the NO_x reduction performance of the LNT catalysts. Frequent desulfation causes durability concerns of the LNT catalysts. SCR involves the reduction of NO_x by ammonia; it can achieve NO_x reductions in excess of 80%. There are concerns over the lack of infrastructure for distributing ammonia or a suitable precursor, urea. SCR also raises concerns relating to the ammonia slip into the environment.

Figure 2.24 shows a schematic of a hybrid LNT/SCR aftertreatment system. The system is comprised of a diesel fuel reformer followed by a LNT catalyst and an NH_3-SCR catalyst. The fuel reformer generates reformate to improve the efficiency of the regeneration cycle as compared to injecting fuel directly into the exhaust system. The functions of LNT include: 1) adsorb NO_x when engine is running under normal (lean) conditions; 2) reduce a portion of stored NO_x to N_2 under regeneration (rich) conditions; and 3) reduce the rest of stored NO_x to ammonia under regeneration conditions. This system uses a fuel reformer to generate H_2 and CO to improve LNT NO_x reduction efficiency and the conversion efficiency of NO_x to NH_3. The NH_3 species will be stored and utilized to reduce NO_x further by the subsequent SCR catalysts. This approach converts the drawbacks of the single-leg LNT approach (low conversion during regeneration and NH_3 slip) into an advantage, while capitalizing on the fact that a urea infrastructure is not required because diesel fuel is the only reductant.

Figure 2.25 shows test results on a 14-L Detroit Diesel Series 60 engine. The engine was rated at 500 hp @ 2100 rpm, 1650 ft-lb at 1200 rpm. The cycle time between LNT regenerations is 100 seconds. As a result, the engine-out NO_x is 12.9 grams per cycle, while the LNT-out NO_x is 3.4 grams per cycle. This accounts for a 74% reduction in NO_x using the LNT catalyst. The regeneration shown by the fuel reformer pulse creates a rich atmosphere in the LNT catalyst that converts stored NO_x to nitrogen, ammonia, and other species. For this case, a large amount of the stored NO_x is converted to ammonia, as depicted by the high ammonia peak (approximately 1500 ppm). This ammonia is stored in the SCR catalyst and later used to further reduce NO_x in the SCR catalyst such that the NO_x at tailpipe out is 1.9 grams per cycle or approximately 25 ppm.

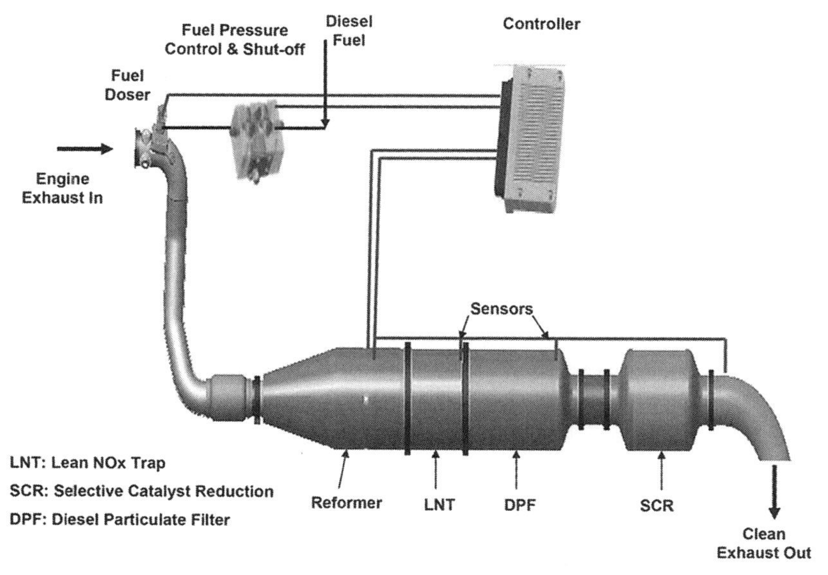

FIG. 2.24 Aftertreatment system schematic [2.18].

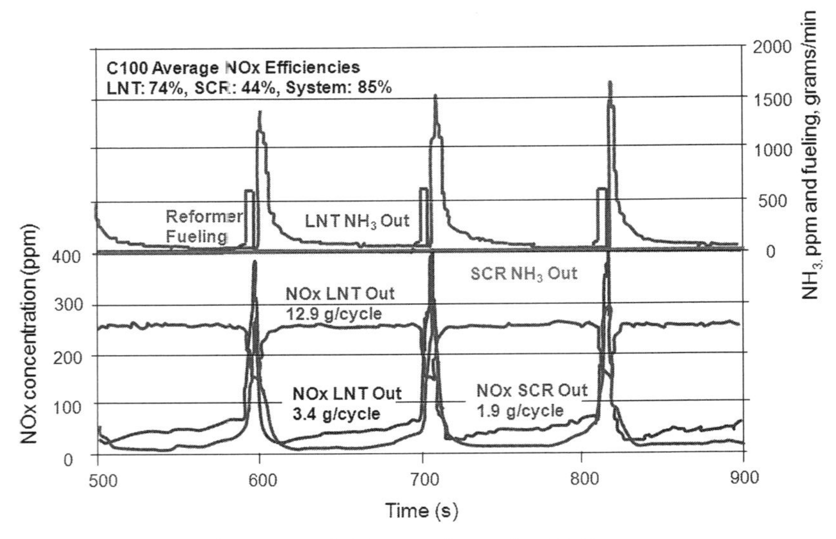

FIG. 2.25 System efficiency at C100 operating point [2.18].

Although the NO_x efficiency in the SCR is only 44% because a large amount of the NO_x was mitigated in the LNT catalyst, the SCR contribution increases the system NO_x efficiency to 85%.

The major advantages of the hybrid LNT/SCR aftertreatment system include: a) the system is independent of urea solution and supply infrastructure, and b) the system eliminates urea sensors and compliance-related penalties. However, because LNT performance is sensitive to fuel sulfur content, periodic desulfation at elevated temperature is needed to maintain the LNT NO_x reduction efficiency, although the interval of desulfation is increased compared to the LNT only aftertreatment system.

2.5.5 Variable Valve Actuation Technology for Heavy-duty Diesel Engines

Variable valve actuation (VVA) allows real-time adjustments to valve openings and closings, maintaining precise control of valve motion. A VVA, or lost motion system, creates a hydraulic link between the cam and the valve, precisely tuning the engine across its operating range. A lost motion system provides a cost-effective means of achieving the desired valve movement, integrating with minimal impact to engine overhead designs for heavy-duty diesel engines. Many systems have been proposed and developed to accomplish the VVA objectives, one of which is described next.

A schematic of a lost motion engine VVA system is shown in Fig. 2.26. Lost motion is the term applied to a class of technical solutions for providing variable engine valve timing strategies [2.21]. These solutions all involve modifying the valve motion prescribed by a conventional camshaft by mechanical,

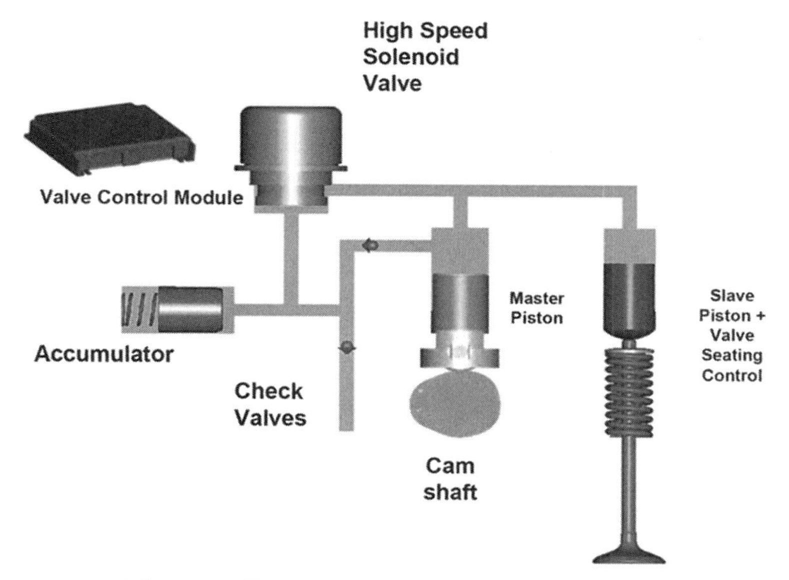

FIG. 2.26 Schematic of lost motion variable valve actuation system [2.21].

hydraulic, or other means. The cam lobe involved is equipped with the "maximum" motion needed over a range of engine operating conditions. A variable-length element in the valve train subtracts, or "loses," motion which is not needed during other operation modes. The system consists of actuator housing, slave piston, master piston, an accumulator, a high-speed solenoid valve, and a control. Valve lift is controlled by the electronically controlled solenoid. The master piston, driven by a camshaft, displaces fluid from its hydraulic chamber and transmits hydraulic force to the engine valve. The system also includes a normally open solenoid valve and a check valve. When the solenoid valve is closed, the engine valve lift profile follows the contour of the cam lobe. If the solenoid valve is open, the working fluid in the hydraulic chamber flows to an accumulator, which modifies the valve lift profile and the duration of opening. The valve lift profile is shown in Fig. 2.27.

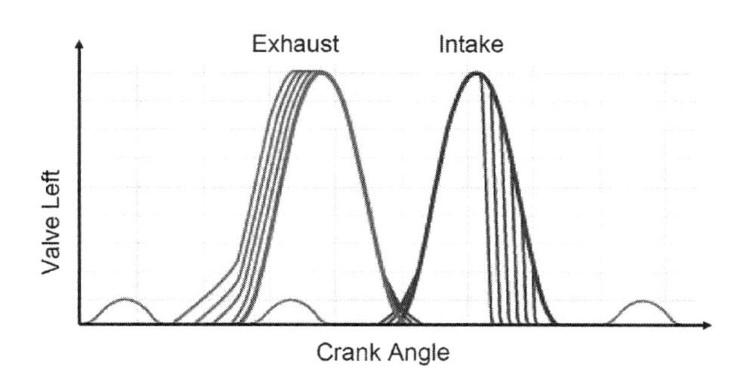

FIG. 2.27 Valve lift profile of lost motion VVA system [2.27].

The lost motion VVA system enables internal exhaust gas recirculation for emissions reduction, flexible timing and lift for the HCCI process at a wide operating range, and compact and powerful engine retarding features for commercial vehicles [2.21].

2.6 References

2.1 "Greenhouse Gas Emissions Standards and Fuel Efficiency Standards for Medium- and Heavy-Duty Engines and Vehicles," http://www.epa.gov/otaq/climate/regulations.htm, Assessed, Sept. 2001.

2.2 Heywood, J. B., *Internal Combustion Engine Fundamentals*, McGraw Hill, 1988.

2.3 Obert, E. F., *Internal Combustion Engines and Air Pollution*, Harper & Row. 1973.

2.4 Ferguson, C. R., *Internal Combustion Engines Applied Thermociences*, Wiley, 1986.

2.5 Taylor, C. F., *The Internal Combustion Engine in Theory and Practice*, MIT Press, 1987.

2.6 Bosch, *Diesel Engine Management*, 3rd Edition, SAE International, Warrendale, PA, 2004.

2.7 Brady, R. N., *Modern Diesel Technology*, Prentice Hall, 1996.

2.8 Majewski, W. A. and Khair, M. K., *Diesel Emissions and Their Control*, SAE International, Warrendale, PA, 2006.

2.9 Merrion, D. F., *Diesel Engine Design for the 1990s*, SP-1011 (940130), SAE International, Warrendale, PA, 1994.

2.10 Heisler, H., *Advanced Engine Technology*, SAE International, Warrendale, PA, 1995.

2.11 Future Powertrain Technologies: Advanced Alternative Fuel Technology Outlook, 2008–2020, TIAX.

2.12 "EPA07 DD13 OPERATOR'S MANUAL," DDC-SVC-MAN-0051, Detroit Diesel Corporation, 2009.

2.13 Zhao, F. et al., *Homogeneous Charge Compression Ignition (HCCI) Engines: Key Research and Development Issues*, SAE International, Warrendale, PA, 2003, pp. 11–12.

2.14 Maier, J., "VARIABLE CHARGE MOTION FOR 2007–2010 DIESEL ENGINES," Diesel Engine Emissions Reduction (DEER) Conference, August 24–28, 2003.

2.15 Arnold, S. et al., "Advances in Turbocharging Technology and its Impact on Meeting Proposed California GHG Emission Regulations," SAE Paper No. 2005-01-1852, SAE International, Warrendale, PA, 2005.

2.16 Pallotti, P. et al., "Application of an Electric Boosting System to a Small, Four-Cylinder S.I. Engine," SAE Paper No. 2003-32-0039, SAE International, Warrendale, PA, 2004.

2.17 George, S. et al., "Optimal Boost Control for an Electrical Supercharging Application," SAE Paper No. 2004-01-0523, SAE International, Warrendale, PA, 2004.
2.18 Hu, H. et al., "Advanced NO_x Aftertreatment System And Controls For On-Highway Heavy Duty Diesel Engines," SAE Paper No. 2006-01-3552, SAE International, Warrendale, PA, 2006.
2.19 Hopmann, U. et al., "Diesel Engine Electric Turbo Compound Technology," SAE Paper No. 2003-01-2294, SAE International, Warrendale, PA, 2003.
2.20 Stobart, R., "Heat Recovery and Bottoming Cycles for SI and CI Engines—A Perspective," SAE Paper No. 2006-01-0662, SAE International, Warrendale, PA, 2006.
2.21 Hu, H. et al., "The Integrated Lost Motion VVT Diesel Engine Retarder," SAE Paper No. 973180, SAE International, Warrendale, PA, 1997.
2.22 "EPA Finalizes Regulations Requiring Onboard Diagnostic Systems on 2010 and Later Heavy-Duty Engines Used in Highway Applications Over 14,000 Pounds; Revisions to Onboard Diagnostic Requirements for Diesel Highway Heavy-duty Applications Under 14,000 Pounds," US EPA, Office of Transportation and Air Quality EPA-420-F-08-032, December 2008.
2.23 Average Fuel Economy Standards Passenger Cars and Light Trucks Model Year 2011, Docket No. NHTSA-2009-0062. http://www.nhtsa.gov/DOT/NHTSA/Rulemaking/Rules/Associated%20Files/CAFE_Updated_Final_Rule_MY2011.pdf.
2.24 "EcoMotors International's Opposed-Piston, Opposed-Cylinder Engine Promises to Revolutionize Commercial Vehicle Design with Powerful, Lightweight, Fuel Efficient, Low Emissions Engines," http://www.ecomotors.com/technology, Feb. 2011.
2.25 Scuderi, S. O.,"Split-cycle Air Hybrid Engine," WO 2007/081445, July 2007.
2.26 "HD Diesel Supercharger Down Speeding with Performance," Eaton internal test report, Oct. 2011.
2.27 "VVA by Jacobs," http://www.jacobsvehiclesystems.com/files/media/Product/VVA_Brochure_Final.pdf, assessed, January 2012.

Chapter 3

Introduction of Clutches and Transmissions for Commercial Vehicles

3.1 Background of Transmissions

Transmissions convert the engine output speed and torque to meet the demand of the vehicle for starting, stopping, slower travel, overdrive, and other operations. A transmission of a commercial vehicle provides speed and torque conversions from the output of the prime mover, such as the internal combustion engine, to the drive wheels.

The transmission will generally be connected to the crankshaft of the engine. The output of the transmission is transmitted via the driveshaft to one or more differentials, which in turn drive the wheels. Unlike transmissions in passenger cars, transmissions for commercial vehicles usually are multi-stage, with long-haul truck designs such as a front splitter unit, main gearbox, and rear range unit, as shown in Fig. 3.1. The transmissions have power take-off capability for on-road utility vehicles and for off-road vehicles.

Early multiple-ratio transmissions were developed for vehicles powered with steam engines to climb gradient roads. In 1784, James Watt patented the constant-mesh gear with constantly meshing gearwheel, which is shown in Fig. 3.2a. Figure 3.2b shows an 1821 disclosure of a two-speed transmission with sliding gear by Griffith [3.1].

Starting at the turn of the last century, the transmission development effort was focused on improving comfort and ease of use, with enhancements such as the development of helical cut spur gears to reduce engine speed to reduce noise and to make the changing gear easier. In 1915, ZF developed the first constant-mesh gearwheels, preselector, and synchronizing mechanisms. The ZF Soden transmission (Fig. 3.3a) provided preselection, whereby the driver set a knob on the steering wheel to the required gear and pressed on the pedal. The clutch disengaged. When the shift pedal released, the preselected gear engaged automatically. Figure 3.3b shows a ZF K45 commercial vehicle transmission with spur toothed sliding gears.

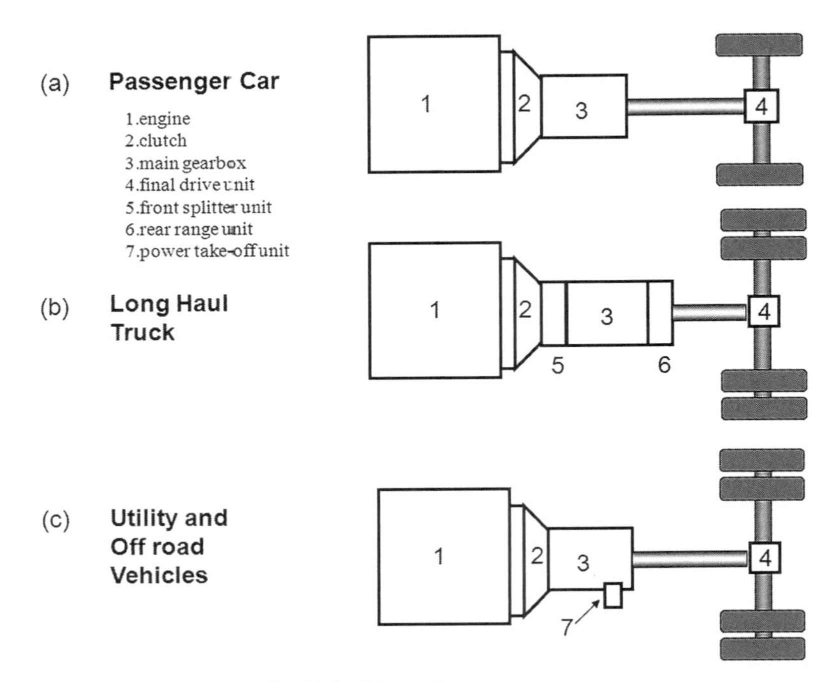

FIG. 3.1 Comparison of vehicle drivetrains.

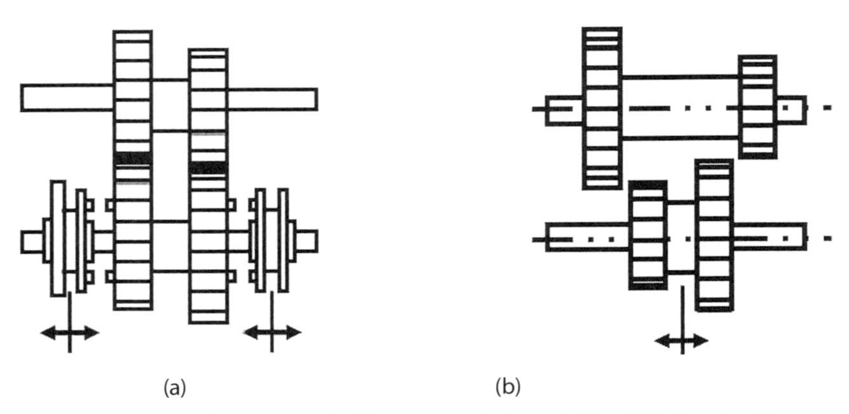

FIG. 3.2 Early developments of vehicle transmissions. (a) 1784 Watt patent two-speed gearbox with dog clutch engagement, (b) 1821 Griffith two-speed gearbox with sliding gears.

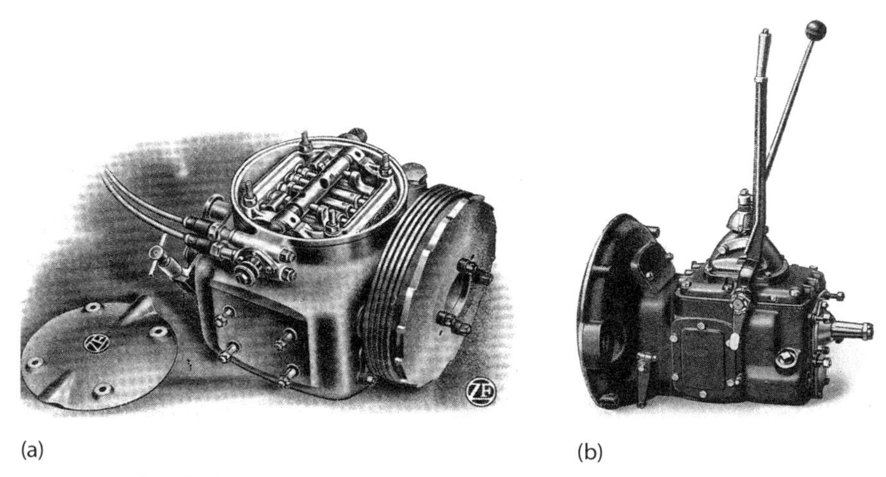

(a) (b)

FIG. 3.3 Early vehicle transmission development [3.2]. (a) ZF Soden (1915); (b) ZF K45 (1925)

Transmissions for passenger cars and commercial vehicles were the same except for their sizes until the Second World War. Since then, there have been fundamental changes. Payloads increased as tires with greater load-bearing capacity were developed, trucks moved into long-distance as opposite to local haulage, and the motorway network was expanded, all of which meant a greater range of ratios and thus higher numbers of speeds.

The development goals for mechanical geared transmissions for commercial vehicles were low weight, reduced noise, and improved ease of use with the introduction of synchronizers. One particular requirement was long service life of up to 1 million kilometers. Initially, five to six speeds were adequate, although the transmissions were fitted with front-mounted splitter units to give finer grading of the overall gear ratio. The six-speed gearbox became a 12-speed gearbox. The increase in specific power output (kW/t) with commercial vehicles gave rise to the requirement for an increased overall gear ratio. To provide greater fuel economy or better performance, transmissions with twelve to sixteen speeds become established in the early seventies for commercial long haul trucks.

Eaton Corporation, one of the primary transmission developers in the North American commercial vehicle market, introduced a nine-speed commercial vehicle transmission with power split to two countershafts (Fig. 3.4) for a short overall design length in 1962; a twin-splitter 12-speed commercial vehicle transmission with four-speed main gearbox and two rear-mounted splitter units in 1983; and an automated manual transmission with the AutoSelect feature, which was designed to increase fuel efficiency, provide safer and easier driving, and drastically lower training costs, in 1993.

The newest to the growing range is the Eaton Fuller UltraShift LEP (Line haul Efficient Performance) transmission, which is designed to optimize shifting to keep engine speeds as low as possible while maintaining performance and drivability. In 2003, Eaton introduced a ten-speed UltraShift automated manual transmission. Its key features include an automatic start with no clutch pedal, automatic shifting, computer-controlled shifts, "skip-shifting" provided as operating and load conditions allow, a high-capacity inertia brake that speeds automated up-shifts and serves as a clutch brake, and clutch-less "float shifting" between gears. The UltraShift ten-speed allows drivers to use the transmission in full automatic mode or with manual intervention, depending on conditions or driver preference.

FIG. 3.4 Eaton Fuller twin countershaft transmission. (Courtesy of Eaton Corporation)

FIG. 3.5 Eaton Fuller ten-speed UltraShift automated manual transmission. (Courtesy of Eaton Corporation

3.2 Key Characteristics of Vehicle Transmissions

The internal combustion engine, a prime mover for almost all today's commercial vehicles, has many advantages, including high power-to-weight ratio, high thermal conversion efficiency, compact size, and good durability. However, it also has fundamental disadvantages as a vehicle prime mover:

1. Unlike electric motors, the internal combustion engine (diesel or gasoline engine) is incapable of producing torque at zero engine speed.
2. The internal combustion engine can only produce maximum power at a certain speed.
3. The fuel efficiency of the engine can only be optimized within a specific operating range.

Therefore, an output converter must be used to match the torque and speed requirements of the vehicle and to minimize the fuel consumption of the IC engines. The primary functions of vehicle transmissions of connecting the engine and drive wheels are:

1. Enable the vehicle to move from rest
2. Adapt power flow
   a. Convert output torque and speed
   b. Enable reverse motion
3. Maximize engine power transmission efficiency
4. Control power matching

3.2.1 Power Requirement

To keep the vehicle moving, the engine has to develop sufficient power to overcome the opposing road resistance, and to pull away from a standstill and to accelerate to the desired operating speed. The driving resistances consist of:

1. Rolling Resistance F_r
2. Air resistance F_f
3. Gradient resistance F_g
4. Acceleration Resistance F_a

The traction F required to drive a vehicle is

$$F = F_r + F_f + F_g + F_a \tag{3.1}$$

or

$$F = f_R m_f g + \tfrac{1}{2} A \rho C_w v^2 + m_f g \sin \alpha_{st} + \lambda m_f a \tag{3.2}$$

where: f_R is the rolling resistance coefficient, $f_R = 0.01 - 0014$ for paved road; m_f is the vehicle mass; g is gravitational acceleration; A is the maximum vehicle cross-section area; ρ represents air density; C_w is the dimensionless drag coefficient for most buses and trucks, $C_w = 0.45 - 0.8$; v is the vehicle traveling speed; α_{st} is the road ascent angle; λ is the rotational inertia coefficient; and a is the acceleration.

For steady-state operation, the road resistance force and power can be simplified for a typical medium size vehicle as

$$F = f_R m_f g + \tfrac{1}{2} A \rho C_w v^2 + m_f g \sin \alpha_{st} \tag{3.3}$$

The power required for driving a vehicle is

$$P = Fv \tag{3.4}$$

3.2.2 Matching the Engine and Transmission

Matching the engine's performance characteristics to suit a vehicle's operating requirement is provided by choosing a final drive gear reduction and then selecting a range of gear ratios for maximum performance in terms of the ability to climb gradients, achieve good acceleration through the gears, and reach some predetermined maximum speed on a level road.

3.2.2.1 *Transmission Ratio*

The transmission ratio I is the relationship between the angular velocity ω_1 of the input shaft of a gearbox to angular velocity ω_2 of the output shaft:

$$I = \omega_1/\omega_2 = n_1/n_2 \tag{3.5}$$

where n_1 and n_2 are the input speed and output speed, respectively.

When: $I > 0$ transmission input and output shaft rotate in the same direction;

$I < 0$ transmission input and output shaft rotate in opposite directions;

$I > 1$ speed reducing ratio;

$I < 1$ speed increasing ratio.

In case of a continuously variable transmission and with transmission combination:

$I = \infty$ stationary output with rotating input;

$I = 0$ stationary input with rotating output.

3.2.2.2 *Top Gear Ratio*

To determine the maximum vehicle speed, the engine brake power should be larger than the vehicle resistance power at the entire vehicle operating speed range. The total resistance F at any speed is given by

$$F = F_r + F_f = f_R m_f g + \tfrac{1}{2} A \rho C_w v^2 \tag{3.6}$$

The top gear ratio G_F is chosen so that the maximum road speed corresponds to the engine speed at which maximum brake power is obtained. Thus,

$$\begin{gathered}\text{Linear wheel speed} = \text{Linear road speed}\\ G_F = 60\pi dN/(1000\text{V})\end{gathered} \tag{3.7}$$

where G_F is the final drive gear ratio, N is the engine speed (rev/min), d is the effective wheel diameter (m), and V is the road speed at which peak power is developed (km/h).

3.2.2.3 *Minimum Gear Ratio*

The maximum payload and gradient determine the minimum required gear ratio of the vehicle. The greatest gradient normally means 20 to 25%. The minimum tractive force effort necessary to propel a vehicle up the steepest slope may be assumed to be approximately equivalent to the sum of both of rolling and gradient resistance opposing motion.

$$F = F_r + F_g = f_R m_f g + m_f g \sin \alpha_{st} \tag{3.8}$$

Minimum gear ratio G_B can be calculated as the maximum required torque equals the rated torque

$$G_B = Fr/(TG_F\eta_m) \tag{3.9}$$

where r is the effective road wheel radius, T is the rated engine torque, and η_m the mechanical efficiency.

3.2.2.4 *Intermediate Gear Ratio*

Intermediate gear ratio can be best selected as a first approximation by using a geometric progression. This method of obtaining the gear ratio requires the engine to operate within the same speed range in each gear, which is normally selected to provide the best fuel economy.

If the ratio of the highest gear G_F and the lowest gear G_B have been determined, and the number of speed of the gearbox n is known, the gear step K between the individual gears can be determined by

$$K = (G_F/G_B)^{1/(n-1)} \tag{3.10}$$

The ratio of the individual gears $n = 1$ to z is then given by

$$G_n = G_z K^{z-n} \tag{3.11}$$

3.2.2.5 *Matching Engine and Transmission*

The criteria of matching transmission to the engine are to optimize the following:

1. Performance
2. Fuel consumption
3. Emissions
4. Comfort

The vehicle performance is defined by its maximum speed and its climbing and acceleration capability. The performance of a vehicle can be determined by comparing the traction available and the traction required at any point. Figure 3.6 shows a traction diagram of the propelling thrust to overcome the resistances of rolling resistance, air resistance, gradient resistance, as well as the acceleration for a vehicle with an eight-speed transmission [3.3].

Fuel consumption and exhaust emissions are major factors for determining the efficiency of a motor vehicle. Transmission design and the operation by the vehicle driver have major impact on the fuel consumption by selecting the operating point of the engine and thus its fuel consumption. Figure 3.7 shows that a transmission design should allow the vehicle to operate at the fuel economy speed range to minimize the fuel consumption and engine exhaust emissions.

Commercial vehicle engines used to pull large loads are normally designed to have a positive torque rise curve; that is, from maximum speed to peak torque with reducing engine speed, the available torque increases. The amount of engine torque rises as a percentage of peak torque from the maximum speed (rated power) to peak torque as:

$$\%\ \text{torque rise} = \text{Max. speed torque/peak torque} \times 100$$

The most common values of torque rise for commercial vehicles range from 15 to 30%. A driver comfort can be determined by less-frequent gear changes and smooth gear changes.

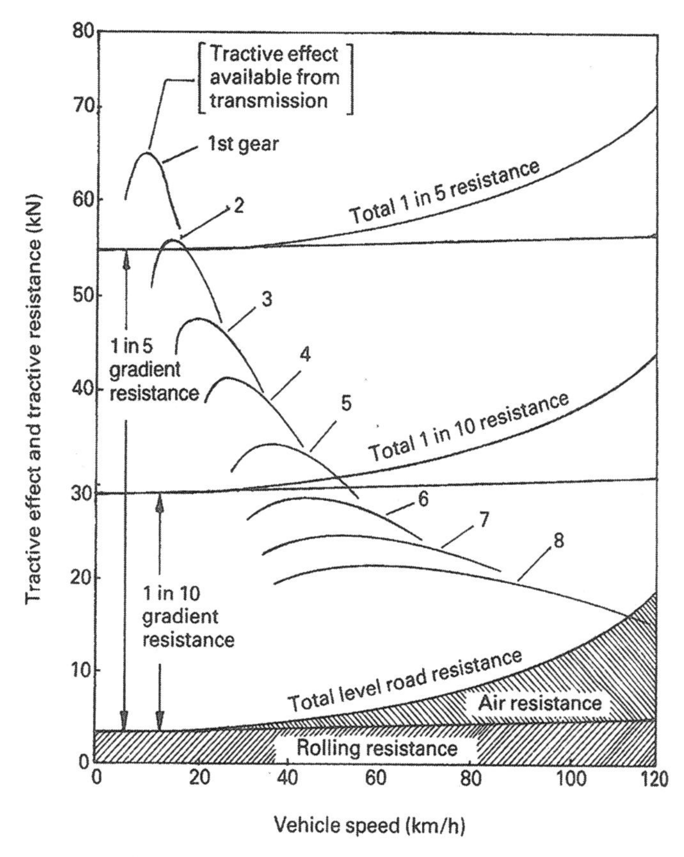

FIG. 3.6 Vehicle tractive resistance and effort performance [3.3].

3.3 Key Components of Commercial Vehicle Drivetrains

There are several key components that are an integral part of commercial vehicle transmissions. In this chapter, friction clutch, torque converter, and planetary gearset will be discussed.

3.3.1 Friction Clutch

Friction clutch is a mechanism for transmitting torque, which can be engaged and disengaged from the engine flywheel to the transmission during vehicle startup or during gear shifting when the vehicle is equipped with a manual transmission.

There are two types of clutches, a wet clutch and a dry clutch. A wet clutch is immersed in a cooling lubricating fluid, which also keeps the surfaces clean and gives a smoother performance and longer life. Wet clutches, however, tend to lose some energy to the liquid. A dry clutch is not bathed in fluid, and can have one or multiple clutch disks based on the clutch design and applications.

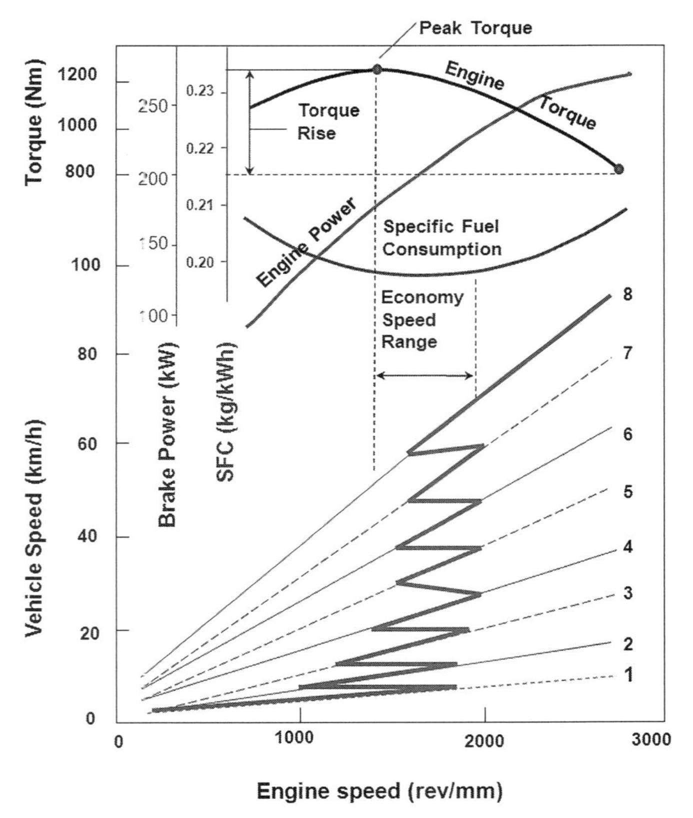

FIG. 3.7 Engine performance and gear split chart [3.3].

Figure 3.8 illustrates an angle spring clutch assembly with two spring-dampened discs. Its components consist of two basic groups, the driving and driven members. The driving members include the cover assembly, the pressure plate, and the intermediate plate if the clutch has two friction discs. The clutch cover is bolted to the flywheel to rotate with it. The pressure plate is machined smooth on the side facing the driven disc. The pressure plate is mounted on the clutch cover and is free to slide back and forth when pressure is applied to the plate. The driven members of a clutch consist of clutch friction discs. The clutch discs are lined on both sides with friction material and mounted on the transmission input shaft. When the discs are clamped between the pressure plate and flywheel, torque is transferred from the engine output shaft to the input shaft of the transmission.

Friction facings are critical to clutch life and performance because they directly receive the torque of the engine any time the clutch is engaged. There are two types of friction facings: organic and Cerametallic. Most organic friction facings are made from non-asbestos materials, such as glass, mineral wool,

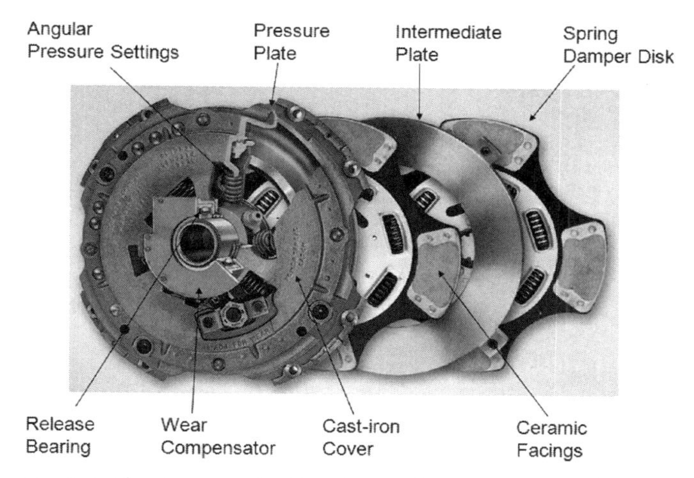

FIG. 3.8 Clutch assembly. (Courtesy of Eaton Corporation)

and carbon. Organic friction facings are usually molded to the full surface of the disc and used in light-duty commercial vehicles.

Cerametallic friction facings are becoming increasingly popular for heavy-duty commercial vehicle applications. It is made from a powder mainly of ceramic and copper, which is compressed into buttons and heated to form a strong metal-ceramic interface bond. These buttons are then riveted to the clutch driven plate and then finally ground flat. The advantages of the Cerametallic-lined driven plates include:

1. Low inertia (about 10% lower than the organic disc)
2. A relatively high and stable coefficient of friction
3. Capable of operating at temperatures up to 752°F (440°C)

For commercial vehicle applications, clutches are specified by driven disc diameter, facing type, number of driven disc, and pressure plate load. Table 3.1 shows typical torque ratings by the size of driven discs.

3.3.2 Torque Converter

A torque converter is a hydrodynamic device that transmits or multiplies torque from the engine flywheel to the transmission input shaft. H. Fottinger applied a torque converter patent for using one in ships in 1905. In 1928, Trilok Consortium in Karlsruhe, Germany developed the Trilok converter. The Trilok converter is a single-stage two-phase converter that has a stator in between the impeller and turbine. The first phase is until the lockup point, and functions as a torque converter. In the second phase, the stator is released from the housing by means of a freewheel; no torque can be transferred between the pump and the turbine, rather it functions like a clutch.

TABLE 3.1 Clutch Torque Rating

Cutch Diameter	Organic	Ceramic
12 Inch Single	300 lb-ft	400 lb-ft
13 Inch Single	400 lb-ft	550 lb-ft
13 Inch Two	700 lb-ft	900 lb-ft
14 Inch Single	550 lb-ft	600 lb-ft
14 Inch Two	1000 lb-ft	1250 lb-ft
15 Inch Single	650 lb-ft	900 lb-ft
15 Inch Two	1300 lb-ft	1600 lb-ft
15-½ Inch Single	700 lb-ft	900 lb-ft
15-½ Inch Two	1450 lb-ft	650 lb-ft
16 Inch Single	900 lb-ft	1350 lb-ft
16 Inch Single	1200 lb-ft	1850 lb-ft

Most torque converters (Fig. 3.9) used in the commercial vehicle industry are Trilok converters and consist of an impeller, a turbine, and a stator, in which

- The impeller is directly mounted on the flywheel and rotates at engine speed. As the impeller rotates, centrifugal force throws the fluid outward through the blades and around the split-guided ring, to propel the turbine.
- The turbine is coupled with the transmission input shaft. When the turbine is rotated by the propelling action of the impeller, the turbine provides rotational input to the transmission input shaft.
- The stator is located between the impeller and the turbine and locked to the turbine shaft by a one-way clutch. The fluid leaves the turbine and is re-directed to the impeller at an accelerated rate to cause the torque to multiply. As the turbine speed increases to close to the speed of the impeller, flow through the stator reduces and eventually stops.

A torque converter first converts the mechanical energy into fluid energy, and then transforms it back into mechanical energy. The input torque valve can be determined using Euler's law and can be calculated as follows:

$$T_P = \lambda \rho_f \omega_p^2 D^5 \qquad (3.12)$$

where T_P is the impeller input torque; λ, the performance coefficient, is a function of speed ratio; ρ_f is the density of transmission fluids (800–900 kg/m^3); ω_p is the angular velocity of the impeller in rad/s; and D is the impeller diameter in meters.

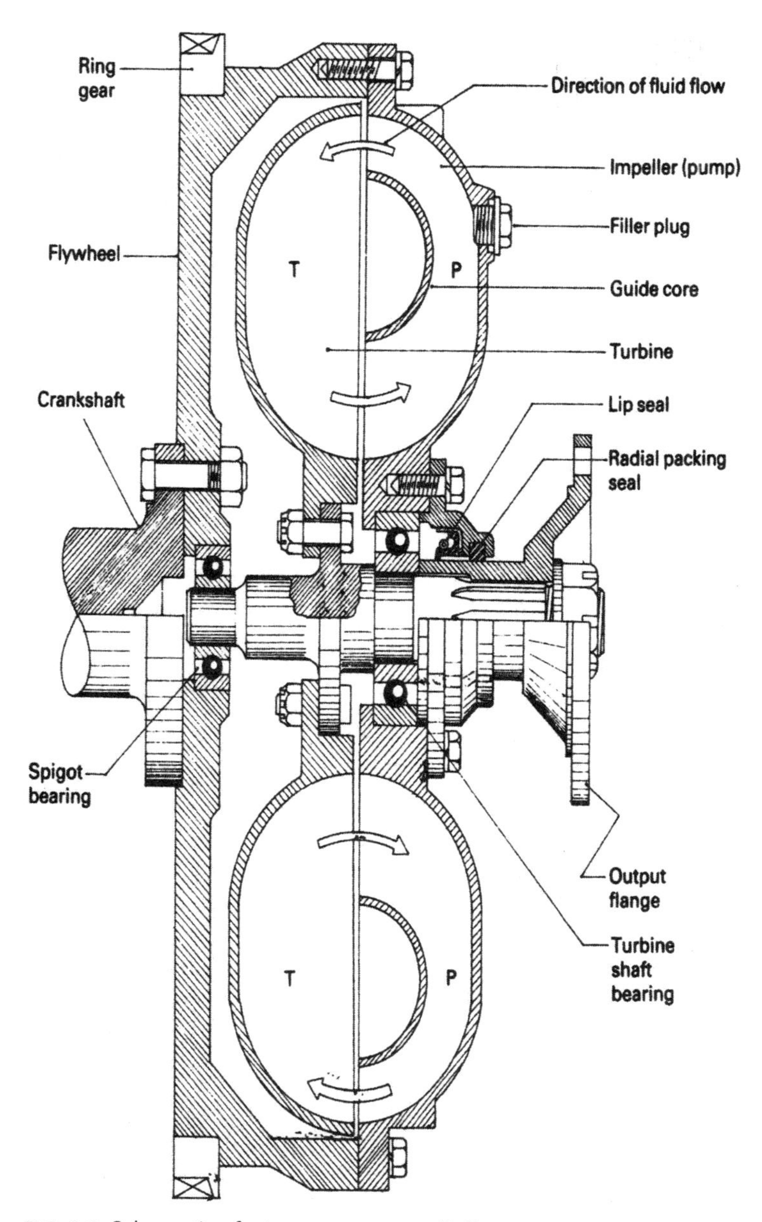

FIG. 3.9 Schematic of a torque converter [3.6].

The ratio of turbine torque T_T to the impeller torque T_P can be determined by the torque conversion factor as

$$\mu = \frac{T_T}{T_P} \tag{3.13}$$

and the hydraulic efficiency η can be defined as

$$\eta = \mu \frac{\omega_T}{\omega_P} \tag{3.14}$$

3.3.3 Planetary Gear Set

Planetary gear sets as shown in Fig. 3.10 contain three main elements: (i) a sun gear A, (ii) a carrier C that hosts planet gears, and (iii) a ring gear B. Either the sun or ring gears may be fixed. The planetary carrier performs rotation about its axis. In 1834, Johann Georg Bodmer, a Swiss inventor, designed a partial power-shift planetary transmission [3.1]. The change in gear ratio is achieved by disengaging the shifting dogs and tightening a brake band.

There are three main configurations for application of a planetary gear. The gear train may have

- One input, one output, and one fixed element. The input shaft may be a sun or the carrier, the output is the remaining element. In this case the mechanism is a speed reducer (or multiplier).
- One input, two outputs, and no fixed element. The input shaft may be the carrier, the outputs the sun and ring gears. In this case the mechanism is a differential. The output torque is divided among the two output shafts.
- Two inputs, one output, and no fixed element. This mechanism is a speed combiner, since the output speed is a linear combination of the speed of the two input shafts.

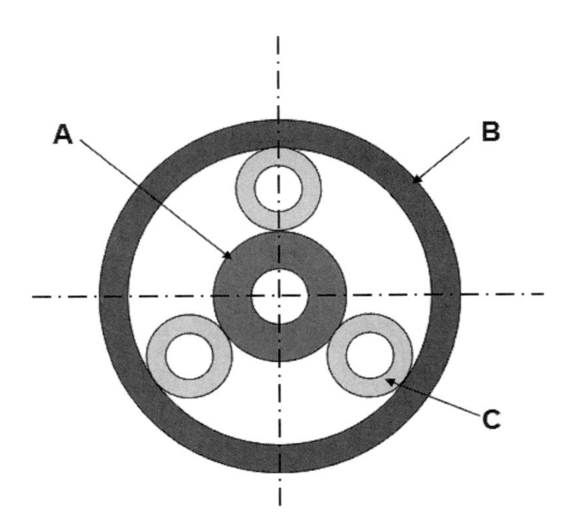

FIG. 3.10 Planetary gearset.

There are even some more applications where the configuration of the system may be changed during the operation. The input shaft is the central sun gear *A*, and the output shaft is the carrier *C*. The first speed is obtained keeping gear *B* at rest. Then, the second speed is obtained connecting gear *B* to the input shaft (direct drive).

For a simple planetary system, the following equation applies:

$$\omega_A Z_A + \omega_B Z_A = \omega_C (Z_A + Z_B) \tag{3.15}$$

where ω_A, ω_B, and ω_C are the rotation speeds of sun, ring, and carrier respectively; Z_A and Z_B are the number of teeth in the sun and ring gears, respectively.

If the input shaft is gear *A* and the output shaft is the carrier with the ring gear locked, the planetary gear ratio will be

$$i = \omega_A / \omega_A = 1 + Z_B / Z_A \tag{3.16}$$

For a simple planetary system, the basic equations and operation can be summarized in Table 3.2.

3.3.4 Compound Planetary Gear Sets

Compound planetary gear sets use multiple planetary components that are a variation on the simple planetary gear set. Since the inception of the simple planetary gear set, there have been numerous compound gear sets introduced. The following gear sets are commonly used for vehicle transmissions [3.23]:

- Simpson Gear Set
- Ravigneax Gear Set
- Lepelletier Gear Set

TABLE 3.2 Planetary gear set conversion ratios [3.9, 3.10].

Input	Output	Locked	Ratio	Remark
A	C	B	$i = 1 + Z_B/Z_A$	$2.5 \le i \le 5$
B	C	A	$i = 1 + Z_A/Z_B$	$1.25 \le i \le 1.67$
C	A	B	$i = 1/(1 + Z_A/Z_B)$ $0.2 \le i \le 0.4$ Upward conversion	
C	B	A	$i = 1/(1 + Z_A/Z_B)$ $0.6 \le i \le 0.8$	
A	B	C	$i = -Z_B/Z_A$	Stationary transmission with direction reversal $-4 \le i \le -1.5$
B	A	C	$i = -Z_A/Z_B$	Stationary transmission with direction reversal $-0.25 \le i \le -0.67$

Simpson Gear Set: Howard Woodworth Simpson patented the two identical planetary gear sets in series, linked by a common sun gear in 1950 for a three-speed automatic transmission. The use of largely identical parts for the front and rear gear set made the transmission substantially cheaper to manufacture by reducing tooling costs. This three-speed gearing arrangement, patented in 1950, is commonly known as the "Simpson Gearset."

Ravigneaux Gear Set: The Ravigneaux gear set is a double planetary gear set commonly used in automatic transmissions. It has two sun gear wheels, a large sun and a small sun; and a single carrier gear with two independent planetary gear wheels connected to it, an inner planet and an outer planet. The carrier is one wheel but has two radii to couple with the inner and outer planets, respectively. The two planet gears rotate independently of the carrier but co-rotate with a fixed gear ratio with respect to each other. The inner planet couples with the small sun gear and co-rotates at a fixed gear ratio with respect to it. The outer planet couples with the large sun gear and co-rotates with a fixed gear ratio with respect to it. Finally, the ring gear also couples and co-rotates with the outer planet in a fixed gear ratio with respect to it. This gear set is capable of four forward gears and one reverse; the overall dimensions of such planetary gears are relatively short. The selection of ratios is restricted; however, depending upon application it may be used with an auxiliary gear set.

Lepelletier Gear Set: The Lepelletier gear set allows for six forward speeds and one reverse gear using a lightweight design. The planetary gear train consists of a single-carrier planetary gear train and a downstream double planetary gear train.

The characteristics of the compound planetary gear sets are listed in Fig. 3.11, and the operations of the planetary will be further discussed in the section on automatic transmissions.

3.4 Manual Transmission

There are two basic types of manual transmissions in today's commercial vehicles market:

- non-synchronized constant-meshed transmission
- synchronized constant-meshed transmission

The earliest form of a manual transmission developed in the late 19th Century engaged the gears by sliding them on their shafts. It required a lot of careful timing and throttle manipulation when shifting, so that the

Gearset Name	Layout	Characteristics
Simpson Gearset[1] The use of largely identical parts for the front and rear gearset made the transmission substantially cheaper to manufacture by reducing tooling costs.		• Two internal ring gears, one rear input ring and one attached to the rear planetary carrier. • Two planetary carriers, each containing three planetary pinions. • One common sun gear, which meshes with both sets of planetary pinions.
Ravigneaux Gearset[2] It is called reduced planetary gear since parts of the individual simple planetary gears are the same size and can therefore be grouped together.		• One planetary carrier which is common to both sets of planetary pinions. • Two sets of planetary pinions, one long set with small diameter and one short set with large diameter. • Two sun gears, one input sun gear and one reaction sun gear. • One common ring gear.
Lepelletier Gearset[3] It allows for 6 forward speeds and one reverse gear using a light weight design.		• A single planetary gearset • A Ravigneaux planetary gearset • Combined along with five shifts.

1. http://web.ncf.ca/ch865/englishdescr/Simpson.html
2. http://www.mathworks.com/help/toolbox/physmod/sdl/ref/ravigneauxgear.html
3. http://www.ingendi.de/index.php?cat=c16_Gear-technique.html

FIG. 3.11 Compound planetary gear sets for automatic transmission applications [3.11].

gears would be spinning at roughly the same speed when engaged; otherwise, the teeth would refuse to mesh. These transmissions are called "sliding mesh" transmissions. Most current transmissions have all gears mesh at all times but allow some gears to rotate freely on their shafts; gears are engaged using sliding-collar dog clutches; these are referred to as "constant-mesh" transmissions.

Manual transmissions for commercial vehicles can also be classified as single-range transmissions or multi-range transmissions. In the case of four- to six-speed transmissions, the single-range design with input constant gear is standard. For more than six gears, a multi-range transmission is preferred to provide as many gear steps with as few gear pairs as possible.

Multi-range transmissions are constructed by combining single-stage, two-stage, or multi-stage transmissions. The system can be arranged with a splitter unit, a main gear box, and a range change unit, as shown in Figs. 3.1(a) and (b). The splitter unit can perform speed-reducing and speed-increasing; however, the range-change unit is always speed-reducing. The appropriate design must always be selected for each range unit.

Nonsynchronized constant-mesh transmissions are common in long-distance trucks in North America because of their robustness. However, synchromesh transmissions are standard in Europe.

Figure 3.12 shows a five-speed constant-mesh gearbox with a synchromesh two-speed, rear-mounted range box. The main gear box has five forward gears and one reverse gear. The rear-mounted two-speed range box extends the application of the gearbox into ten speeds.

There is a countershaft on either side of the mainshaft. Each of the countershafts is provided with an identical grouping of countershaft gears, which are in constant meshing with a group of mainshaft gears. Power flows to the main gearbox through the input first motion shaft and gear wheel, and is divided between the two countershaft gear wheels and the corresponding second-stage gears of the mainshaft. Three dog clutches can be used to clutch the input gear to the mainshaft. Torque will be transferred from the input shaft to the mainshaft when the selected dog clutch hub is slid in to mesh with the designed gear dog teeth.

The basic idea behind the two countershaft gear wheels is to reduce the gear contact face width. These two countershafts are located in the same plane as the mainshaft for reasons of symmetrical loading. The gear wheel on the mainshaft can be meshed with gears in the two countershafts. In theory, the gear face width of countershafts can be reduced by 50%; however, face width can be reduced by 40% compared with the gearbox using a single countershaft.

The auxiliary transmission range section includes two substantially identical auxiliary countershaft assemblies, each containing an auxiliary countershaft carrying two auxiliary section countershaft gears. The auxiliary countershaft gears are constantly meshed with and support the range/output gear while auxiliary section countershaft gears are constantly meshed with output gear, which is fixed with the transmission output shaft. A two-position clutch assembly is provided for clutching either for low-range operation or for direct or high-range operation of the transmission.

To prevent gears from grinding or clashing during engagement, a constant-mesh, fully "synchronized" manual transmission is equipped with synchronizers. A synchronizer typically consists of an inner-splined hub, an outer sleeve, shifter plates, lock rings (or springs), and blocking rings. The hub is splined onto the mainshaft between a pair of main drive gears. Held in place by the lock rings, the shifter plates position the sleeve over the hub while also holding the floating blocking rings in proper alignment [3.3].

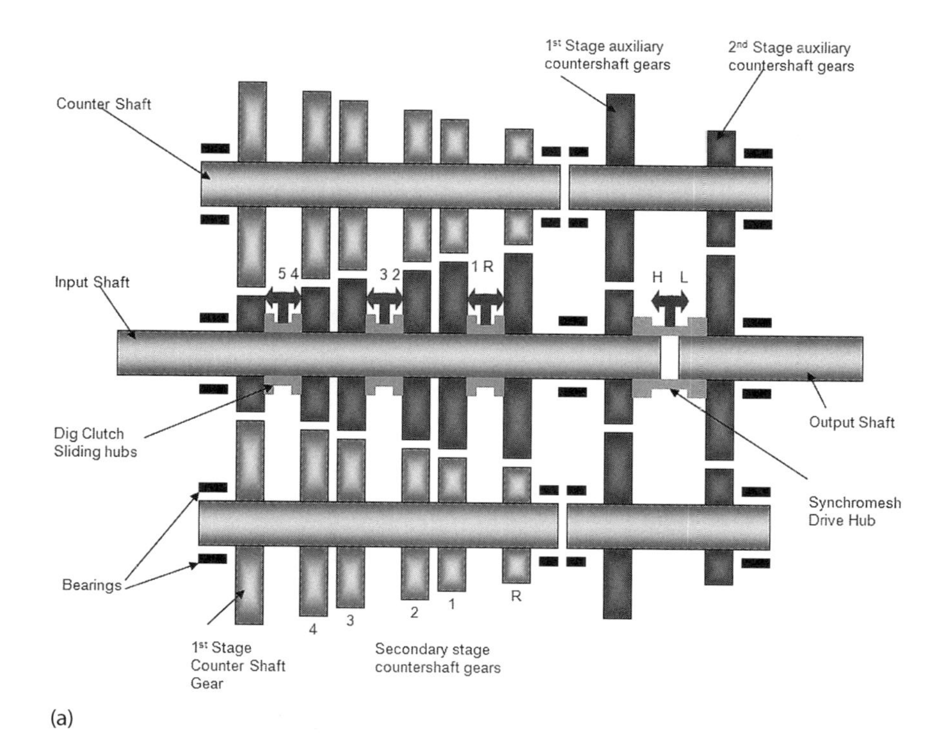

(a)

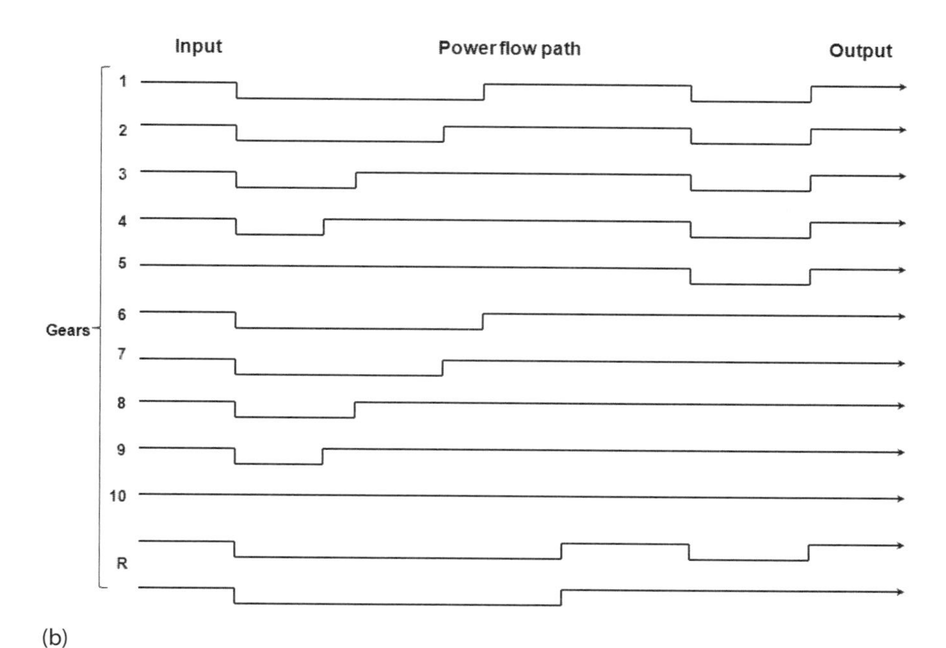

(b)

FIG. 3.12 Eaton Fuller 10 twin countershaft ten-speed constant-mesh gear box with two-speed range box. (Courtesy of Eaton Corporation)

(a)

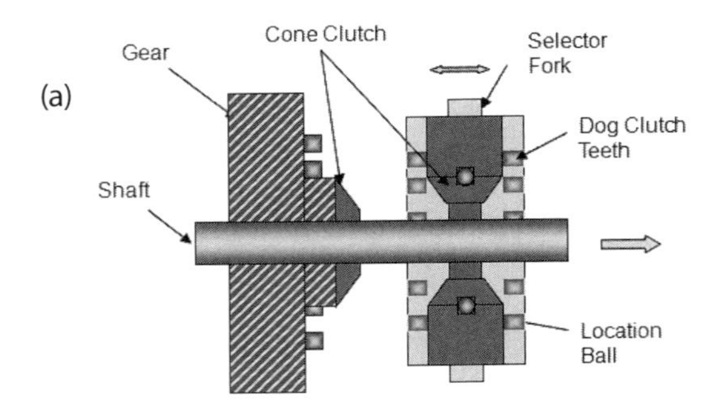

(b)

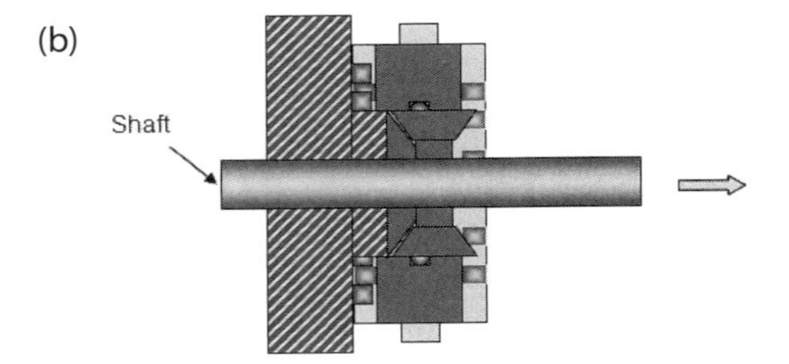

FIG. 3.13 Nonpositive constant-load synchromesh unit.

Figure 3.13 shows a synchromesh unit with a dog clutch along with a cone clutch. The dog clutch is positively splined to the mainshaft shaft and an outer ring that can slide forward against each gear. Both the mainshaft gear and the ring of the dog clutch have a row of teeth (Fig. 3.13 a). Moving the shift linkage moves the dog clutch against the adjacent mainshaft gear, causing the teeth to interlock and solidly lock the gear to the mainshaft. The cone clutch is usually made of a softer material, such as brass. When the pair of conical faces contact, friction torque will be generated due to the combination of the axial thrust and the difference of input and output shaft members. It allows the teeth on the dog clutch to engage without a hub sleeve (Fig. 3.13 b).

3.5 Automated Manual Transmission

An automated Manual Transmission (AMT) automates clutch actuation or shifting of a manual transmission. The automated shifting is controlled electronically (shift-by-wire) and performed by electric motors or a hydraulic system. AMT provides the convenience of automatic clutch actuation and gear shifting while retaining the excellent fuel efficiency of the manual transmission.

Figure 3.14 shows a schematic of an automated manual transmission for long-haul commercial vehicles. The AMT is matched with an electronically controlled prime mover such as a diesel engine or a CNG engine. The ECU of

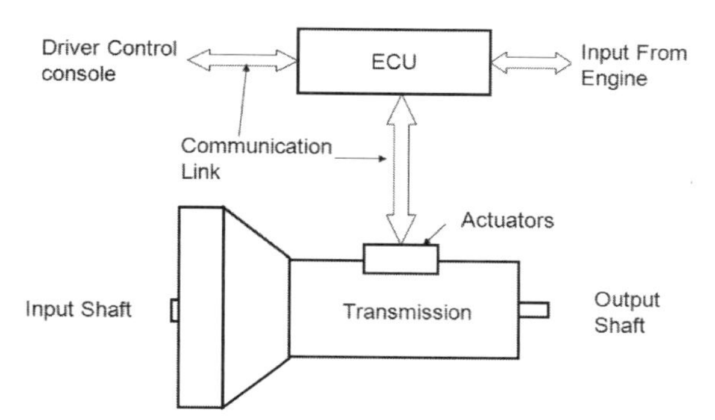

FIG. 3.14 Schematic of AMT and controls.

the AMT communicates to the engine electronic control module through a date link as defined in SAE J1922 and J1939 or ISO 11898 protocols. Electric control with automatic clutch activation and engine management will be the focus of this section.

An AMT control system includes an ECU for receiving input signals from the engine speed sensor, input shaft speed sensor, output shaft speed sensor, and driver control console (Fig. 3.15). The ECU may also receive inputs from an auxiliary section position sensor, and to the data links as defined in SAE J1922 and J1939 or ISO 11898 protocols. The ECU processes the inputs in accordance with predetermined logic rules to issue command output signals to control the main section shift actuator and the auxiliary section shift actuator, and to the driver control console.

The driver control console, shown in Fig. 3.15, includes three buttons that indicate the transmission is in a forward drive, neutral, or reverse drive condition, respectively. The console also includes three selectively lighted pushbuttons, which allow the operator to select a manual mode, upshift, or downshift. A selection made by depressing or pushing any one of button R, N, D, and

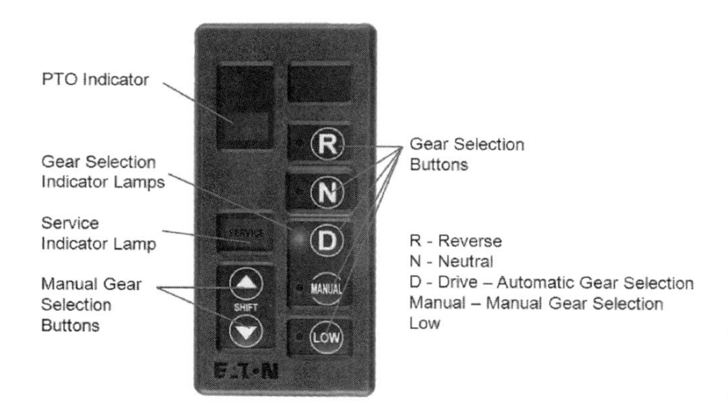

FIG. 3.15 Driver control console. (Courtesy of Eaton Corporation)

Manual and may be canceled by redepressing the buttons. As an alternative, multiple depressions of buttons may be used as commands for skip shifts. In operation, to select upshifts and downshifts manually, the operator will depress either button ▲ or button ▼ as appropriate. The selected button will then be lighted until the selected shift is implemented or until the selection is canceled.

The driver control console allows the operator to select a manual or hold mode of operation for manually selecting a shift in a given direction (i.e., upshifts or downshifts) or to neutral from the currently engaged ratio, or to select a semi-automatic preselect mode of operation, and provides a display for informing the operator of the current mode of operation (automatic or manual preselection of shifting), the current transmission operation condition (forward, reverse, or neutral), and of any ratio change or shift (upshift, downshift, or shift to neutral) that has been preselected but not yet implemented.

To implement a selected shift, the ECU commands the XY Shifter actuator to shift main transmission section into neutral. In the manual controlled transmission, this is accomplished by the operator causing a torque break or reversal by manually momentarily decreasing and/or increasing the supply of fuel to the engine and/or manually disengaging the master clutch. In an automated transmission, a torque break can be achieved by manipulating engine fueling through the communication between transmission ECU and the engine control module.

As the transmission is shifted into neutral, and neutral is verified by the ECU (neutral sensed for a period of time such as 1.5 seconds), the neutral condition button N is lighted. If the selected shift is a compound shift, i.e., a shift of both the main section and of the range section, such as a shift from fourth to fifth speeds, the ECU will issue command output signals to cause the auxiliary section actuator to complete the range shift after neutral is sensed in the front box.

When the range auxiliary section is engaged in the proper ratio, the ECU will calculate an enabling range of input shaft speeds based upon sensed output shaft (vehicle) speed and the ratio to be engaged, which will result in an acceptably synchronous engagement of the ratio to be engaged. When the input shaft speed falls within the acceptable range by throttle manipulation and/or use of the input shaft brake, the ECU will issue command output signals to cause the actuator to engage the main section ratio.

To select a shift into transmission neutral, selection button N is pushed. The indicating light will flash until the ECU confirms that neutral is obtained, at which time the light will assume a continuously lighted condition while the transmission remains in neutral.

In the automatic preselection mode of operation, selected by pressing pushbutton D, the ECU will, based upon stored logic rules, currently engaged ratio (which may be calculated by comparing input shaft to output shaft speed), and output shaft speed, determine if an upshift or a downshift is required and preselect the same. The operator is informed that an upshift or downshift is preselected and will be semi-automatically implemented by a command output signal from the ECU causing either lighted pushbutton ▲ or lighted pushbutton ▼ to flash and/or an audible shift alert signal. The operator or controller may initiate semi-automatic implementation of the automatically preselected shift as indicated above or may cancel the automatic mode by depression of pushbutton D.

The computer-controlled shift feature optimizes the engine operating condition during the shifts, which improves engine operating performance and fuel economy and improves the clutch lift. Its clutchless "float shifting" between gears eliminates hundreds of clutch actuations per day and provides driver comfort and extended clutch life.

3.6 Dual-clutch Transmission

In a conventional manual transmission or in an automated manual transmission, there is not a continuous flow of power from the engine to the wheels during gear shifting. The power delivery changes from on to off to on during gearshift, causing a phenomenon known as "torque interrupt." The torque interruption can become an issue of driver comfort and driveline durability for light- and medium-duty commercial vehicles and for stop-and-go heavy-duty commercial vehicles. Dual-clutch transmissions (DCT) offer the advantages of smooth gear shift without torque interruption, as with the fully automatic transmission and the high drivetrain efficiency of the standard manual transmission.

The dual-clutch transmission concept was invented by Adolphe Kégresse, a pioneer in automotive engineering, in 1939. However, the development of the DCT technologies was not until the 1980s by Audi and Porsche for car racing. Since the first series production for the 2003 Volkswagen Golf Mk4 R32, the DCT technology has attracted the attention of automakers and commercial vehicle manufacturers. In 2008, there were about 600,000 cars and light commercial vehicles with DCT transmissions being sold in Europe [3.21].

A dual-clutch transmission offers the function of two manual gearboxes in one, with two clutches (Fig. 3.16). Unlike a manual, the two clutches in a DCT are linked to two input shafts, and the shift and clutch actuation is controlled through a transmission control module. The input shaft from the engine is connected through the damper with the outer plates of both clutches. One clutch operates with one shaft with odd gears (first, third, fifth, and reverse), while the other clutch operates the shaft with even gears (second, fourth, and sixth) and a shared transmission output shaft.

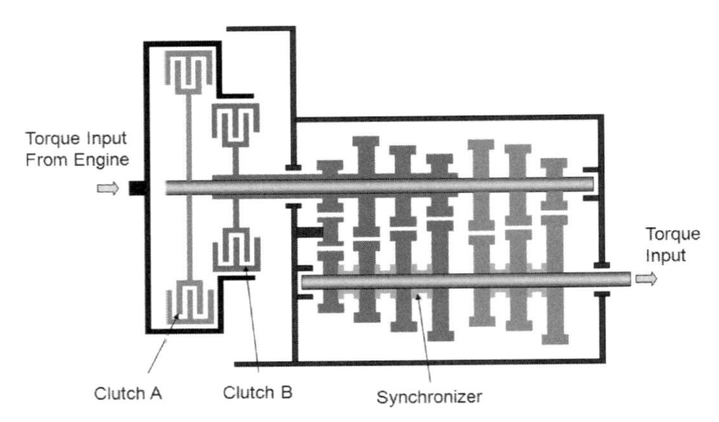

FIG. 3.16 Schematic of dual-clutch transmission. (Courtesy of Eaton Corporation)

A transmission control unit commands the clutch engagement and gear shifting. When a gear shift is commanded, the transmission control unit preselects the appropriate gear in the unused shaft to transfer power before a shift is made. Using this arrangement, gears can be changed without interrupting the power flow from the engine to the transmission.

Figure 3.17 shows the simulated shifting performance of a medium-duty commercial vehicle (vehicle weight 30,000 lb) with both conventional manual transmission and dual-clutch transmission [3.15]. It shows that the DCT shifting performance over a single-clutch transmission is not as good as for passenger cars. Cost and durability are also concerns for DCT applications in the commercial vehicle market.

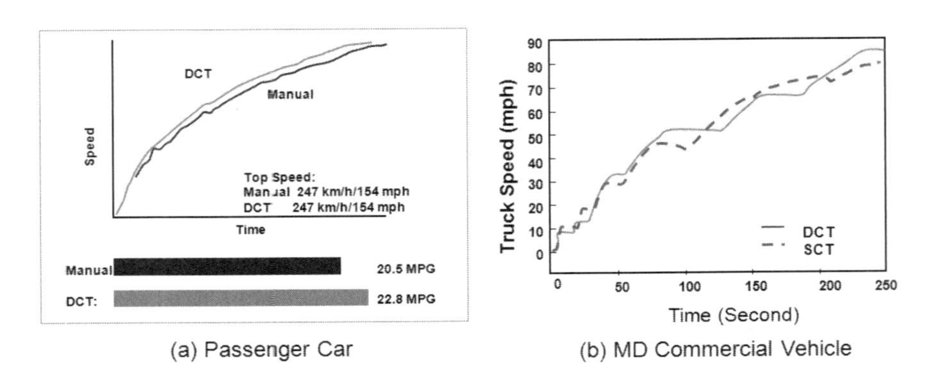

FIG. 3.17 DCT shifting performance for passenger cars and for medium-duty commercial vehicles [3.15].

3.7 Fully Automatic Transmission

A fully automatic transmission for heavy-duty commercial and industrial vehicles and equipment is a gearbox that can free the driver from having to engage the clutch and to shift gears manually. There are three types of fully automatic transmissions:

1. Conventional automatic transmissions, which consist of a torque convertor and rear-mounted planetary gear sets.
2. Fully automatic countershaft transmissions such as the automated manual transmission and dual-clutch transmission, as discussed in the previous chapter.
3. Continuously variable transmissions such as the belt-driven transmission and hydrostatic transmissions, which will be discussed in the next chapter.

Planetary gear sets have a theoretical advantage in weight savings when compared with countershaft units because the gear separating forces are counterbalanced within the ring gear. However, planetary automatics also have relatively fixed relationships between the ratios, especially if a compact, lightweight design is required. This reduces the resolution available between ratios. The recent adoption of the Lepelletier planetary gear set has helped to reduce planetary transmission size, weight, and control complexity. Parasitic losses in the actuating elements are reduced due to a decrease in the number of clutches required. These benefits are offset by spin and gear mesh losses and rotating inertia, which are higher in Lepelletier gear sets compared with Ravigneaux, Simpson, and other layouts.

Although the planetary transmission was invented in the 1880s, the world's first mass-produced automatic transmission was not introduced until 1939 by General Motors for its automotive applications. In 1947, Allison Transmission, then a division of General Motors, introduced a fully automatic transmission for commercial vehicles. Although the fully automatic transmissions have a higher unit cost compared to standard transmissions, the advantages of requiring a low level of driving skills are especially suited for vocational applications such as garbage packers, fire trucks, utility vehicles, and pickup and delivery applications. Using a fully automatic transmission in a frequent stop-and-go operation such as picking up garbage is not only kinder to the driver but also reduces the drivetrain shock load because of its fluid coupling. With the electric controls and more-precise range shifting, the transmission is more protected from driver abuse.

Figure 3.18 shows a schematic of a planetary gear set arrangement from a fully automatic transmission for heavy-duty vocational applications such as buses, fire trucks, and garbage packers [3.31]. It uses two rotating clutches, three planetary gear sets, and three stationary clutches to provide six forward

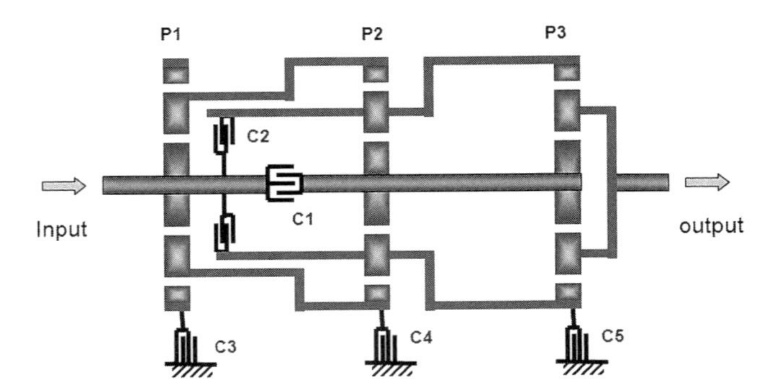

FIG. 3.18 Schematic of fully automatic transmission.

ratios, a neutral, and a reverse, as shown in Table 3.3. The fully automatic transmission consists of major subsystems called modules. Each module is an integral subassembly that may be removed for service, repair, or replacement as separate unit, providing significant service advantage. The modules may be grouped into a) input modules, b) gear box modules; c) output modules, and d) electronic control modules.

The input modules consist of torque converter module, converter housing modules, and charging pump modules. The input modules are located between the engine and the transmission gearing. The gear box modules have five clutch assemblies and three planetary gear sets. The output modules consists of transfer case and output retarder. The electronic control module is located outside of the main transmission housing. This unit is responsible for processing input signals and issuing command signals.

3.8 Continuously Variable Transmissions for Commercial Vehicles

A continuously variable transmission (CVT) is a transmission which can change steplessly through an infinite number of effective gear ratios between maximum and minimum values. This contrasts with other mechanical

TABLE 3.3 Clutch Application and Speed Status of a Fully Automatic Transmission

	C1	C2	C3	C4	C5	Gear Ratio
FWD 1	x				x	3.49
FWD 2	x			x		1.86
FWD 3	x		x			1.41
FWD 4	x	x				1.00
FWD 5		x	x			0.75
FWD 6		x				0.65
REV 1			x		x	–5.03

transmissions that only allow a few different distinct gear ratios to be selected. The flexibility of a CVT allows the driving shaft to maintain a constant angular velocity over a range of output velocities. This can provide better fuel economy than other transmissions by enabling the engine to run at its most efficient revolutions per minute (RPM) for a range of vehicle speeds. Alternatively, it can be used to maximize the performance of a vehicle by allowing the engine to turn at the RPM at which it produces peak power. This is typically higher than the RPM that achieves peak efficiency.

Leonardo da Vinci conceptualized the CVT more than 500 years ago, and the first toroidal CVT patent was filed in 1886. Today, several car manufacturers, including General Motors, Toyota, Audi, Honda, and Nissan, are designing their drivetrains around CVTs.

There are three types of CVTs:

1. Mechanical Continuously Variable Transmission
2. Hydrostatic Transmission
3. Electrically Variable Transmission

The mechanical CVTs such as pulley-based or toroidal-based are frictional CVTs that work by varying the radius of the contact point between two rotating objects. The pulley-based CVTs have been used in passenger cars for torque up to 350 N·m. In 2009, Allison Transmission signed a joint development agreement with Torotrac to develop a CVT with toroidal variators for light- and medium-duty commercial vehicles [3.26, 3.27]. Currently, the CVTs in commercial applications include electrically variable transmissions for on-road applications [3.32], such as for city buses and delivery vehicles, and hydrostatic transmissions for off-road vehicles [3.28].

3.8.1 Mechanical Continuously Variable Transmission

The pulley-based CVT was the first commercially successful continuously variable transmission. The first successful automotive CVTs were offered by DAF in 1955. The variable-diameter pulleys, which consist of two pulleys connected with a V-shaped belt, are the heart of a CVT. The pulley that is connected to the engine is called the driving or primary pulley; the other pulley is connected to the wheels and is called the driven or secondary pulley [3.16]. By changing the axial position of the moveable sheave of each pulley, the pitch radius of the belt is changed and, in turn, the transmission ratio is modified, as shown in Fig. 3.19.

CVTs may use hydraulic pressure, centrifugal force, or spring tension to create the force necessary to adjust the pulley halves. The pulley-based CVT has the advantages over the conventional transmission design of acceleration

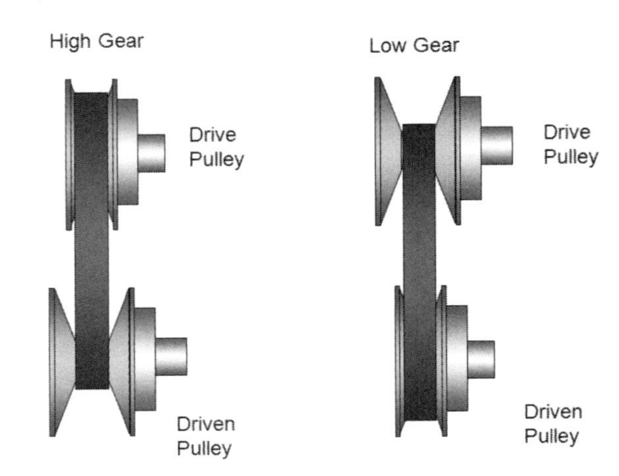

FIG. 3.19 Schematic of a V-shaped pulley.

without the jerk or jolt from changing gears. However, one of the major limitations of the belt CVTs is packaging. The two offset but parallel shafts that carry the sheaves are spaced much farther apart than the shafts in a countershaft transmission. This wide spacing makes it so difficult to package a belt-type CVT in a rear-drive car or light truck that the general consensus is that these will be limited to front-drive applications. Another limitation of the belt CVT is that the system lacks durability for the long-haul commercial vehicle application. Eaton looked into a pulley-based CVT for commercial vehicle applications in early 2000. The project was stopped because the system lacked durability for the long-haul commercial vehicle industry.

Another version of the mechanical CVT—the toroidal CVT system—replaces the belts and pulleys with discs and power rollers. Although such a system seems drastically different, all of the components are analogous to a belt-and-pulley system and lead to the same results—a continuously variable transmission [3.26]. Figure 3.20 shows the schematics of a toroidal-based CVT system. The function of the CVT system is as follows:

- One disc connects to the engine. This is equivalent to the driving pulley.
- Another disc connects to the drive shaft. This is equivalent to the driven pulley.
- Rollers or wheels located between the discs act as the belt, transmitting power from one disc to the other.

The wheels can rotate along two axes. They spin around the horizontal axis and tilt in or out around the vertical axis, which allows the wheels to touch the discs in different areas. When the wheels are in contact with the driving disc near the center, they must contact the driven disc near the rim, resulting in a

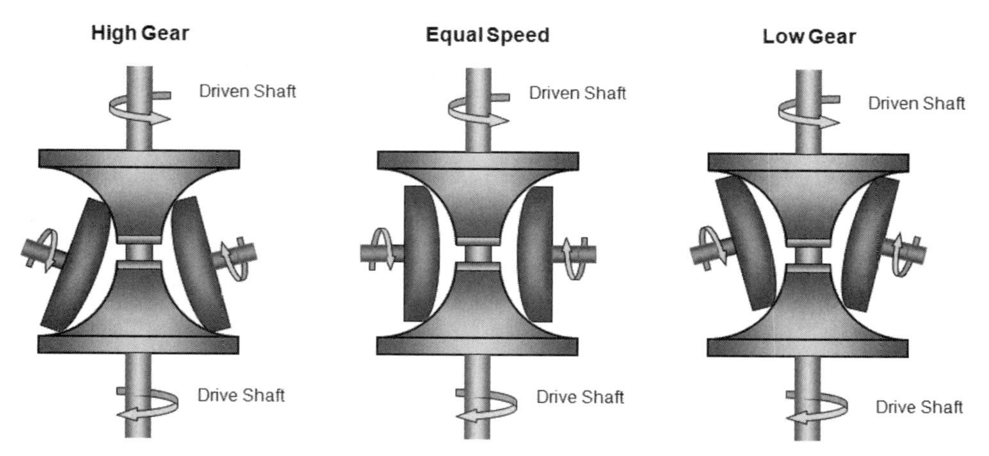

FIG. 3.20 Schematic of a toroidal variator.

reduction in speed and an increase in torque (i.e., low gear). When the wheels touch the driving disc near the rim, they must contact the driven disc near the center, resulting in an increase in speed and a decrease in torque (i.e., overdrive gear). A simple tilt of the wheels, then, incrementally changes the gear ratio, providing for smooth, nearly instantaneous ratio changes.

The variator alone cannot provide neutral and reverse drive, nor can it provide the ratio spread to achieve high overdrive ratios. Figure 3.21 shows schematics of a toroidal CVT. The output from the variator is connected to both the sun gear of the planetary gear set and the output shaft of the transmission. The annulus of the planetary is also linked to the output shaft of the transmission through another wet plate clutch (the low regime clutch). It allows the variator to provide forward and reverse rotation, generation of high

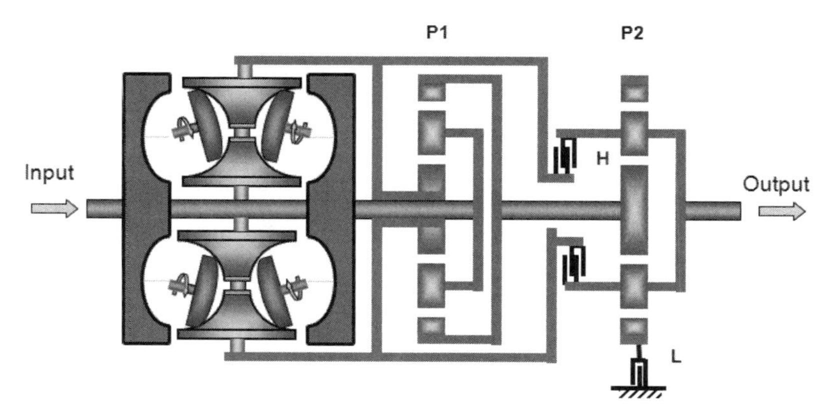

FIG. 3.21 Schematics of a toroidal CVT.

output torques in "low regime," overdrive capabilities in "high regime," and the "geared neutral" function of a zero output speed.

3.8.2 Hydrostatic CVTs

Another type of CVT, known as a hydrostatic CVT, uses variable-displacement pumps to vary the fluid flow into hydrostatic motors. In this type of transmission, the rotational motion of the engine operates a hydrostatic pump on the driving side. The pump converts rotational motion into fluid flow. Then, with a hydrostatic motor located on the driven side, the fluid flow is converted back into rotational motion.

Often, a hydrostatic transmission is combined with a planetary gear set and clutches to create a hybrid system known as a hydromechanical transmission. Hydromechanical transmissions transfer power from the engine to the wheels in three different modes. At a low speed, power is transmitted hydraulically, and at a high speed, power is transmitted mechanically. Between these extremes, the transmission uses both hydraulic and mechanical means to transfer power. Hydromechanical transmissions are ideal for heavy-duty applications, which is why they are common in agricultural tractors and all-terrain vehicles.

The basic structure of the hydromechanical transmission is to use one planetary gear set as the differential mechanism, which includes input split and output split. The input split is the combination of hydraulic motor and output shaft at a constant speed ratio, as shown in Fig. 3.22. Output split is the combination of hydraulic pump and input shaft at a constant speed ratio. Fundamentally, they are the same mechanism. The split type is determined by which shaft is input or output.

The hydromechanical transmission can be linked to a hydraulic accumulator to adjust the amount of power flow through the hydraulic path, as

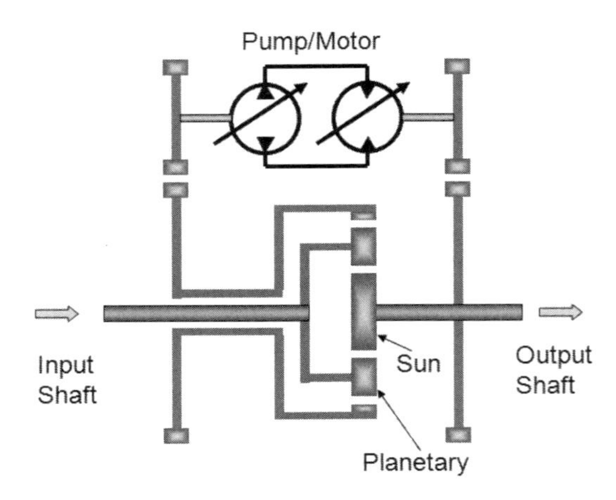

FIG. 3.22 Schematic of input—split hydrostatic transmission.

opposed to the mechanical path. It is especially useful for regenerative hydro-mechanical drivetrains for the application of hydraulic hybrids to enable the engine speed and vehicle velocity to be independent. Various configurations of hydraulic power generation, storage, and usage can be implemented for optimized vehicle drivability and fuel savings.

3.8.3 Electric Variable Transmission

The Electric Variable Transmission (EVT) is a transmission that achieves CVT action by using an electric motor to operate at any speed ratio through the transmission. An EVT is usually designed around a planetary gear system. The planetary gear acts as a differential, performing a "power-split" function; a portion of the mechanical power is carried directly through the gear set (the "mechanical path"). The rest of the power is converted to and from electrical energy by electric motor-generators (the "electrical path").

Figure 3.23 is a schematic cross section of this one-mode EVT arrangement that includes a single set of planetary gears, two electric motors, and the connecting shafts. The input shaft is on the far left and is connected to the ring gear. The smaller of the two motors is connected with a sleeve shaft to the sun gear. The planets are on a carrier, which is connected to the long output shaft. The output shaft extends from the carrier through the hollow sun gear and sleeve shaft to the far right, and the output shaft holds the larger of the two motors.

This one-mode EVT arrangement is known as an input-split EVT because the input is connected by itself to the planetary gearing, and the power flow through the transmission is effectively split by the gearing at the input.

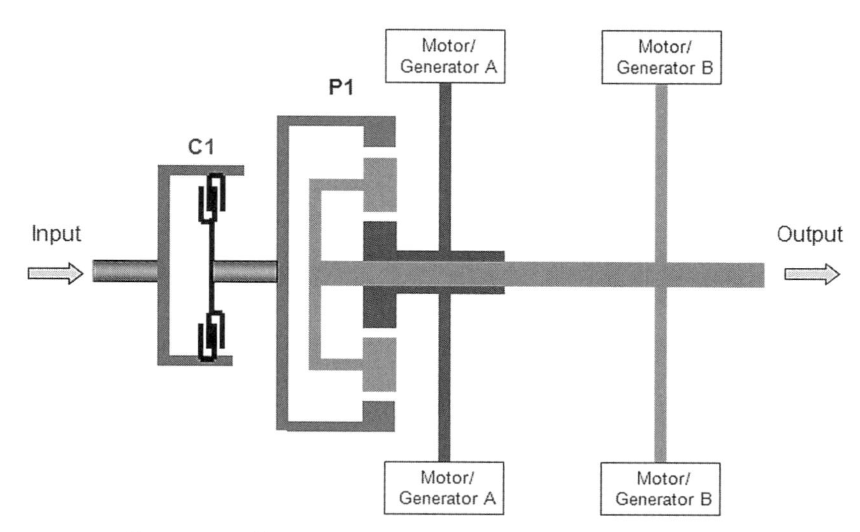

FIG. 3.23 Electric Variable Transmission [3.31, 3.32].

Typically, some of the input power flows to motor A, which acts as a generator and turns that power into electricity. The rest of the input power flows along the output shaft. Output shaft power is added from motor B, which turns the electrical power from motor A back into mechanical power, except for the fraction lost in these conversions. Thus, there are two power paths through the transmission from input to output: an entirely mechanical path from input to gears to output, and an electrical or electromechanical path from input to gears to generator (A) to motor (B) to output.

In an input-split EVT, the motor A controls the speed ratio through the transmission using the sun gear and typically generates electricity. The motor B is connected directly to the output shaft and does not affect the speed ratio. For the planetary gear set, the speed of the carrier is the weighted average of the speed of the ring and the speed of the sun. So, for this particular EVT arrangement, which maximizes output torque, the speed of the output is the weighted average of the speed of the input and the speed of motor A. Motor B has the same speed as the output.

3.9 Characteristics and Efficiency of Commercial Vehicle Transmissions

There is a wide range of transmissions for commercial vehicles. Table 3.4 shows the transmission efficiency under the U.S. Federal Test Procedure Test Cycle. Figure 3.24 shows a study on comfort rating vs. transmission efficiency. The characteristics of major transmissions discussed in this chapter can be summarized as follows:

The conventional manual transmission offers the best fuel efficiency, lowest cost, and best durability for heavy-duty long-haul commercial vehicles. However, AT and AMT are more popular at the light- and medium-duty commercial vehicle market and the vocational heavy-duty market when driver comfort becomes an important factor for frequent shifting.

TABLE 3.4 Average Operating Efficiencies of Advanced Transmission Technologies [3.23] Over the U.S. Federal Test Procedure Drive Cycle

Type	Start-Up Device	Efficiency over the FTP
Manual or robotized countershaft	Dry clutch	96–97%
Automatic countershaft	Dual wet clutch	94–95%
Toroidal CVT	Lock-up torque converter	88–90%
Belt CVT	Lock-up torque converter	83–85%
Planetary automatic	Lock-up torque converter	85–87%

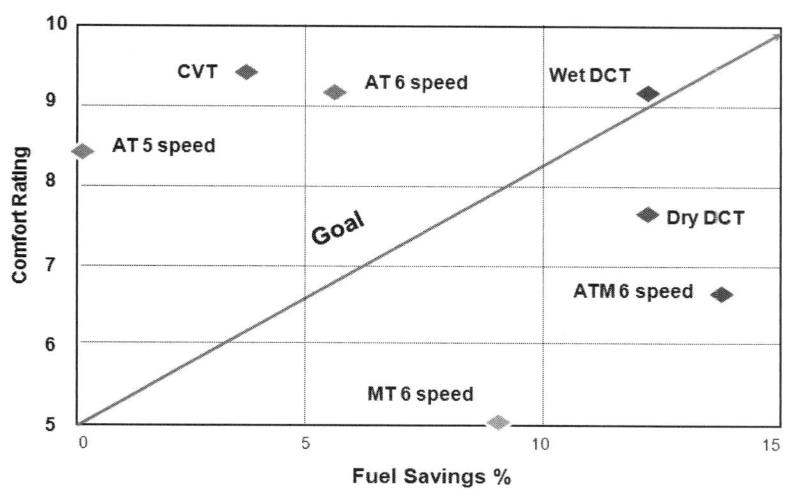

FIG. 3.24 Fuel efficiency vs. engine displacement comparison for automatic transmissions [3.14].

The AMT, using an automated single dry clutch for shifting, provides the best fuel efficiency and the lowest cost, but does not meet customer expectations due to the uncomfortable torque interrupt when shifting, in particular in the lower gears.

The fully automatic transmission (torque converter) offers drive comfort. However, it has lower drivetrain efficiency compared with manual transmissions and AMT.

CVTs have the disadvantage of low transmission efficiency. The pulley-type CVT has limited application for commercial vehicles due to its limited torque range. The toroidal CVT has the potential for light and medium commercial vehicles, and the hydrostatic transmission has broad applications for off-road vehicles.

DCT provides excellent fuel efficiency, responsiveness, and the shift and launch comfort of an automatic transmission. It may have broad potential for light and medium commercial vehicles. For long-haul heavy-duty commercial vehicles, the cost and durability may outweigh the benefit of smoothness of shifting.

3.10 References

3.1 Lechner, G. and Naunheimer, H., *Automotive Transmission—Fundamentals, selection, design and application*, Springer, 1999, pp. 7–17.

3.2 http://www.zf.com/corporate/en/company/tradition/company_history/theme_downloads/theme_downloads.html.

3.3 Heisler, H., *Advanced Vehicle Technology*, Second Edition, SAE International, Warrendale, PA, 2002, pp. 60–62.
3.4 Heisler, H., *Advanced Vehicle Technology*, Second Edition, SAE International, Warrendale, PA, 2002, p. 43.
3.5 *Truck System Design Handbook*, PT-41, SAE International, Warrendale, PA. 1992, p. 458.
3.6 Heisler, H., *Advanced Vehicle Technology*, Second Edition, SAE International, Warrendale, PA, 2002, p. 99.
3.7 Lechner, G. and Naunheimer, H., *Automotive Transmission—Fundamentals, selection, design and application*, Springer, 1999, p. 267.
3.8 Bosch, *Automotive Handbook*, Third Edition, 1993, SAE International, Warrendale, PA.
3.9 Fitch, J. W., *Motor Truck Engineering Handbook*, Fourth Edition, SAE International, Warrendale, PA, 1994, p. 553.
3.10 Bennett, S. and Norman, I. A., *Heavy Duty Truck System*, 4th Edition, Delmar Cengage Learning, 2006, p. 531.
3.11 "Transmission Fundamentals," shttp://alphacars.com/Documents/Tech%20BMW%20Transm%202.pdf.
3.12 "Automatic-shifting dual-clutch transmissions are poised to grab share from traditional transmissions thanks to their combination of efficiency and convenience" (PDF), *AEI-online.org* (DCTfacts.com), June 2009.
3.13 Amendola, C. and Alves, M., "Gear shift strategies analysis of the automatic transmission in comparison with the double clutch transmission," SAE Paper No. 2006-01-2872, SAE International, Warrendale, PA, 2006.
3.14 Matthes, B., "Dual Clutch Transmissions—Lessons Learned and Future Potential," SAE Paper No. 2005-01-1021, SAE International, Warrendale, PA, 2006.
3.15 Song, X., Liu, J., and Smedley, D., "Simulation Study of Dual Clutch Transmission for Medium Duty Truck Applications," SAE Paper No. 2005-01-3590, SAE International, Warrendale, PA 2005.
3.16 Harris, W., *How CVT works*, http://auto.howstuffworks.com/cvt.htm.
3.17 Amsallem, M., "Control system/method for automated mechanical transmission systems," US patent: 5,729,454.
3.18 Markyvech, R. and Genise, T., "Control for transmission system utilizing centrifugal clutch," US patent: 6,561,948.
3.19 Eaton Fuller Ultrashift Plus, Installation Guide, TRIG1110 2009.
3.20 Nice, M., "How Automatic Transmissions Work," http://auto.howstuffworks.com/automatic-transmission12.htm.
3.21 "How the Dual Clutch Transmission Works," http://www.dctfacts.com/widc_pg3a.asp.
3.22 Harris, W., "How Dual-clutch Transmissions Work," http://auto.howstuffworks.com/dual-clutch-transmission3.htm.
3.23 "Future Transmission Technologies," Copyright 2001 by DRI..WEFA, Inc.

3.24 Scherer, H. and Bek, M., "ZF New 8-speed Automatic Transmission 8HP70—Basic Design and Hybridization," SAE Paper No. 2009-01-0510, SAE International, Warrendale, PA, 2009.
3.25 Bender, J. G. and Struthers, K. D., "Advanced Controls for Heavy Duty Transmission Applications," SAE Paper No. 901157, SAE International, Warrendale, PA, 1990.
3.26 Brockbank, C. and Greenwood, C., "Developments in Full-Toroidal Traction Drive Infinitely & Continuously Variable Transmissions," www.torotrac.com.
3.27 Brockbank, C., "Application of a Variable Drive to Supercharger & Turbo Compounder Applications," SAE Paper No. 2009-01-1465, SAE International, Warrendale, PA, 2009.
3.28 "Hydrostatic Transmissions," http://www.hydraulicspneumatics.com/200/TechZone/HydraulicPumpsM/Article/True/6450/TechZone-HydraulicPumpsM, assessed, Sept. 2011.
3.29 "Facts about gear: contribution by Bodmer," http://www.britannica.com/facts/5/287646/gear-as-discussed-in-Johann-Georg-Bodmer-Swiss-inventor, assessed, Sept. 2011.
3.30 "Transmission Fundamentals," http://alphacars.com/Documents/Tech%20BMW%20Transm%202.pdf, assessed, Sept. 2011.
3.31 Bullock, B. L., "The Allison MD3066 Transmission," SAE Paper No. 982792, SAE International, Warrendale, PA, 1998.
3.32 Grewe, T. M. et al.," Defining the General Motors 2-Mode Hybrid Transmission," SAE Paper No. 2007-01-0273, SAE International, Warrendale, PA, 2007.

3.11 Appendix

Dry and Wet Clutch: a coupling device is used to separate the engine and transmission when necessary. A "wet clutch" is immersed in a cooling lubricating fluid, which also keeps the surfaces clean and gives smoother performance and longer life. A "dry clutch," as the name implies, is not bathed in fluid.

Dog clutch: a type of clutch that couples two rotating shafts or other rotating components, not by friction but by interference. The two parts of the clutch are designed such that one will push the other, causing both to rotate at the same speed and never slip.

Constant-mesh transmissions: also called sliding-mesh gearbox, individual gears are mounted so they always engage the shaft, but gears on one shaft can be moved axially. To engage a particular pair of gears, one gear is slid axially until it fully engages a gear on the other shaft.

Synchromesh: a synchronizing mechanism to slow down the fast gear, so that engagement of the shifting mechanism can be made quickly and noiselessly without double clutching.

Splitter: a gear set for splitting the input power into two or more output powers with different gear ratios.

Range Box: a gear set for multiplying the range ratio of the main gearbox. The range box is always used for speed reducing and to increase the torque output.

Transfer Gearbox: a device to split the input power into more than one powered axle.

Power Take-Off (PTO): a device for splitting power into an auxiliary unit.

Torque Convertor: a fluid coupling device used to transfer rotating power from a prime mover, such as an internal combustion engine or electric motor, to a rotating driven load.

Ratio Span: Ratio Span = (Road speed in highest gear)/(Road speed in lowest gear).

Chapter 4

Energy Storage Systems: Battery, Ultracapacitor, Accumulator, and Flywheel

4.1 Energy Storage Systems for Commercial Hybrid Applications

Powertrain hybridization of commercial vehicles improves overall drivetrain efficiency over a conventional drivetrain by capturing the vehicle kinetic energy during braking, storing the energy in an energy storage system (ESS), and supplying the energy to assist the main power source during high power demands. A high-quality ESS is one of the most crucial components; it affects vehicle performance characteristics, as well as vehicle safety, reliability, and durability.

There are several types of energy storage systems that are being developed for hybrid applications of commercial vehicles. The energy storage systems to be discussed in this chapter include:

- Electrical chemical batteries
- Supercapacitor
- Hydraulic/pneumatic accumulator
- Flywheel energy storage system

Other types of energy storage systems, such as fuel cells as energy storage systems for commercial vehicles, are discussed in Chapter 9.

4.1.1 Hybrid Vehicle Energy Storage System Operating Requirements

Hybrid vehicles for commercial applications can be classified according to the energy storage system as Hybrid-Electric Vehicles (HEV) when the electrical chemical battery or supercapacitor, or a combination of the battery and supercapacitor, are used as the energy storage system; or as Hybrid Hydraulic Vehicles (HHV) when a hydraulic accumulator is used as the energy storage system. The ESS stores the energy recuperated during regenerative braking and depletes for assisted starting or acceleration for a parallel hybrid powertrain because the internal combustion engine provides the primary propulsion. For a series hybrid powertrain, the internal combustion engine provides energy to ESS, and the EES provides all required energy for vehicle propulsion. The ESS can be optimized for powertrain efficiency, performance, and cost [4.1, 4.2, 4.3]. Some examples of hybrid powertrain design goals that affect the specification of ESS are:

Efficiency Optimization—The ESS provides power for load-leveling and the engine runs at its most efficient operating load and speed to improve vehicle fuel economy. The improved efficiency reduces the fuel consumption, which in turn automatically reduces exhaust emissions.

Efficiency Boost—The energy of the ESS can be used to provide power boost for acceleration and hill climbing.

Stop/Start Mode—When the vehicle is temporarily stationary at traffic lights or in traffic jams etc., the engine can be switched off to save fuel and the vehicle can be powered by the ESS alone. The engine is restarted when a vehicle predetermined speed is reached.

Plug-in HEV Mode—The plug-in hybrid-electric commercial vehicle (PHEV) is designed to be used both as pure electric vehicle or output electric power for utility application and as HEV when the charge is depleted for highway driving. Battery ESS can be charged from the electric grid and be optimized for high energy capacity to allow the vehicle to be powered by the energy from ESS only or the pure EV mode.

Because of the wide range of operating requirements, the ESS can be designed to meet the specific requirements of the hybrid vehicle for commercial applications. Figure 4.1, which shows the performance comparison of various energy storage devices, is called the Ragone plot [4.4]. The Ragone plot is useful for characterizing the tradeoff between effective energy capacity and power of the ESS. The sloping lines on the Ragone plots indicate the relative time to charge or discharge of the ESS. For example, supercapacitors can be charged or discharged in microseconds, which makes them ideal for capturing regenerative braking energy in HEV applications. On the other hand, lead acid batteries have poor dynamic performance for energy capture and delivery. Lithium batteries are somewhere in between and provide a reasonable compromise between the two.

Table 4.1 lists typical properties of the ESS available in the market to meet the requirements for the intended applications in fuel saving, cost, performance, reliability, and durability, and the industry specifications developed by USABC [4.5]. As specified by USABC, the basic requirements for high-power electrical chemical batteries are low cost, long life (more than 1000 cycles), low self-discharge rates (less than 5% per month), and low maintenance.

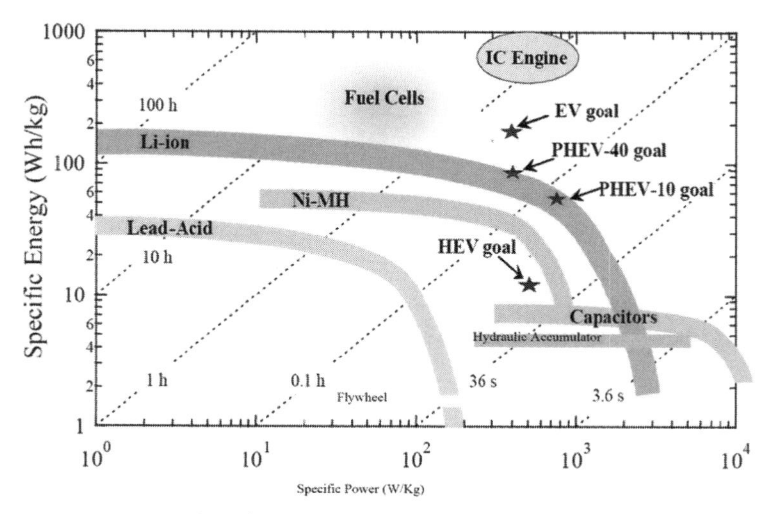

FIG. 4.1 Ragone plot of selected energy storage systems [4.4].

TABLE 4.1 Properties of Selected Energy Storage Systems [4.5]

	USABC Specification for Hybrid	Pb-Acid	NiMH	Li-Ion	Super Cap	Bladder Accumulator	Metal Bellows Accumulator
Energy/weight (Wh/kg)	150	30–40	30–80	70	5	2.8	4
Energy/size (Wh/L)	230	60–75	140–300	270		2	4.1
Power/weight (W/kg)	300	180	250–1000	1800	2500	2006	3600
Charge/discharge efficiency		70%–92%	66%	99.90%		90%	95%
Energy specific cost ($/Wh)	<0.15	0.12	0.65	0.5	16	1.98	1.32
Power specific cost ($/kW)		80	75	35	12	36	24
Self-discharge rate (/ month)		3%–20%	15%–30%	5%–10%			
Cycle durability	300	500–800 cycles	500–1000	1200 cycles		>1E6	1.00E+06

Note: data from USABC Conference and John Miller, SAE Webcast, 11/29/2007

4.2 Electrical Chemical Battery

For over a century lead acid batteries have been the prime source of energy storage for traction applications such as for fork lift trucks and similar applications, because they are both robust and relatively inexpensive. However, their low power density and weight limit applications for on-road vehicles.

In the 1980s, the advancement of high-power Nickel Metal Hydride (NiMH) cells has encouraged several vehicle and powertrain manufacturers of HEVs using NiMH batteries for commercial applications. NiMH batteries have a higher energy and power density than lead acid batteries. The NiHM batteries use hydrogen as the active element at a hydrogen-absorbing negative electrode (anode). This electrode is made from a metal hydride, usually the alloys of Lanthanum and rare earth materials that serve as a solid source of reduced hydrogen, which can be oxidized to form protons. The electrolyte is alkaline potassium hydroxide.

High-power lithium-ion batteries have been tested for commercial hybrid applications since 2000. They have an even higher energy density than NiMH batteries and are being introduced into new hybrid commercial vehicle applications. A comparison of the electrical chemical batteries is listed in Table 4.1.

4.2.1 Fundamentals of Electrochemical Cells

An electrical chemical battery for an HEV contains electrochemical cells in which an electrochemical reaction is reversible through a discharging and recharging process. Each cell consists of the following components, as shown in Fig. 4.2.

The **Anode** or **negative electrode** is the fuel electrode, and is generally a metal or an alloy. It releases electrons to the external circuit during the electrochemical (discharge) reaction.

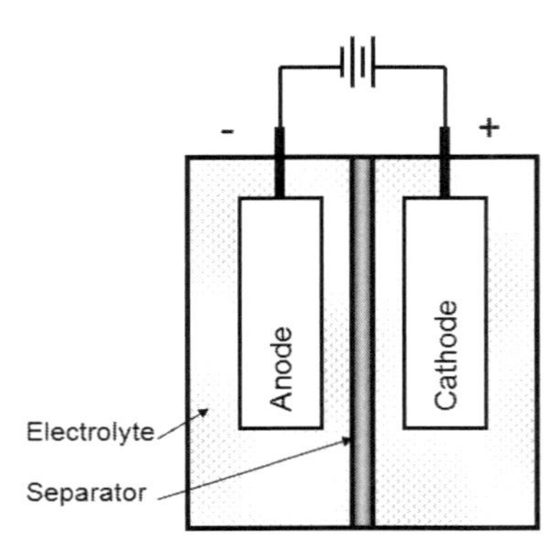

FIG. 4.2 Schematic of an electrochemical battery cell.

The **Cathode** or **positive electrode** is the oxidizing electrode, and is usually a metallic oxide or a sulfide. It accepts electrons from the external circuit during the electrochemical (discharge) reaction.

The **Electrolyte** provides the medium for transfer of charge as ions inside the cell between the positive and negative electrodes. The electrolyte is typically a solvent containing dissolved chemicals providing ionic conductivity, but it must be a non-conductor of electrons to avoid self-discharge of the cell.

The **Separator** is the electrically isolating layer of material that physically separates the positive electrodes from the negative electrodes, but is permeable to the ions of the electrolyte.

The internal chemical reaction within the battery between the electrolyte and the electrodes builds up free electrons with negative charges at the anode and positive ions at the cathode of the battery. The electrical potential difference between the anode and cathode is illustrated in Fig. 4.2.

Different metals have different affinities for electrons. When two dissimilar metals (or metal compounds) are in contact through a conducting medium, there is a tendency for electrons to pass from the metal with the smaller affinity to the metal with the greater affinity. A potential difference between the metals will therefore build up until it just balances the tendency of the electron transfer between the metals. At this point the "equilibrium potential" is that which balances the difference between the propensity of the two metals to gain or lose electrons.

When the circuit is completed, the surplus electrons flow in the external circuit from the negatively charged anode, which loses all its charge, to the positively charged cathode, which accepts it, neutralizing its positive charge. This action reduces the potential difference across the cell to zero. The circuit is completed or balanced by the flow of positive ions in the electrolyte from the anode to the cathode. Because the electrons are negatively charged, the electrical current they represent flows in the opposite direction, from the cathode (positive terminal) to the anode (negative terminal).

The electrochemical series listed in Table 4.2 is a list of metallic elements or ions arranged according to their electrode potentials. The order shows the tendency of one metal to reduce the ions of any other metal below it in the series.

The values for the table entries are reduction potentials. Lithium, which has the most negative number, indicates that it is the strongest reducing agent. The strongest oxidizing agent is fluorine with the largest positive number for standard electrode potential.

4.2.1.1 Thermodynamic Voltage

The energy available in an atom to do external work is called the **Gibbs Free Energy**, and an indication of the magnitude of this potential energy release is

TABLE 4.2 Strengths of Oxidizing and Reducing Agents

Cathode (Reduction) Half-Reaction	Standard Potential V_0 (volts)
Li^+ (aq) + $e^- \rightarrow$ Li(s)	−3.04
K^+ (aq) + $e^- \rightarrow$ K(s)	−2.92
Ca^{2+} (aq) + $2e^- \rightarrow$ Ca(s)	−2.76
Na^+ (aq) + $e^- \rightarrow$ Na(s)	−2.71
Zn^{2+} (aq) + $2e^- \rightarrow$ Zn(s)	−0.76
$2H^+ + 2e^- \rightarrow H_2$	0
Cu^{2+} (aq) + $2e^- \rightarrow$ Cu(s)	0.34
O_3^+ (q) + $2H^+$(aq) +$2e^- \rightarrow O_2$(g)+H_2O(l)	2.07
F_2 (q) + $2e^- \rightarrow 2F^-$(aq)	2.87

given by the electrode potential of the element. For a balanced reaction, this is expressed in the following equation:

$$\Delta G = -V_0 nF \tag{4.1}$$

where ΔG is the change in Gibbs free energy in Joules; V_0 is the standard electrode potential in volts (See Table 4.2); n is the number of moles of electrons transferred in the cell reaction per mole of product; and F is the Faraday constant in Coulombs per mole (the magnitude of electric charge per mole of electrons).

The open-circuit voltage V_0 of a battery cell can be expressed as

$$V_0 = -\frac{\Delta G}{nF} \tag{4.2}$$

And the cell voltage V is expressed in the Nernst relationship, which is named after the German physical chemist, Walther Nernst, who first formulated it as

$$V = V_0 - \frac{RT}{nF} \ln \frac{\Pi_j \alpha_j^{v_j}}{\Pi_i \alpha_i^{v_i}} \tag{4.3}$$

Where R is the universal gas constant and T is the absolute temperature in K. The numerator is a product of reaction product activities, α_j, each raised to the power of a stoichiometric coefficient, v_j, and the denominator is a similar product of reactant activities, α_i.

4.2.1.2 *Battery Capacity*

The amount of free charge generated by the active material at the negative electrodes and consumed by the positive electrodes is called the battery capacity. The capacity is measured by Ah (1 Ah = 3600 C, where 1 C is a Coulomb of charge transferred in 1 second by 1 A current).

The theoretical capacity of a battery (in C) is

$$Q_t = xnF \tag{4.4}$$

where x is the number of moles of limiting reactant associated with complete discharge of the battery, and n is the number of electrons produced by the negative electrode discharge reaction. F is the Faraday constant.

4.2.1.3 *Specific Energy and Specific Power*

Specific energy of a battery is defined as the energy capacity per unit battery weight (W·h/kg). The theoretical specific energy (*SE*) is the maximum energy that can be generated per unit mass of cell reactants.

$$SE = -\frac{\Delta G}{3.6\Sigma M_i} = \frac{nFV}{3.6\Sigma M_i}(\text{Wh/kg}) \tag{4.5}$$

Specific power (*SP*) is defined as the maximum power per unit battery weight that the battery can produce in a short period as

$$SP = \frac{P}{M_b}(\text{Wh/kg}) \tag{4.6}$$

where P is the maximum power delivered by the battery, and M_b is the mass of the battery.

4.2.2 Lead-acid Battery

The rechargeable lead-acid battery was invented in 1859 by Gaston Planté, a French physicist, and by the late nineteenth century its biggest market was in the light electric vehicle (EV) industry. Although its use in EVs declined during the early years of the twentieth century, it remained the dominant battery technology in automotive applications until recently and for some HEV commercial applications.

The typical lead-acid battery has a cell potential of 2.1 V, a gravimetric energy content of 35–50 W·h/kg, and volumetric energy of 100 W·h/L. The overall cell discharging chemical reaction of a lead-acid battery is:

$$Pb + PbO_2 + H_2SO_4 \leftrightarrow 2PbSO_4 + 2H_2O \tag{4.7}$$

Lead-acid batteries are typically characterized at a C/20 discharging rate. Higher discharging rates incur higher internal losses and lower resultant power output. Fig. 4.3 shows the voltage-current discharging characteristics of a lead-acid battery.

The lead-acid battery used in the industry is the valve-regulated lead-acid (VRLA) battery. Unlike the traditional flooded-electrolyte lead-acid battery, the VRLA provides a path for the oxygen, generated at the positive electrodes, to reach the negative electrodes, where it recombines to form lead sulfate. There are two mechanisms to immobilize the sulfuric acid electrolyte, i.e., gel cell battery and AGM (absorptive glass microfiber) battery.

A gel cell battery is a VRLA battery with a gelified electrolyte; the sulfuric acid is mixed with silica fume, which makes the resulting mass gel-like and immobile. Gel batteries reduce the electrolyte evaporation and spillage (and subsequent corrosion issues) common to the wet-cell battery, and provide better resistance to extreme temperatures, shock, and vibration. Chemically, they are the same as wet (nonsealed) batteries except that the antimony in the lead plates is replaced by calcium.

An Absorbed Glass Mat (AGM) battery is just like flooded lead-acid batteries, except the electrolyte is held in the glass mats, as opposed to freely flooding the plates. Very thin glass fibers are woven into a mat to increase surface area enough to hold sufficient electrolyte on the cells for their lifetime.

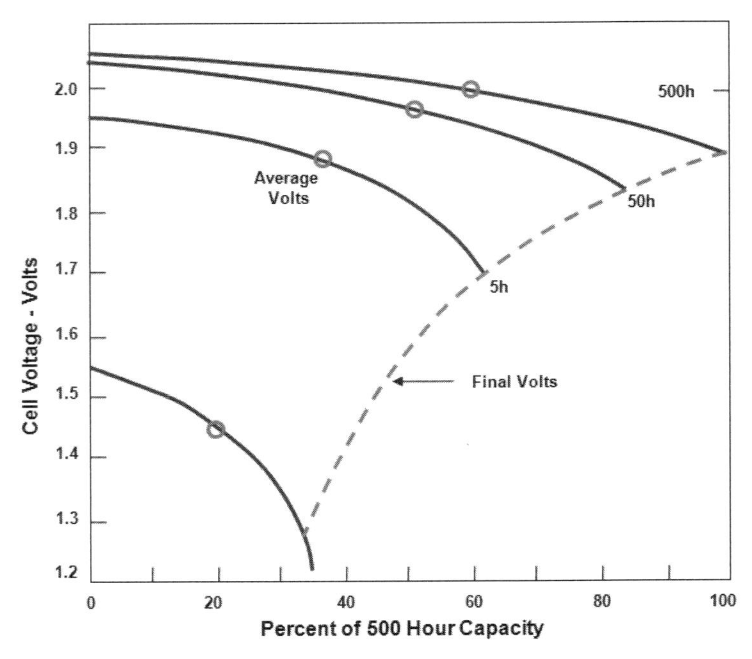

FIG. 4.3 Effect of discharge rate on output at 25°C [4.4].

The fibers that comprise the fine-glass-fiber glass mat do not absorb and are not affected by the acidic electrolyte they reside in. These mats are wrung out 2–5% after being soaked in acids, prior to manufacture completion and sealing. The AGM battery can now accumulate more acid than is available, to prevent spill.

The lead-acid battery has been used in HEVs for commercial applications in recent years. In 1998, New York City Transit (NYCT), part of the Metropolitan Transportation Authority in New York, began operating the first of ten heavy-duty diesel hybrid-electric transit buses (Model VI) from Orion Bus Industries [4.6, 4.7]. All ten buses were in revenue service by mid-2000. The hybrid buses are intended to provide NYCT with increased fuel economy and lower levels of harmful exhaust emissions, compared with NYCT's diesel transit buses. The diesel hybrid-electric transit buses have BAE series hybrid-electric systems, which include a downsized diesel engine running at an optimal controlled speed; a generator that produces electricity for the electric drive motor; and batteries. The electric-drive motor drives the vehicle and acts as a generator to capture energy during regenerative braking. The batteries supply additional power during acceleration and hill climbing and store energy recovered during regenerative braking and idling.

The batteries are sealed gel lead-acid batteries in two separate tubs of 23 batteries each, or 46 battery modules [4.7]. These batteries are expected to last at least three years, and then they will need to be replaced (or three sets of traction batteries during the 12-year life of the bus). A battery management subsystem monitors and maintains the charge of each individual battery. The propulsion control subsystem manages the entire system and optimizes performance for emissions, fuel economy, and power.

4.2.3 Nickel Metal Hydride

Nickel-metal-hydride (NiMH) batteries evolved from the nickel-hydrogen batteries used to power satellites in the 1970s. Such batteries are expensive and have low volumetric energy because they require high-pressure hydrogen-storage tanks, but they offer high energy-density, higher cycle life, and long calendar life [4.8, 4.9, 4.10, 4.11].

The modern nickel-metal-hydride (NiMH) electric vehicle battery was invented by Dr. Masahiko Oshitani, of the GS Yuasa Corporation, and Stanford Ovshinsky, the founder of the Ovonics Battery Company [4.8]. In NiMH batteries, the negative electrode has been replaced with the metal compound to store the hydrogen. The metal hydride is capable of undergoing a reversible hydrogen absorbing-desorbing reaction as the battery is charged and discharged. Both NiMH and nickel hydrogen batteries use the same electrolyte, potassium hydroxide (KOH).

Metal hydride cell chemistry depends on the ability of some metals to absorb large quantities of hydrogen. These metallic alloys, termed hydrides, can provide a storage sink of hydrogen that can reversibly react in battery cell chemistry. Such metals or alloys are used for the negative electrodes. The positive electrode is nickel hydroxide. The electrolyte, which is also a hydrogen absorbent aqueous solution such as potassium hydroxide, takes no part in the reaction but serves to transport the hydrogen between the electrodes.

Positive Electrode (Nickel)

$$Ni(OH)_2 + OH^- \leftrightarrow NiOOH + H_2O + e^- \quad (4.8)$$

Negative Electrode (Metal Hydride)

$$M + H_2O + e^- \leftrightarrow MH + OH^- \quad (4.9)$$

Complete Cell Reaction

$$M + Ni(OH)_2 \leftrightarrow MH + NiOOH \quad (4.10)$$

M stands for metallic alloy, which takes up hydrogen to form the metal hydride MH in negative electrodes; nickel oxyhydroxide (*NiOOH*) is formed on the positive electrode during charging. During the charge reaction, hydroxide from the electrolyte reacts with the nickel oxyhydroxide $Ni(OH)_2$ found on the positive electrode to form *NiOOH* and water, while, on the negative electrode, water reacts with the metal alloys to form metal hydride. The charge reaction is exothermic. The heat produced during the charge process must be released to avoid continuous temperature rise of the cells. When the battery is discharged, metal hydride in the negative electrodes is oxidized to form metal alloy, and nickel oxyhydroxide in the positive electrodes is reduced to nickel hydroxide.

Figure 4.4 shows a discharging rate of a Cobasys NiMH battery at a constant temperature of 25°C. The battery cell capacity diminishes rapidly as the discharging rate increases.

In recent years, developments of NiMH batteries have improved the self-discharging rate by using an improved separator and positive electrode [4.11]. This improved battery reduces self-discharge and, therefore, lengthens shelf life compared to normal NiMH batteries by up to 70 to 85% of their capacity after one year when stored at 20°C (68°F), while standard NiMH batteries may lose half their charge in this time period.

NiMH batteries became commercially viable during the 1980s and offer energy density of 60–120Wh/kg, exhibit minimal charge memory effects,

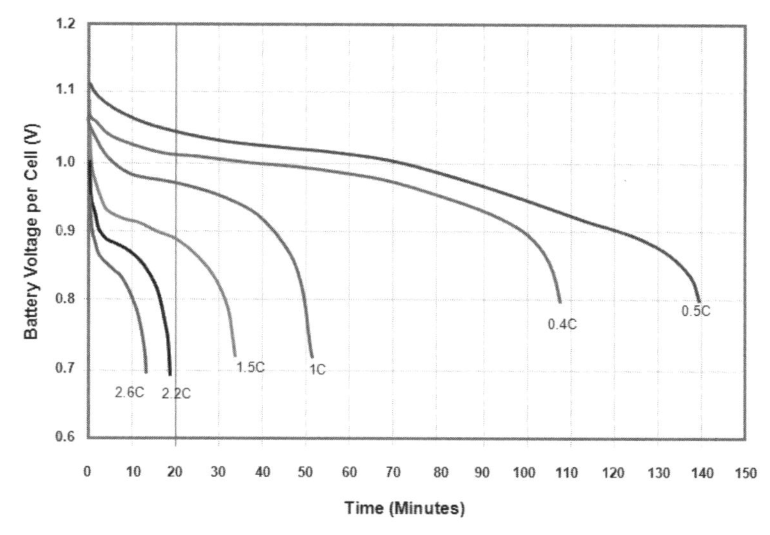

FIG. 4.4 Constant power discharge curves of a NiMH battery at 25°C [4.10].

accept three times as many charge cycles as conventional lead-acid batteries, and are completely recyclable. However, they have a self-discharge rate of up to 30% per month, and service life can be shortened by high discharge rates. The technology had progressed in terms of durability and serviceability for commercial vehicle applications to the extent that Allison has used the NiMH battery for hybrid bus applications since early 2000 [4.9].

As shown in Fig. 4.5, the Allison NiMH ESS contains six NiMH battery sub-packs, each sub-pack containing 40 NiMH battery modules, for a total of 240 battery modules and 437 kg. The sub-packs are located on either side of the ESS; control modules, fuses, and relays are arranged in the center. The

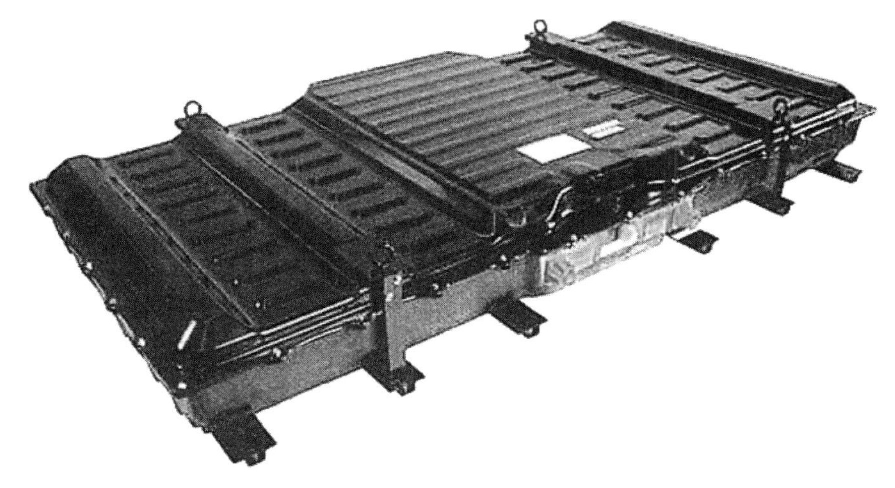

FIG. 4.5 Allison NiMH energy storage system [4.9].

system is air-cooled, and a nominal voltage of 600 V is maintained. Safety features include a High Voltage Interlock Loop to protect individuals from electric shock. Furthermore, the high voltage is not chassis-grounded, and each sub-pack is individually covered to separate it from the low-voltage components in the center of the pack. However, the power density is lower than the comparable Li-ion battery.

4.2.4 Li-ion Battery

Lithium-ion batteries evolved from nonrechargeable lithium batteries, such as those used in watches and hearing aids. The first commercial lithium-ion battery, which was launched by Sony in 1991, used cobalt oxide for the positive electrode, and graphite (carbon) for the negative one.

In 1997, John Goodenough and his colleagues at the University of Texas developed a new material for the positive electrode—iron phosphate—with the potential of being cheaper, safer, and more environmentally friendly than cobalt oxide. However, it had a lower energy-density than cobalt oxide. In 2002, Dr. Yet-Ming Chiang of MIT and his colleagues had dramatically boosted the material's conductivity by doping the electrode with aluminum, niobium, and zirconium and by using iron-phosphate particles less than 100 nanometers to increase the surface area of the electrode and improves the battery's ability to store and deliver energy [4.13].

Rather than the traditional redox galvanic action, Lithium-ion cell chemistry involves the insertion of lithium ions into the crystalline lattice of the host electrode without changing its crystal structure. These electrodes have two key properties: (1) open crystal structures which allow the insertion or extraction of lithium ions; and (2) the ability to accept compensating electrons at the same time.

In a typical Lithium cell, the anode or negative electrode is based on Carbon, and the cathode or positive electrode is made from Lithium Cobalt Dioxide or Lithium Manganese Dioxide. During discharge, Lithium ions are dissociated from the anode and migrate across the electrolyte and are inserted into the crystal structure of the host compound. At the same time, the compensating electrons travel in the external circuit and are accepted by the host to balance the reaction. The general electrochemical reaction can be described as:

$$Li_xC + Li_{1-x}M_yO_z \leftrightarrow C + LiM_yO_z \tag{4.11}$$

The process is completely reversible. Thus, the Lithium ions pass back and forth between the electrodes during charging and discharging.

Since Lithium reacts with water, the electrolyte is composed of nonaqueous organic Lithium salts and acts purely as a conducting medium and does not take part in the chemical action. Figure 4.6 shows the discharge curves for a Lithium-ion cell, which reveals that the effective capacity of the cell is reduced if the cell is discharged at very high rates (or conversely increased with low discharge rates). This is called the capacity offset, and the effect is common to most cell chemistries.

If the discharge takes place over a long period of several hours, as with some high-rate applications such as electric vehicles, the effective capacity of the battery can be as much as double the specified capacity at the C rate. This can be most important when dimensioning an expensive battery for high-power use. The capacity of low-power, consumer electronics batteries is normally specified for discharge at the C rate, whereas SAE International uses the discharge over a period of 20 hours (0.05C) as the standard condition for measuring the Amphour capacity of automotive batteries.

Lithium-polymer batteries have a polymer gel instead of a liquid electrolyte, and each cell has the cathode, separator, and anode laminated together, enabling simpler manufacturing. The technology offers higher energy density, lower manufacturing costs, reduced size, flexible packaging, and greater durability in terms of charge-discharge cycles, thermal robustness, and resistance to physical damage.

The most promising positive electrode material for the Li-polymer battery is vanadium oxide V_6O_{13}. This oxide interlaces up to eight lithium atoms per oxide molecule with the following positive electrode reactions:

$$Li_x + V_6O_{13} + xe^- \leftrightarrow Li_xV_6O_{13} \quad 0 < x < 8 \tag{4.12}$$

Li-polymer has the potential of the highest specific energy and power.

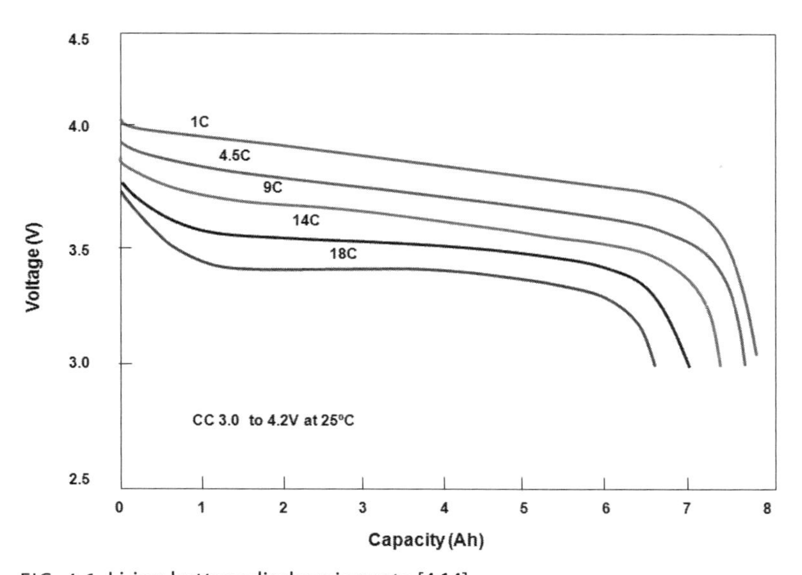

FIG. 4.6 Li-ion battery discharging rate [4.14].

According to the Advanced Lithium Ion Battery industry group, the development of the Li-ion battery can be classified into four broad categories (Table 4.3), based on the formulation contained in the cathode: 1) NCA, or nickel cobalt aluminum; 2) NMC, or nickel manganese cobalt; 3) LMO, or manganese spinel; and 4) LFP, or iron phosphate.

Oxide-based lithium-ion batteries such as lithium cobalt oxide and lithium manganese oxide offer up to 60% higher energy density than phosphate-based alternatives, but they are more vulnerable to overheating and require protection circuitry within each cell to limit peak voltage during charge and low voltage during discharge [4.13, 4.15].

Although lithium batteries with phosphate-based cathodes provide significantly lower energy density than oxide-based batteries, they are stable at temperatures of up to 932°F (500°C) and offer four times the durability in terms of charge-discharge cycles, greater resistance to overcharge, increased life cycle, and increased safety. Furthermore, they can be produced for around 25% the cost of oxide-based batteries because they contain no heavy or rare metals, which are more subject to commodity price fluctuations, and current research promises increasing energy density.

TABLE 4.3 Lithium-ion Cathode Chemistry Comparisons [4.23]

Chemistry	Wh/Kg	Positive	Negatives	Makers
Nickel/Cobalt/ Alum (NCA)	160	Energy density Power	Safety Commodity exposure Life Expectancy Range of Charge	JCI / Ssft PEVE AESC
Manganess Spinel (LMO)	150	Cost Safety Power	Life Expectancy Usable Energy	Hitachi, AESC, Sanyo GS Yuasa, LG Chem Samsung, Toshiba Ener1, Evonik, GS Yuasa
Nickel Manganess Cobalt (NMC)	150	Energy density Range of charge	Safety (better than NCA) Cost / Commodity Exposure	PEVE, Hitachi, Sanyo GS Yuasa, LG Chem Ener1, Evonik, GS Yuasa
Lithium Iron Phosphate (LFP)	140	Safety Life expectancy Range of charge Material cost	Low Temp Performance Processing Costs	A123, BYD GS YUASA, JCI/Saft

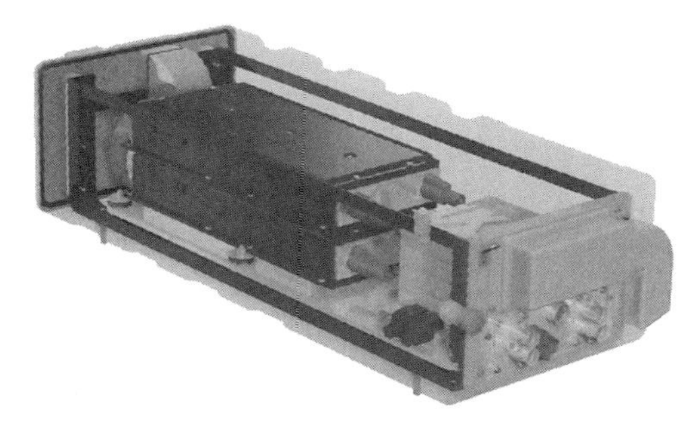

Capacity (Ah):	5.5
Weight (Kg):	23
L x W x H (mm):	675 x 260 x 100
Energy density (Wh/kg):	165
Power density (W/kg):	970
Discharge (C rate):	>25
Charge (C rate):	>22

FIG. 4.7 Eaton Li-ion battery pack and key parameters [4.16]. (Courtesy of Eaton Corporation)

The recent development of using nano-technology to produce nanotubes and other shape-controlled structures for carbon or other electrode materials results in electrodes with higher surface areas that enhance the load current characteristics and charge-discharge cycle life of lithium-ion batteries. The technology also reduces battery weight and size, and the amount of material required. The recent development of Lithium-air, which has oxygen as an unlimited cathode reactant and a battery capacity that is limited by the Li anode, shows the potential of significantly improved battery power density [4.17].

Eaton Corporation pioneered the application of the Li-ion battery for the hybrid-electric vehicle for commercial delivery vehicles starting in early 2000 [4.16]. Lithium-ion batteries were chosen as the energy storage system for their high energy density and low discharging rate. As shown in Fig. 4.7, the battery pack includes two lithium-ion battery modules. The battery module's key parameters are listed in the figure.

4.2.5 Supercapacitor

Supercapacitors, also known as ultra-capacitors, are energy storage devices in that energy is stored via charge separation at the electrode-electrolyte interface. Unlike batteries that store charges chemically, supercapacitors store charges electrostatically and can withstand hundreds of thousands of charge/discharge cycles without degrading.

A supercapacitor, also known as a double-layer capacitor, can be viewed as two nonreactive porous electrodes suspended within an electrolyte, which polarize an electrolytic solution to store energy electrostatically. In an individual supercapacitor cell, the potential on the positive electrode attracts the negative ions in the electrolyte, while the potential on the negative electrode attracts the positive ions. A dielectric separator between the two electrodes prevents the charge from moving between the two electrodes, as shown in Fig. 4.8.

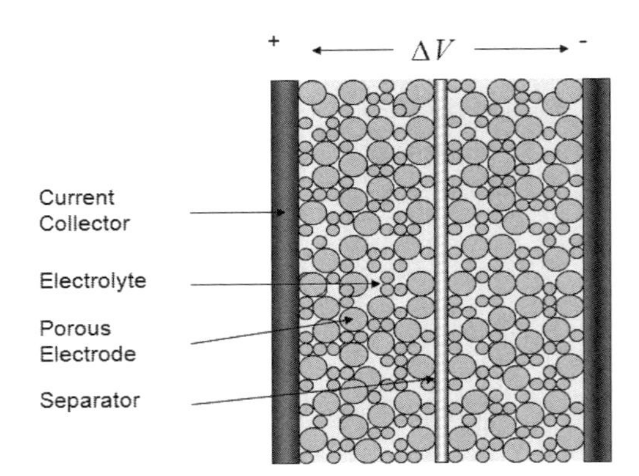

FIG. 4.8 Illustration of an electric double-layer supercapacitor [4.18].

Supercapacitors have extremely rapid charge and discharge rates, making them ideal for accumulating charge from regenerative braking systems and for providing electrically powered acceleration. However, the energy density of supercapacitors is low compared with Li-ion batteries, and they lose voltage level as they discharge, limiting their application as the sole energy storage system to urban operations with special charging facilities or in conjunction with a battery in order to enable the use of a smaller battery pack.

A supercapacitor uses ions and an electrolyte rather than simply relying on the static charges. The ions do not chemically combine with the electrodes as in a battery, and remain on the surfaces of the electrodes. Supercapacitors get their improved capacity through the use of electrodes coated with carbon and etched to produce holes, rather like a sponge, although research with nano-engineering is working toward coating them with a layer of carbon nanotubes, which could increase the supercapacitor to 50% of a battery's storage capacity rather than the current 5%.

An equivalent circuit of a supercapacitor is shown in Fig. 4.9. The terminal voltage of the supercapacitor during discharging can be expressed as

$$V_t = V_c - iR_s \tag{4.13}$$

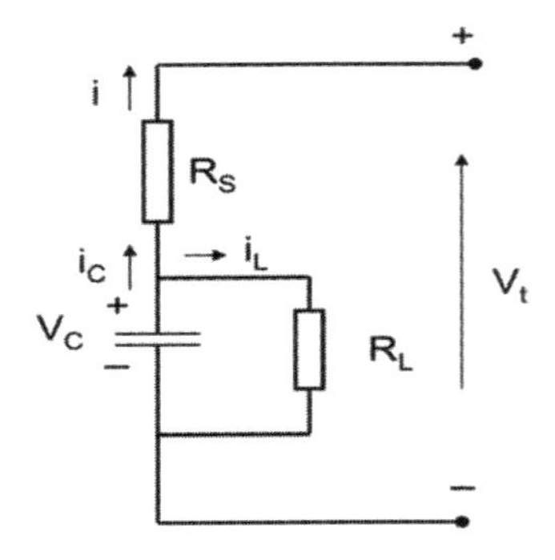

FIG. 4.9 Equivalent circuit of a supercapacitor.

where V_t is the terminal voltage, V_c is the capacitor electric potential, and R_s is the series resistance; the electric potential of the capacitor can be expressed as

$$\frac{dV_c}{dt} = -\left(\frac{i + i_L}{C}\right) \tag{4.14}$$

$$V_c = \left[V_{c_0} - \int_0^t \frac{i}{C} e^{t/CR_L} dt \right] e^{-(t/CR_L)} \tag{4.15}$$

where i is the discharging current, and i_L is the leakage current

Figure 4.10 shows the measured capacitance of the device as a function of charge/discharge test currents ranging from 10 to 500 A, and for a charge/ discharge voltage range of 0.03 to 2.7 V for a Maxwell Technologies Company Model #PC-2700 ultracapacitor. Three consecutive charge/discharge test cycles were used and, thus, permitted the calculation of the capacitance from three tests. As can be seen in the figure, the discharge capacitance at +25°C is ~2700 farads at 10 A, which decreases in a linear manner with increasing discharge current until at 500 A the capacitance is ~2375 farads. The capacitance measured at +50°C is slightly lower that that measured at +25°C. A list of commercially available supercapacitors is presented in Table 4.4.

4.2.6 Battery/Supercapacitor Hybrid Energy Storage Systems

The efficiency of hybrid-electric vehicles depends on their capability to store larger amounts of energy and to quickly extract power from that energy. Currently, HEVs rely on large battery systems, such as Li-ion batteries, to store their onboard electric energy. The development of these batteries has allowed for superior performance with regard to energy density. However,

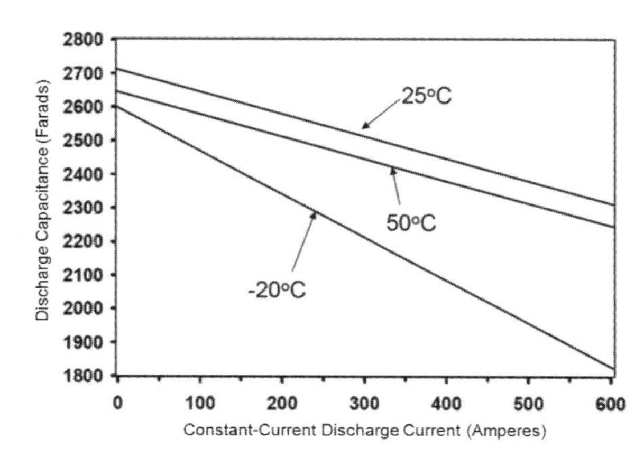

FIG. 4.10 Discharge capacitance as a function of constant-current discharge current [4.10].

TABLE 4.4 Commercially Available Supercapacitor [4.23]

Manufacturer	Technology	Voltage (V)	Capacity (°F)	Power (kW/kg)	Energy (Wh/kg)	Development
EPCOs, Japan & Germany	Organic/ Carbon	2.3	2700	3.04	2.74	On sale
Maxwell, USA & Switzerland	Organic/ Carbon	2.7	3000	11	5.52	On sale
Nesscap, USA & Korea	Carbon power	2.7	5000	5.2	5.8	On sale
Evans Corp, USA	H_2SO_4/ Carbon	1.4	65	2.5	0.35	On sale
Montena, Switzerland	Organic/ Carbon	2.5	1400	3.45	4.34	On sale
SAFT, France	Organic/ Carbon	2.5	3500	3	4.7	Development
Elit. Russia, Aqueous	Aqueous Carbon/ NiO_2	1.17	470	9.5	3.84	Development

these technologies are not so well suited for the dynamic power demand of an HEV. To meet the peak power demands, battery storage systems tend to be oversized and heavy. The supercapacitor, which stories the energy by means of static electronic charge rather than by an electrochemical process of the battery, has higher power density than a battery. It is advantageous to combine these two energy-storage devices to gain better power and energy performances. The supercapacitor can be used to meet the power output demand, while the battery can be used to meet the energy demand. The combination of the high power density of the supercapacitor and the high energy density of the battery allows for a reduction in size and weight of the overall energy storage system [4.20, 4.21].

Two of the most common architectures of combining the battery and the supercapacitor are illustrated in Fig. 4.11. Fig. 4.11(a) shows the battery and supercapacitor connected in parallel, with the load current denoted by I_L defined to flow downward, i.e., positive during acceleration and coasting, and negative during regenerative braking.

For a given load profile, the battery current I_b and the supercapacitor current I_b can be expressed as

$$I_L = I_c + I_b \tag{4.16}$$

$$V = V_c - I_c R_c = V_b - I_b R_b \tag{4.17}$$

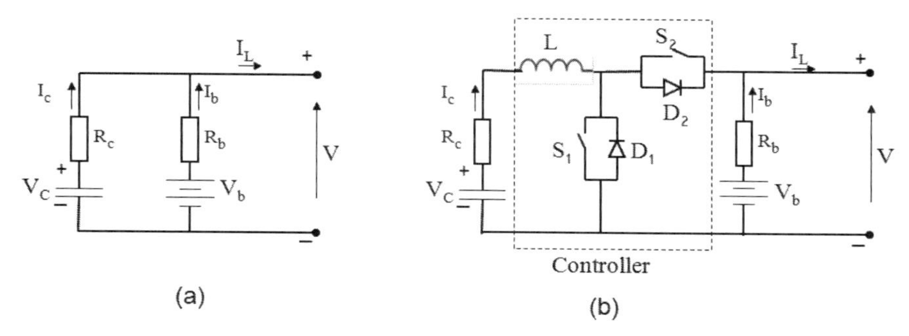

FIG. 4.11 Battery-capacitor combination: (a) Parallel and (b) Independent power converter architecture.

$$I_c = -C\frac{dV_c}{dt} \tag{4.18}$$

where V_c and V_b represent the internal capacitor and battery voltages. By substituting Eqs. (4.16) and (4.18) into Eq. (4.17), the internal voltage of supercapacitor V_c can be expressed as

$$\frac{dV_c}{dt} = \alpha\left((V_b - V_c) + R_b I\right) \tag{4.19}$$

where $\alpha = 1/C(R_b + R_c)$, and the V_c can be solved as

$$V_c = Ke^{-\alpha t} + V_b + R_b\alpha e^{-\alpha t}\int Ie^{\alpha t}dt \tag{4.20}$$

where K is determined by setting the initial values of V_c and V_b.

This configuration limits the control over the charge and discharge of the components. The voltage and the state of charger (*SOC*) of each component and the overall system follow a nonlinear characteristic curve that is defined by the battery. The current division between the battery and the supercapacitor is determined solely by their internal resistances and voltages. Figure 4.12 shows a current profile of a supercapacitor/battery energy storage system without power controller during acceleration [4.19].

Figure 4.11(b) shows that the battery and the supercapacitor are connected through an electronic controller. The controller can regulate the energy transfer to and from each component and the system, which allows for greater flexibility in the design of the battery and supercapacitor system.

The first application of a supercapacitor in a vehicle was on the Nissan Diesel-Capacitor mild hybrid delivery truck launched in 2002. Diesel-supercapacitors

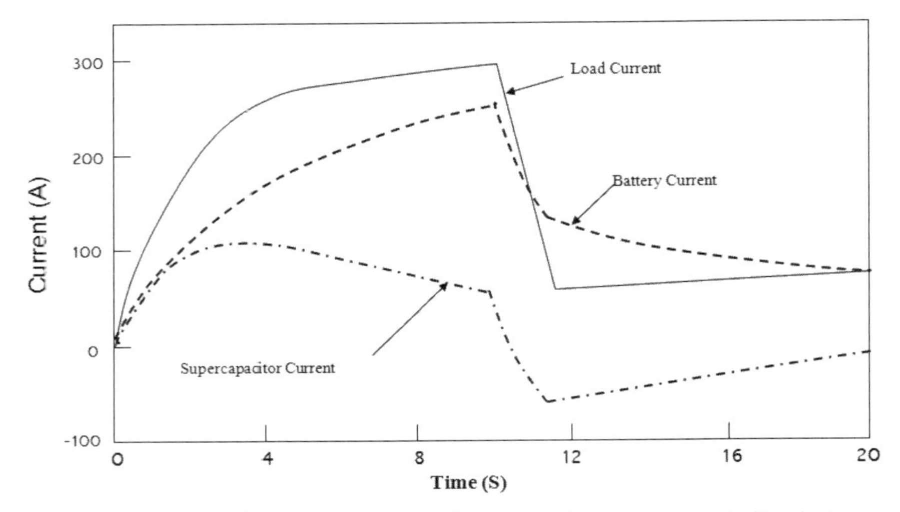

FIG. 4.12 Current profile of supercapacitor/battery without power controller during acceleration [4.19].

for hybrid buses were also used at the 2010 Shanghai World Expo [4.25]. These supercapacitors, made by Shanghai Aowei Technologies, can be charged and discharged quickly compared to standard battery-powered vehicles.

The specification of the supercapacitor is listed in Table 4.5, and the installation of the supercapacitors is shown in Fig. 4.13. The recharging is done by means of a railway type pantograph fitted to the bus roof that makes contact with an overhead power supply. The bus can be quickly recharged in 30 seconds with an overhead electric charger when it stops at the station, just about the time needed for passengers to get on and off the bus, enabling it to run for 3 to 6 km depending on the load and whether air-conditioning is on [4.25].

TABLE 4.5 Specification of Supercapacitor

Operating Voltage (V)	360~600	
Max Charging Voltage (V)	610	
Static Capacitor (F)	200	
Energy Capacity (W · h)	6400	300~600V
Max discharging Current (A)	200	
Internal Resistance (Ω)	0.22	
Weight (kg)	980	Including all sub-components

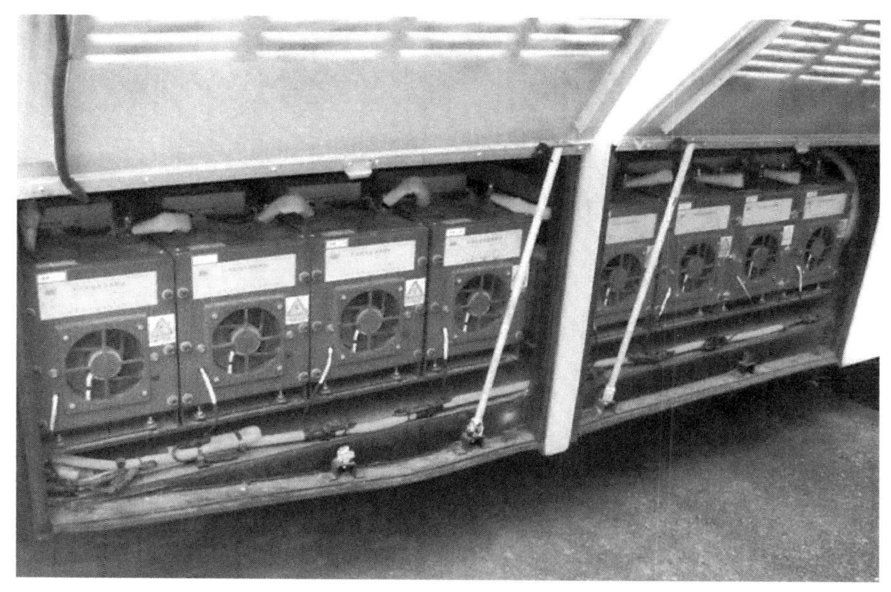

FIG. 4.13 Supercapacitor-powered city bus at the 2010 Shanghai World Expo.

4.3 Battery Management Systems

The battery management system (BMS) controls and optimizes battery packs and protects them from electrical and thermal abuse under varying demands and driving conditions. For commercial hybrid vehicle applications, the battery management system is a component of a much more complex vehicle energy management system and must interface with other onboard systems such as engine management, climate controls, communications, and safety systems, and feed system information to the vehicle controllers. The basic features of the BMS are as follows [4.36]:

- Voltage, current, and temperature monitoring of individual cells and battery pack State-of-Charge (*SOC*) maintenance in accordance with the energy management system algorithm
- State-of-Health (*SOH*) monitoring and power availability to the vehicle powertrain
- Cell balancing
- Safety supervising and over-charging and over-discharging protection
- Communication interfacing
- Interfacing with auxiliaries such as sensors and contactors

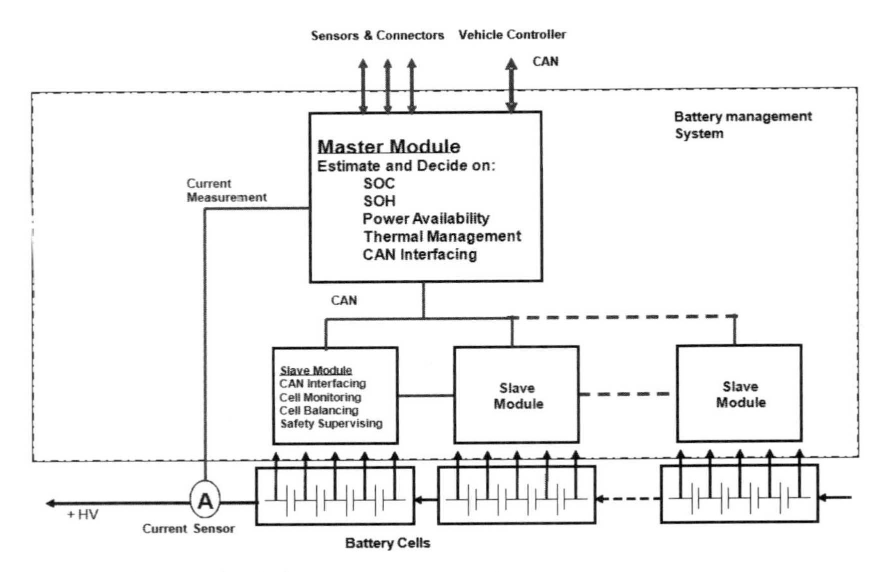

FIG. 4.14 Schematic of a battery management system.

There are many ways of implementing the battery management system. The diagram of Fig. 4.14 is a conceptual representation of the primary BMS functions for a battery made up from Lithium iron phosphate cells.

The BMS consists of two basic types of modules, a master module and slave module. The master module connects to the slave modules and communicates to the vehicle CAN communication link. The functions of the master module and slave module are:

The Slave Module—The slave module connects to each cell through temperature, voltage, and other sensors to monitor the conditions of the cell, implement the cell balancing, and communicate with the master module.

The Master Module—The master module connects to multiple slave modules and calculates the battery *SOC*, *SOH*, and the availability of power to the vehicle powertrain. The master module controls the main battery isolation contactor(s) initiating battery protection and the thermal management system in response to data from the main current sensor or voltage and temperature data from the slave modules. The master module also provides the system communications.

4.3.1 Cell Protection

One of the prime functions of the battery management system is to protect the cells from out of tolerance ambient and operating conditions. This is of particular importance in commercial vehicle applications because of their harsh working environment. Individual cell protection must be designed to respond to external fault conditions by isolating the battery as well as addressing the cause of the fault. For example, cooling fans can be turned on if the battery overheats. If the overheating becomes excessive, then the battery can be disconnected.

A typical BMS can detect the following battery warnings and faults:

- Over and under voltage
- Cell over and under voltage
- Over current
- Over temperature
- Leakage fault
- Cell voltage unbalanced
- CAN communication fault

4.3.2 Battery State of Charge

State of Charge (*SOC*) is defined as the percentage of the maximum possible charge that is present in a rechargeable battery. The BMS monitors and calculates the *SOC* of each individual cell in the battery.

Batteries for HEV applications require both high-power-charge capabilities for regenerative braking and high-power-discharge capabilities for launch assist or boost. For this reason, the batteries must be maintained at an *SOC* that can discharge the required power but also be able to accept the necessary regenerative power for fuel economy improvement without risking overcharging the cells. Over-charging and over-discharging are two of the prime causes of Li-ion battery failure, and the BMS must maintain the cells within the desired Depth of Discharge (*DOD*) operating limits. An accurate estimate of *SOC* provides the benefits of minimizing the number and size of cells needed to provide the range of power and energy required by the propulsion system.

Systems capable of indicating *SOC* of a battery have been around for almost as long as rechargeable batteries have existed. One example is a single-meter device that measures the battery voltage drop, which can be translated as an estimation of *SOC* for lead-acid batteries. A major limitation with this technique is that the battery voltage is also affected by temperature and battery discharging current. Other methods have been developed to take into account the effects of temperature and current to provide more accurate estimation of *SOC*. The following sections discuss three commonly used techniques of estimating *SOC*.

4.3.2.1 Open-Circuit Voltage (OCV) Method

The *OCV* is directly proportional to the battery *SOC* and can be calculated using the following equations [4.27]:

$$V_{bat} \approx OCV\,(SOC) - I \cdot R \tag{4.21}$$

$$SOC \approx OCV^{-1}(V_{bat} + I \cdot R) \tag{4.22}$$

where V_{bat} is the battery terminal voltage, I is the actual battery current—regarded as a positive value during discharge and a negative value during charge—and R is the internal resistance. Note that $OCV = V_{bat}$ when $I = 0$, but after current interruption this takes a while due to several relaxation processes occurring inside a battery.

For a lithium-ion battery, the *SOC–OCV* relationship is obtained by defining the battery charge amount when the *OCV* of the cell is 3.9 V as *SOC* = 100% and defining the battery charge amount when the *OCV* of the cell is 3.5 V as *SOC* = 0%. By defining *SOC* = 100% and *SOC* = 0%, the *SOC* can be calculated and displayed.

This method does not take into account hysteresis, polarization voltages, etc., especially at low temperatures where resistance can be quite high and where hysteresis can also be quite large. It is also poor at extreme values of *SOC* where the resistance is much larger than at moderate *SOC* values.

4.3.2.2 Coulomb Counting Method

The Coulomb counting method is based on battery current measurement and integration as

$$SOC\ (\%) = \frac{\text{Estimated Capacity} - \text{Where capacity removed}}{\text{Estimated Capacity}} \tag{4.23}$$

Where capacity removed = $\int_0^t I(t)dt$.

For constant-current discharge, capacity removed = $I \times t$, where I = discharge current in Amperes and t = time in hours. The charge and discharge current of a battery is measured in C-rate. 1C is often referred to as a one-hour discharge; a 0.5C would be a two-hour, and a 0.1C a ten-hour.

Integrating the current into or out of a battery gives the relative value of its charge. The "Coulomb Counting" method needs a starting point. If the initial charge in the battery is known, from then on "Coulomb Counting" can be used to calculate its *SOC*. Another limitation of Coulomb Counting is drift. In any integration, any small, constant error in the variable being integrated results in a drift in the result. In the case of Coulomb Counting, any small offset in the measurement of battery current will result in the *SOC* drifting up (or down) over time.

Depending on the battery chemistry, Coulomb Counting can be a very accurate technique.

Coulomb Counting does not work as well with lead-acid batteries, because the significant leakage current within lead-acid batteries does not go through the battery current sensor and is therefore not taken into account. The Coulomb Counting method provides better estimation of *SOC* with Li-ion batteries, because of their low leakage and less discharging current effect.

4.3.2.3 Adaptive Methods

The adaptive method is based on a comparison of the estimated values with observed battery behavior. Several adaptive methods are being developed to estimate the *SOC* of the rechargeable batteries. One of the most widely used adaptive methods is the Kalman filter, which will be discussed in more detail.

The Kalman filter is a set of mathematical equations that provides an efficient computational (recursive) means to estimate the state of a process, such as *SOC*, in a way that minimizes the mean of the squared error [4.26]. The basis of the filter is a numeric battery model description, as shown in Fig. 4.15. The battery voltage is estimated on the basis of the current and temperature measurements, and the results are compared with the measured battery voltage value. The Kalman filter estimates the system's state on the basis of statistical knowledge of parameters and measurements. The battery model, which is expressed by differential equations, will be corrected by the statistical knowledge. The main advantage of this method in comparison to the Coulomb Counting method is that the state of charge calculation is not influenced by the accumulation of measurement errors.

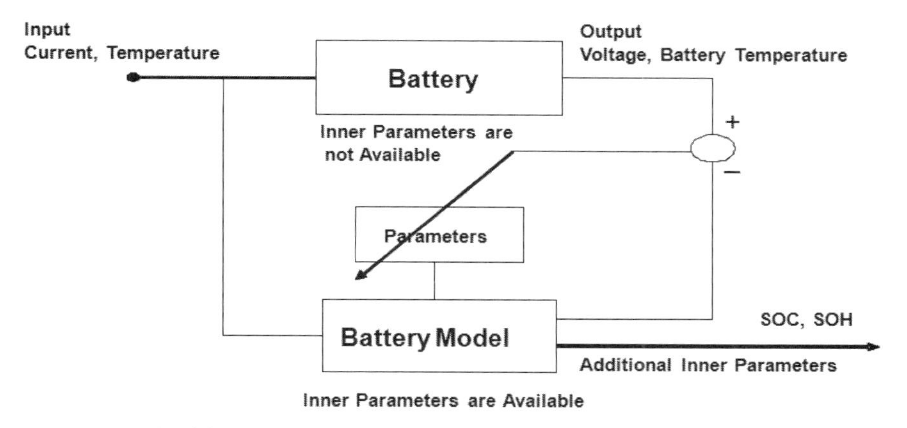

FIG. 4.15 Method for *SOC* and SOH determination using a Kalman Filter [4.27].

Different forms of Kalman filter, such as the EKF (Extended Kalman Filter) and SPKF (Sigma Point Kalman Filter) can be used to adaptively identify unknown parameters for modeling Lithium battery cell dynamics [4.28]. Compared with the Kalman filter, which is used for a system governed by a set of linear differential equations, the EKF approach is used for a system that can be described by a set of nonlinear stochastic differential equations, which are linearized at each sample point using a Taylor-series expansion. The SPKF is an alternate approach of linearization by using a small fixed number of function evaluations instead. The nonlinear model includes terms that describe the dynamic contributions due to open-circuit voltage, polarization time constants, electrochemical hysteresis, ohmic loss, and the effects of temperature. EKF provides a reasonable solution for long-term *SOC* estimation, at some cost in complexity. The SPKF *SOC* estimator has a built-in mechanism for determining whether a measurement is statistically valid to produce better solutions.

4.3.3 Cell Balancing

Li-ion batteries used for HEV applications are made up of long strings of cells in series to achieve higher operating voltages of 200 V to 500 V and in parallel to achieve the desired capacity or power levels.

Because lead-acid and NiMH cells can withstand a level of over-voltage without sustaining permanent damage, a degree of cell balancing can occur naturally with these technologies simply by prolonging the charging time since the fully charged cells will release energy by gassing until the weaker cells reach their full charge. This is not possible with Lithium cells, which cannot tolerate over-voltages. Any capacity mismatch between series-connected cells reduces the overall pack capacity because the Li-ion cells must each maintain a voltage with strict limits. If the voltage on any cell goes too high, there is a danger that once it has reached its full charge it will be subject to overcharging until the rest of the cells in the chain reach their full charge. The result is temperature and pressure build up and possible damage to the cell. During discharging, the weakest cell will have the greatest depth of discharge and will tend to fail before the others. It is even possible for the voltage on the weaker cells to be reversed as they become fully discharged before the rest of the cells, resulting in early failure of the cell.

The most common mismatch of Li-ion cells is the *SOC* mismatch, which is caused by tiny imperfections in construction that can result in soft shorts inside the cells [4.29, 4.30]. The soft shorts can increase the self-discharge rate or affect the charge acceptance rate. The *SOC* mismatch for the series cells also can be cumulative and exaggerated when the cells are subject to the rapid charge and discharge cycles found in HEV applications. Therefore, cell balancing techniques are required for Li-ion batteries for HEV applications to increase the battery pack capacity and to prevent premature battery failures. Three most commonly used cell balancing techniques are discussed next.

4.3.3.1 *Charge Shunting*

The charge-shunting cell balancing method selectively shunts the charging current around each cell as it becomes fully charged (Fig. 4.16). This method is most efficiently employed on systems with known charge rates. The shunt resistor R is sized to shunt exactly the charging current I when the fully charged cell voltage V is reached. If the charging current decreases, resistor R will discharge the shunted cell. To avoid extremely large power dissipations due to R, this method is best used with stepped-current chargers with a small end-of-charge current.

Disadvantages of the charge shunting method are the requirement for large power dissipating resistors, high current switches, and thermal management requirements. This method is best suited for systems that are charged often with small charge currents.

4.3.3.2 *Charge Shuttling*

Charge shuttling cell balancing mechanisms consist of a device that removes charge from a selected cell, stores that charge, and then delivers it to another cell. There are several embodiments of charge shuttling schemes, the most notable being a "flying capacitor" (Fig. 4.17).

The control electronics close the proper switches to charge capacitor C across cell B_1. Once the capacitor is charged, the switches are opened. The switches are then closed to connect capacitor C across cell B_2. The capacitor then delivers charge to B_2 based on the differential of voltage between B_1 and B_2.

$$\text{Charge} = \frac{1}{2}C\left(V_{B_1}^2 - V_{B_2}^2\right) \tag{4.24}$$

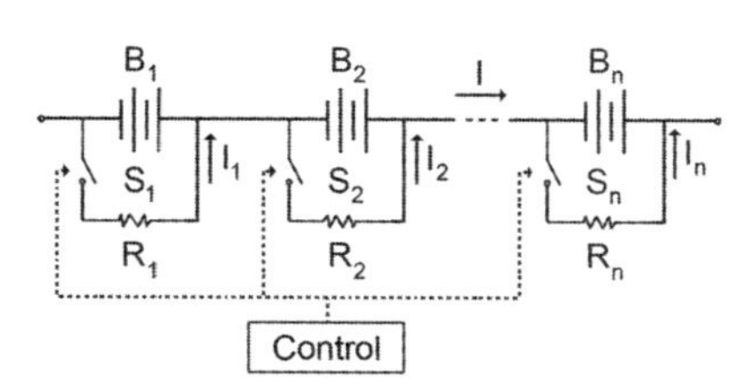

FIG. 4.16 Charge shunting [4.29].

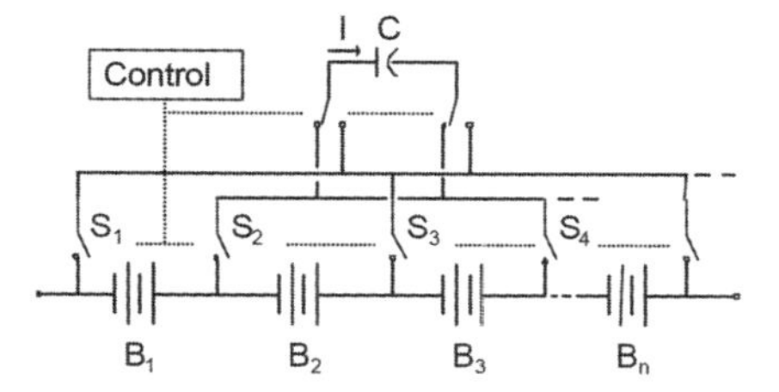

FIG. 4.17 Flying capacitor charge shuttling [4.29].

The capacitor is then connected in the same manner across B_3, B_4, . . . , B_n. The highest charged cells will charge C and the lowest charged cells will take charge from C. In this way, the charges of the most charged cells are distributed to the least charged cells. The only electronic control needed for this method is a fixed switching sequence to open and close the proper switches.

Alternatively, the process can be speeded up by programming the capacitor to repeatedly transfer charge from the highest-voltage cell to the lowest-voltage cell. Efficiency is reduced as the cell voltage differences are reduced. The method is fairly complex with expensive electronics.

4.3.3.3 Energy Converters

Cell-balancing utilizing energy conversion devices employ inductors or transformers to move energy from one cell or group of cells to another cell or group of cells. Two active energy converter methods are the switched transformer and the shared transformer.

A shared transformer has a single magnetic core with secondary taps for each cell (Fig. 4.18). Current I from the cell stack is switched into the transformer primary and induces currents in each of the secondaries. The secondary with the least reactance (due to a low terminal voltage on B_n) will have the most induced current. In this way, each cell receives charging current inversely proportional its relative *SOC*. The only active component in the shared transformer is the switching transistor for the transformer primary. No closed-loop controls are required. The shared transformer can rapidly balance a multicell pack with minimal losses [4.30].

Disadvantages of this cell-balancing method include complex magnetics and high parts count due to each secondary's rectifier. The balancing circuit would have to be designed for the maximum expected number of cells; additional secondary taps could not be easily added.

The cell balancing technologies for the Li-ion battery are still in the development phase. The charge shunting method works well but is limited by the amount of current that must be dissipated. The charge shuttling method would be expensive due to the switches required to handle the large peak capacitor

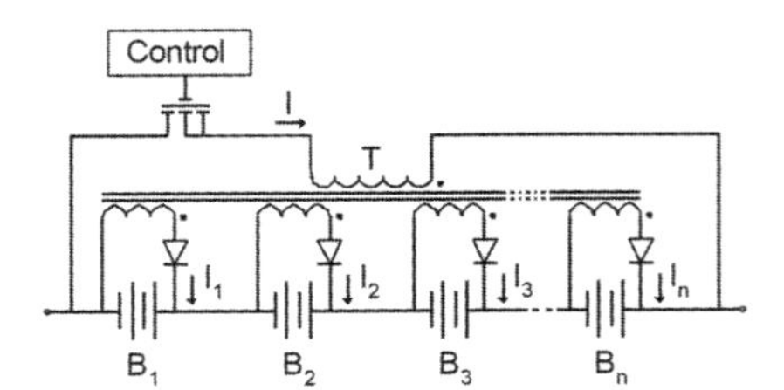

FIG. 4.18 Shared transformer [4.29].

charging currents. The shared transformer method is costly in terms of magnetics and parts count. Other methods, such as the passive dissipative method, which shunts selected cells with high value resistors to remove charge from the highest cells until they match the charge of the lowest cells, also has been used in the industry because of its low complexity and low cost of implementation.

4.4 Hydraulic Energy Storage

The energy storage device of hydraulic hybrid vehicles is a gas-charged accumulator. Energy is stored by forcing hydraulic fluid into the accumulator, occupying some of the volume that is otherwise occupied by a fixed charge of inert gas that is also located inside the accumulator. Conversely, when hydraulic fluid is withdrawn from the accumulator, the gas expands, delivering Work (energy) back to the hydraulic fluid. Thus, the gas is the energy storage medium; the hydraulic fluid is the means to transfer that energy.

Fig. 4.19 shows a typical accumulator configuration as (a) Bladder, (b) Diaphragm, and (c) Piston, etc. The major mechanical components of an accumulator include (1) a strong, rigid pressure vessel that confines the gas and fluid and carries the internal pressure loads, (2) a flexible or movable barrier, such as bladder, diaphragm, or piston that keeps the hydraulic fluid and gas charge separate, (3) a fill and maintenance port and valve at the gas side of the accumulator, and (4) a port and shut-off valve at the fluid side of the accumulator.

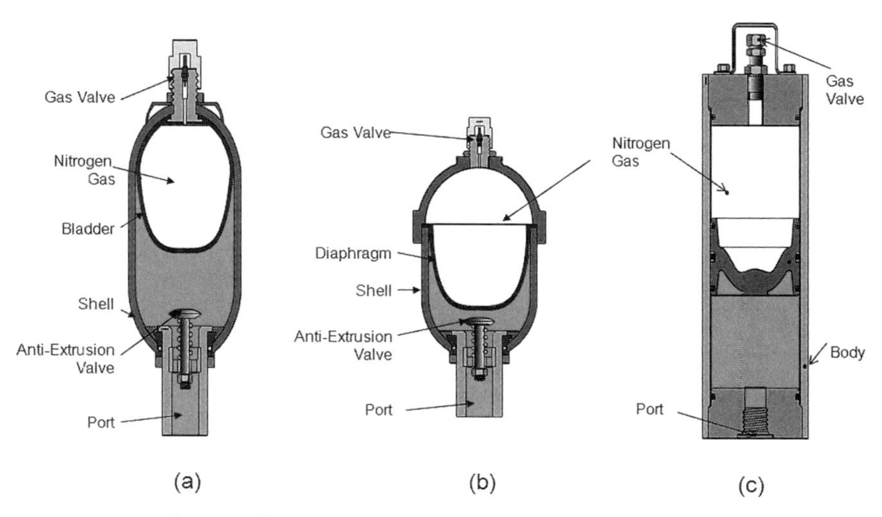

FIG. 4.19 Typical types of accumulators: (a) bladder, (b) diaphragm, and (c) piston accumulator.

Although a hydraulic accumulator has the cost advantage over the energy-storage devices of a hybrid-electric system, such as the Li-ion battery, the low energy density is one of the key limiting factors for the hydraulic hybrid powertrain for broad applications in the commercial vehicle market. However, high energy density and long life make these components cost effective, particularly where frequent braking and other transients feature in the vehicle use pattern.

4.4.1 Background of Hydraulic Accumulators

In a hydraulic accumulator, incoming fluid displaces a nearly equal volume of gas contained in the accumulator, thus compressing it and storing energy in the process. During discharge, the gas expands as it expels fluid from the accumulator, transferring this energy back to the fluid. Mathematically, the energy stored during compression of an enclosed gas can be described by the sum or integral of the product of pressure and volume change:

$$\text{Energy} = \int P dV \tag{4.25}$$

where P is the instantaneous pressure and V is the gas volume.

An important aspect of the equation above is that it is path dependent and that accumulated work (energy) depends on the relationship between pressure and volume. Assuming ideal gas behavior and that P and V are related in a simple polytropic process,

$$P(V) = P_2 \cdot \left(\frac{V_2}{V}\right)^n \tag{4.26}$$

The relationships for the amount of energy (work) stored in the compressed gas can be written as

$$W = \frac{P_1 \cdot V_1 - P_2 \cdot V_2}{n-1} \qquad n \neq 1 \tag{4.27}$$

$$W = P_2 \cdot V_2 \cdot \ln\left(\frac{V_2}{V_1}\right) \qquad n = 1 \tag{4.28}$$

In the above relationships, certain values of n have special physical significance, as shown in Fig. 4.20. For $n = 0$, the process proceeds at constant pressure and is called an "isobaric" process.

For $n = 1$, the process proceeds at constant temperature (by extracting or adding heat energy during the compression or expansion process). This is said to be an "isothermal" process.

For $n = k = C_p/C_v$ the process proceeds reversibly (no friction losses) and without heat exchange. This is referred to as an isentropic process.

For reasonably fast compression or expansion, such as for a vehicle undergoing many starts and stops, the accumulator tends towards isentropic behavior, with $n = k$ generally lying between 1.6 and 2.0 for real gases. Comparing this to the amount of energy that could be stored if a lower value of n were used shows that isentropic compression is actually the least desirable from an energy density standpoint. Thus, real and significant gains in energy density, on the order of 30–50% and in the limit as high as 100%, may be possible if the process can be shifted to lower values of n, that is, closer to isobaric. This behavior is shown in Fig. 4.20 as a plot of total stored energy vs. the volume change ratio for the gas.

4.4.2 Types of Hydraulic Accumulators

While the hydraulic system provides a means of transferring energy between the vehicle and accumulator, it is compression of a gas in the high-pressure accumulator that provides the energy storage. Each accumulator contains a fixed mass of inert gas (nitrogen) that changes in occupied volume and pressure in proportion to the instantaneous amount of hydraulic fluid contained. Several types of accumulator designs have been developed and are used in hydraulic systems today. The most common designs of hydraulic accumulators in commercial vehicle applications are (a) Bladder, (b) Impermeable Bag; (c) Diaphragm; (d) Metal bellows; and (e) Piston.

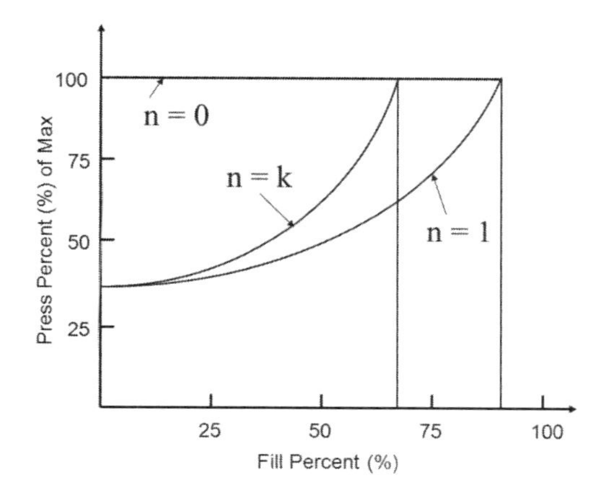

FIG. 4.20 Stored energy density as a function of the polytropic exponent and the gas volume change ratio.

4.4.2.1 *Bladder Accumulators*

Bladder accumulators feature a nonpleated, flexible rubber bladder housed within a steel shell. These bladders are made from an elastomer, such as rubber, and inhibit the mixing of the gas and liquid inside the accumulator by effectively sealing the gas inside the bladder. The bladder accommodates pressure deviations by expanding and contracting like a balloon inside the accumulator; decompressing or compressing the gas inside the bladder to match the pressure outside the bladder. The open end of the bladder is attached to the precharging valve at the gas end of the shell. A poppet valve, normally held open by spring pressure, prevents extrusion of the bladder into the exhaust passage when the accumulator empties of fluid.

An accumulator with an elastic bladder is advantageous for many reasons. First, it is a low-cost solution; the elastomer wall itself is often formed out of a simple rubber compound, and the only additional component is a seal or port that allows for the addition or removal of gas during maintenance. This leads directly to its second strength, simplicity. The bladder normally contains only two parts, the elastic wall and a port, which are easy to design, manufacture, and maintain. The bladder also possesses an inherently high durability. The elastic wall, the bladder's only moving part, is extremely resilient in resisting fatigue, due to the simple elastic nature of its expansion and contraction. It also experiences almost no friction with the walls of the accumulator or the exterior liquid. As a further development, the elastic bladder can be filled with low-density compressible foam, along with gas. This foam has been shown to increase the efficiency of the bladder by reducing losses due to heat flow [4.31].

An elastic bladder has one inherent weakness that has not yet been overcome: significant leakage of the nitrogen charge gas into the hydraulic fluid leading to loss of pressure over a period of months. This is a weakness in the long molecular chains that allow an elastomer to flex and expand without permanent damage. Unlike materials with a crystalline structure, gas molecules are able to squeeze through the gaps between these chains in elastomers. Several actions can be taken to minimize this leakage, such as increasing the thickness of the bladder walls beyond that necessary merely to withstand applied stresses. This extra thickness can provide improvements in leakage rates at the cost of flexibility.

4.4.2.2 *Diaphragm Accumulators*

Diaphragm accumulators feature a one-piece molded diaphragm that is mechanically sealed to the high-strength metal shell, as shown in Fig. 4.19(b).

The flexible diaphragm provides excellent gas and fluid separation. A button molded to the bottom of the diaphragm prevents the diaphragm from being extruded out the hydraulic port. The non-repairable electron beam-welded construction reduces size, weight, and ultimately cost.

The bladder/diaphragm is charged with a dry inert gas, such as nitrogen, to a set precharge pressure determined by the system requirements. As system pressure fluctuates, the bladder/ diaphragm expands and contracts to discharge fluid from, or allow fluid into, the accumulator shell.

4.4.2.3 Piston Accumulators

Piston accumulators consist of a cylindrical body, sealed by a gas cap and charging valve at the gas end, and by a hydraulic cap at the hydraulic end.

A lightweight piston separates the gas side of the accumulator from the hydraulic side.

As with the bladder/diaphragm accumulator, the gas side is charged with nitrogen to a predetermined pressure. Changes in system pressure cause the piston to rise and fall, allowing fluid to enter or forcing it to be discharged from the accumulator body.

Piston accumulators function using a metal piston that slides back and forth, compressing the gas. The piston is snug against the sides of the accumulator, with a seal or seals preventing leakage of gas past the piston and into the liquid. The piston often has guide rings along its sides to maintain alignment within the accumulator and may also have scrapers along its walls to prevent debris from damaging the seals, as shown in Fig. 4.19(c). The pistons are also almost perfectly impermeable to gas leakage. The only way for gas to leak in a piston is for the gas to pass through the thin gap between the piston and the accumulator wall, and past the seal(s) along the piston's edge. These seals themselves are manufactured from materials with very low permeation rates and are very resistant to gas flow.

There are ways for the gas to make it past the seal and into the hydraulic fluid. First, the surface finish on the accumulator walls must be extremely precise; any minute cracks or crevices can allow gas to slowly leak past a seal when the piston cycles back and forth. The piston and accumulator must also be extremely round; any deviations cause irregularities in the pressure of the piston seal against the accumulator wall. Low pressure against the wall can allow gas to seep past. High pressure against the wall causes increased friction between the wall and the piston, increasing pressure imbalances between gas and liquid, which accelerate wear. The piston seals generate a significant amount of pressure against the accumulator wall, causing friction, and will eventually wear out. This wear will initially occur as small abrasions along the surface of the seals, causing a small amount of leakage, but it will ultimately result in a catastrophic failure of the system due to fracture of the seal.

4.4.2.4 *Metal Bellows Accumulators*

In a bellows accumulator, a thin, collapsible, convoluted, metal foil or welded metal disk separates the gas and hydraulic fluid [4.37]. The metal foil provides an absolute barrier between the gas and fluid yet allows for significant volume change (approximately three-fold) of the gas space. The bellows would be filled with nitrogen, with hydraulic fluid existing between the bellows and the accumulator. A bellows system may also feature rails or other objects along the walls of the accumulator to guide its expansion and contraction.

The bellows are generally made in one of two ways, by edge welding of individual stamped diaphragms, or plastic forming of a seam welded tube. Forming is a high-volume manufacturing approach and leads to a cost significantly below that of welding due to drastically reduced scrap (the inner portion of each disk represents about 80% of the total foil and is discarded) and inherent process efficiency (all convolutes are formed simultaneously). The forming process itself uses a combination of hydro and then mechanical forming to fully generate the convolute shape.

The primary advantage of a bellows is that it is entirely made of metal. Thus, it is initially impermeable to gas transmission, perfectly eliminating the saturation of the fluid with nitrogen gas.

However, a bellows accumulator of sufficient size would be extremely large and complex. It would contain several hundred small metal rings and welds, resulting in large manufacturing costs. It would also be extremely vulnerable to fatigue; if only a small portion of any of these welds fails during any of the hundreds of thousands of expansions and compressions it is expected to experience during its lifetime, it would result in a catastrophic failure of the system. Therefore, different technologies such as hydroformed bellows are required to improve the system durability and performance [4.37].

4.4.3 Application of Hydraulic Accumulators

A lightweight composite shell is required for the hydraulic accumulator to meet weight requirements of commercial vehicle applications. Advanced composites, specifically a filament-wound carbon composite, can be used to significantly reduce weight—by a factor of three to four over high-strength steel—while accommodating higher system pressure. Such a structure can be achieved by filament winding over a metal liner. However, high cycle fatigue life of the liner is difficult to achieve because of the relatively lower cyclic strain capacity of metals (steel) relative to carbon fiber. A fundamentally better match is to use a polymer liner with a full composite shell enclosing the liner.

A 21-gallon hydraulic accumulator, which has been in production in Eaton's Hydraulic Launch Assist (HLA) system since 2009, is shown in

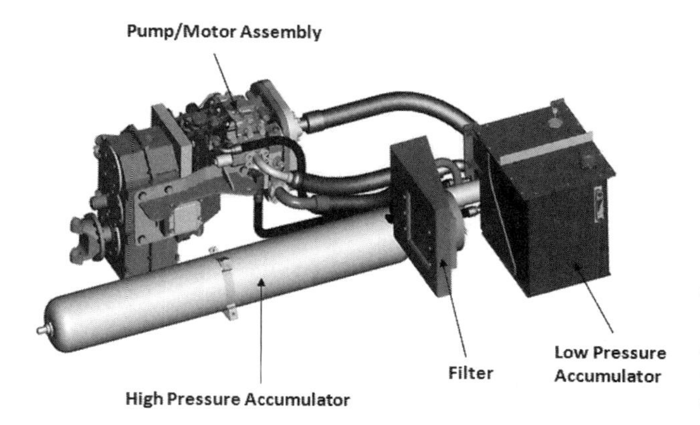

FIG. 4.21 Hydraulic Accumulator for Hydraulic Launch Assist (HLA). (Courtesy of Eaton Corporation)

Fig. 4.21. In the HLA, a hydraulic pump-motor is mounted to the vehicle drivetrain through electronically controlled clutches and transfers hydraulic fluid between a low- and high-pressure accumulator. During braking, the pump-motor generates a braking torque (torque that would otherwise come from friction brakes) as it drives fluid from the low-pressure accumulator to the high-pressure accumulator. At acceleration (launch), the flow is reversed and hydraulic fluid flows from the high-pressure accumulator back through the piston pump-motor to the low-pressure accumulator, generating a driving torque.

4.5 Flywheel Energy Storage (FES)

A flywheel is a mechanical device that can be used as a storage device for rotational energy, as shown in Fig. 4.22. A flywheel system consists of rotors made of high-strength carbon filaments, a shaft that is suspended by magnetic bearings, and an electric motor/generator. The rotor can spin at speeds from 20,000 to up to 100,000 rpm in a vacuum enclosure.

The stored energy depends on the moment of inertia of the rotor and the square of the rotational velocity of the flywheel, as shown in Eq. (4.29). Energy is transferred to the flywheel when the machine operates as a motor (the flywheel accelerates), charging the energy storage device. The flywheel is discharged when the electric machine regenerates through the drive (slowing the flywheel):

$$E = \frac{1}{2} I \cdot \omega^2 \tag{4.29}$$

where ω is the angular velocity of the flywheel and I is the moment of inertia of the rotor, which can be expressed as

$$I = \frac{r^2 \cdot m \cdot h}{2} \tag{4.30}$$

where r is the radius of the rotor, m is the mass of the rotor, and h is the height (length) of the rotor.

The energy storage capability of flywheels can be improved either by increasing the moment of inertia of the flywheel or by turning it at higher rotational velocities, or both. For the rotor, some designs use hollow cylinders that allow the mass to be concentrated at the outer radius of the flywheel, which improve storage capability with a smaller weight increase, utilizing compact carbon fiber composite rotors on magnetic bearings. The rotor turns in a vacuum at very high rotational velocities (up to 100,000 rpm). This approach results in compact, lightweight energy storage devices. Modular designs are possible, with a large number of small flywheels used as an alternative to a few large flywheels.

One example of a more recent development is the magnetic loaded composite (MLC) flywheel by Williams Hybrid Power [4.33]. Instead of using discrete permanent magnets to form the rotor of a flywheel's integrated motor/generator, magnetic powder is mixed into the composite matrix of the rotor. After the flywheel has been manufactured using filament winding, flash magnetization of the integrated magnetic particles generates the required field

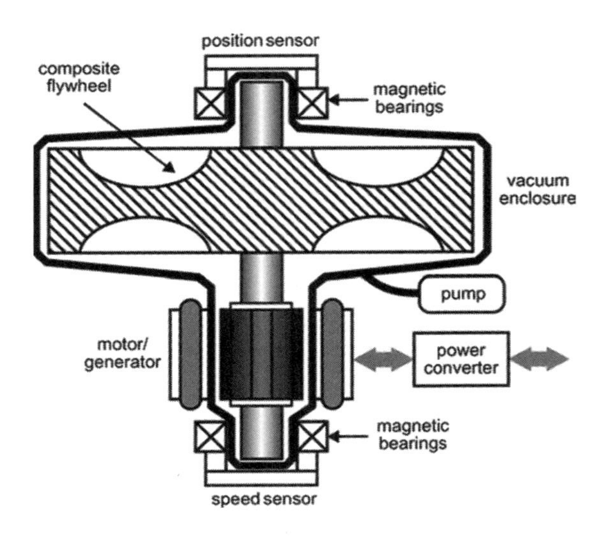

FIG. 4.22 Schematic of flywheel energy storage device [4.38].

configuration forming the rotor. With no large metallic structures in the flywheel rotor, eddy current losses and heating are negligible, resulting in very high electrical efficiencies. The lack of rotor heating gives MLC flywheels a unique advantage over other composite flywheel designs: they can be continuously deep-cycled at high power with no detriment to performance or reduction in life.

The advantages of the flywheel system include:

1. Compact weight and size—The entire system (the flywheel and the housing) is roughly half the weight and packaging of a battery hybrid system.
2. Higher efficiency—The flywheel system has no chemical-electric conversion losses like the batteries or supercapacitors of the engine storage system.
3. Lower cost—Smaller size and weight and reduced complexity make these arrangements about one-quarter the cost of a battery-electric system.

4.6 References

4.1 Zuercher, J. and Briggs, R., "Energy Storage Technology Study," Eaton Internal report, Project #2534, August 2007.

4.2 Ehsani, M. et al., *Modern Electric, Hybrid Electric, and Fuel Cell Vehicles*, Second Edition, CRC Press, 2010, pp. 375–410.

4.3 Miller, J. M., *Propulsion Systems for Hybrid Vehicles*, Institution of Electrical Engineers, 2004, pp. 359–410.

4.4 Kromer, M. A. and Heywood, J. B., " Electric Powertrains: Opportunities and Challenges in the U.S. Light-duty vehicle fleet," LEEE 2007-03 RP, pp. 30–47, Sloan Automotive Laboratory MIT, May 2007.

4.5 FreedomCAR Energy Storage System Performance Goals for Power-Assist Hybrid Electric Vehicles (11/2002), http://www.uscar.org/guest/view_team.php?teams_id=11, assessed Sept. 2011.

4.6 Chandler, K. et al., "New York City Transit Diesel Hybrid-Electric Buses: Final Results," DOE/NREL Transit Bus Evaluation Project, NREL/BR-540-32427, 2002.

4.7 Barnitt, R., "BAE/Orion Hybrid Electric Buses at New York City Transit," Technical Report NREL/TP-540-42217 Revised March 2008.

4.8 Fritz, G., et al., "Ovonic NiMH Batteries: The Enabling Technology for Heavy-Duty Electric & Hybrid Electric Vehicles," SAE Paper No. 2000-01-3108, SAE International, Warrendale, PA, 2000.

4.9 Chiang, P. K., "Two-Mode Urban Transit Hybrid Bus In-Use Fuel Economy Results from 20 Million Fleet Miles," SAE Paper No. 2007-01-0272, SAE International, Warrendale, PA, 2007.

4.10 Gibbard, H. F., "Nickel metal hydride battery technology," WESCON/'93, Conference Record, 1993, pp. 215–219.

4.11 Moorthi, M. M., "Field experience with large Nickel Metal Hydride (NiMH) batteries in stationary applications," 1-4244-0431-2/06/2006 IEEE.

4.12 "Basic Construction of a Ni-MH Battery," http://www.eneloop.info/home/technology.html, assessed, Sept. 2011.

4.13 "In search of the perfect battery," http://www.mitenergyclub.org/assets/2009/9/25/Economist_Batteries2008.pdf. Assessed, Sept. 2011.

4.14 Barsacq, F. "Lithium-Ion Battery Overview for PHEV Applications," EPRI Workshop, April 3, 2005.

4.15 Chen, C. H. et al., "Aluminum-doped lithium nickel cobalt oxide electrodes for high-power lithium-ion batteries," Journal of Power Sources 128, pp. 278–285, 2004.

4.16 Nellums, R. et al. "Class 4 Hybrid Electric Truck for Pick Up and Delivery Applications," SAE Paper No. 2003-01-3368, SAE International, Warrendale, PA, 2003.

4.17 Zhang, Z. et al. "Lithium-Air Rechargeable Battery," SAE 2011 Hybrid Vehicle Technologies Symposium, Anaheim, California, February 10, 2011.

4.18 "Ultracapacitors," http://www.nrel.gov/vehiclesandfuels/energystorage/ultracapacitors.html, assessd, April 2011.

4.19 Wright, R. B. et al., "Performance Testing Of Selected Commercial Ultracapacitors," www.electrochem.org/dl/ma/201/pdfs/0230.pdf, assessed April 2011.

4.20 Pay, S. and Baghzouz, Y., "Effectiveness of Battery-Supercapacitor Combination in Electric Vehicles," IEEE Bologna Power Tech Conference, June 23th–26th, Bologna, Italy, 2003.

4.21 Lohner, A. and Evers, W., "Intelligent Power Management of a Supercapacitor based Hybrid Power Train for Light-rail Vehicles and City Busses," 2004 35th Annual 1EEE Power Electronics Specialists Conference, Aachen, Germany, 2004.

4.22 Hoelscher, D., Skorcz, A., Yimin, Gao, and Ehsani, M., "Hybridized electric energy storage systems for hybrid electric vehicle," Vehicle Power and Propulsion Conference, VPPC '06. IEEE, 2006.

4.23 Hadartz, M. and Julander, M., "Battery-Supercapacitor Energy Storage," Master Thesis, Chalmers University of Technology, Göteborg, Sweden, 2008.

4.24 "UNEP Environmental Assessment: Expo2010—Shanghai, China," http://www.unep.org/publications/contents/pub_details_search.asp?ID=4040, assessed, Sept. 2011.

4.25 Welch, G. and Bishop, G., "An introduction to the Kalman Filter," www.cs.unc.edu/~welch/kalman/kalmanIntro.html, assessed, April 2011.

4.26 Garche, J. and Jossen, A., "Battery management systems (BMS) for increasing battery life time," Telecommunications Energy Special, 3, pp. 81–84, 2000.

4.27 Plett, G. L. and Klein, M., "Advances in HEV Battery Management System," SAE Paper No. 2006-21-0060, SAE International, Warrendale, PA, 2006.

4.28 Pop, V. et al., Battery Management Systems. Accurate State-of-Charge Indication for Battery-Powered Applications. doi: 10.1007/978-1-4020-6945-1_2, © Springer Science + Business Media B.V. 2008.

4.29 Moore, S. W. and Schneider, P. J., "A Review of Cell Equalization Methods for Lithium Ion and Lithium Polymer Battery Systems," SAE Paper No. 2001-01-0959, SAE International, Warrendale, PA, 2001.

4.30 Pourmovahed, A. et al., "Experimental Evaluation of Hydraulic Accumulator Efficiency with and Without Elastomeric Foam," *J. Propulsion*, Vol. 4, No. 2, March–April 1988.

4.31 "Ricardo Kinergy delivers breakthrough technology for effective, ultra-efficient and low cost hybridisation," www. Recardo.com, 24 November 2009. assessed, April 2011.

4.32 "WHP's Flywheel Technology," http://www.williamshybridpower.com/technology/whps-flywheel-technology., assessed, April 2011.

4.33 Foley, I., "Flywheel Energy Storage," SAE 2011 Hybrid Vehicle Technologies Symposium + Electric Vehicle Technologies Day, February 2011.

4.34 Andrea, D., *Battery Management System for large Lithium-Ion Battery Packs*, Artech House, 2010.

4.35 "Assessment of Needs and Research Roadmaps for Rechargeable Energy Storage System Onboard Electric Drive Buses," Http://Www.Fta.Dot.Gov/Research, Report No. Fta-Tri-Ma-26-7125-2011.1, December 2010.

4.36 Suzuki, K., "Metal bellows hydraulic accumulator," US Patent 7013923, Mar 21, 2006.

4.37 Briat, O. et al., "Principle, design and experimental validation of a flywheel-battery hybrid source for heavy-duty electric vehicles," IET Electr. Power Appl., 2007, 1, (5), pp. 665–674, 2007.

Chapter 5

Hybrid-electric System Design and Optimization

5.1 Characteristics of Hybrid-electric Powertrains

Hybrid-electric powertrains can be used to optimize energy flow in a vehicle system by selecting the appropriate amount of power to be supplied from the internal combustion engine, and combining it with power from the electric motor/generator(s) to meet the needs of vehicle operation. Electric motor/generators can provide positive or negative torque, with a positive or negative direction of rotation. A typical hybrid system configuration includes one or more electric traction motor/generators, a traction motor/generator controller, an energy storage system such as a high-voltage battery- or ultracapacitor-based system, and appropriate thermal management and power distribution. The electrical power path can provide energy independent of the internal combustion engine, or store energy to be used at another time. This additional flexibility in function of the powertrain provides opportunities for substantial improvement in fuel economy and emission performance, as demonstrated by the thousands of hybrid commercial vehicles and buses in use around the United States.

5.1.1 User Requirements

Hybrid system design and optimization starts with the consideration of customer and end user requirements. At the highest level, most users express their requirements in terms of fuel savings and reduced operational cost. However, a deeper investigation of user needs and requirements reveals a richer set of requirements. Included in this set are requirements that have emerged from early use of hybrid systems that revealed unexpected benefits.

5.1.1.1 *Economics*

Hybrid truck economics are clearly a driving factor for many individual and fleet end users. Economics include the initial purchase price of the hybrid vehicle, the amount of fuel saved from hybrid operation, maintenance requirements, and warranty period for the hybrid vehicle. Typically, end users expect

30 to 50% fuel savings from their hybrid trucks and a payback of the higher initial cost of the vehicle in less than five years. Especially in the early days of hybrid systems for commercial vehicles, economics of hybrid trucks were not very favorable. Early adopters of these systems were primarily fleets purchasing these vehicles for evaluation purposes or simply to build a "green" image.

5.1.1.2 *Performance*

Performance of hybrid trucks is generally desired to be similar or better than conventional vehicles. Acceleration performance, for instance, should be similar even if the conventional vehicle's internal combustion engine is replaced with a downsized version in the hybrid equivalent vehicle. Acceleration should be smooth, with a shift feel similar to that of conventional vehicles with automatic transmissions. The latter has proven to be a challenge for hybrid systems incorporating an Automated Manual Transmission (AMT). These systems can suffer from very distinct torque interrupts during shifting, which has all but prevented their adoption into the U.S. bus hybrid market, where smooth acceleration is strongly desired. Brake feel should be similar, even if retarding torque is derived from the foundation brakes blended with electric motor/generator retarding torque. Powertrain reliability and durability should be equivalent to those features for a conventional vehicle powertrain.

5.1.1.3 *Auxiliary Features*

Hybrid systems offer additional benefits that contribute to fuel savings indirectly, but also provide additional value to the end users of these vehicles. Over the last decade, the automotive industry has witnessed a significant shift from largely mechanical and hydraulic subsystems to electrically driven versions. This shift has been driven largely by the need for efficiency improvements as well as improved packaging and reliability of these subsystems. Examples of these are electric steering and braking, electrically driven cooling fans and water pumps, and so on. More generally, this trend is termed X-by-wire. While this same need for efficiency improvements is desired for commercial vehicles, the much higher power demands for these subsystems make their implementation impractical with conventional 12-V vehicle electrical power systems. Hybrid systems, however, offer both a high-voltage power bus as well as generally sufficient energy storage to sustain electric accessories, even during vehicle idle conditions.

Commercial vehicles are manufactured and sold for a wide range of vocational applications. Hybrid system manufacturers such as Eaton and Azure Dynamics have identified a number of these vocational applications to be particularly suited for hybridization, along with complementary electrical subsystems that provide significant additional value to the end users:

Utility vehicles with aerial booms: These trucks, as shown in Fig. 5.1, are typically used to install or service electric transmission and distribution infrastructure. In service, these trucks have demonstrated their value with overall fuel consumption reduction up to 60%. A significant portion, up to 75%, of the fuel savings is derived from stationary operation of the truck. In standard aerial trucks, the engine idles while the truck is stationary to allow for operation of the aerial device. Operation of the aerial device is achieved via the transmission Power Take Off (PTO), which drives a hydraulic pump that provides pressure and flow to the hydraulic system of the aerial device. In the hybrid version, the engine is shut off, and the aerial device is operated using electrical energy from the hybrid battery system. In this case it is not the engine, but the electric traction motor that drives the PTO, or in this case ePTO to denote electric operation of the PTO. Further demonstrated benefits of the system are 30–60% reduction in criteria pollutant emissions (HC, CO, NO_x, and PM) and silent operation at the work site.

The typical hybrid system on these trucks also integrates a DC/AC converter, providing 110-V AC power to operate any standard electrical tools, and obviating the need to have a separate generator on board the vehicle. While these benefits were designed into the system in response to customer needs and the understanding of vocational vehicle mission profiles, there have also been unanticipated benefits. One has been the fact that while utility

FIG. 5.1 Hybrid utility truck. (Courtesy of Eaton Corporation)

trucks are operating on site, they are very quiet, allowing the crew to communicate easily without having to shout over an idling engine. It has also allowed crews to work in residential areas at times that would normally not be acceptable due to the noise the vehicle generates. These emergent benefits have quickly made the few hybrid trucks in any particular fleet the most desired ones to operate.

Refrigeration trucks: Refrigerated city delivery trucks are another subset of vocational vehicles that can benefit significantly from hybridization. Typically these vehicles use a small auxiliary engine to power the cooling system. However, this cooling system can be powered electrically from the hybrid system's high-voltage energy storage system. This would allow for the elimination of the auxiliary engine that normally powers the cooling system.

5.1.2 Driving Cycles

Vehicle fuel economy improvement and emission reduction benefits from hybridizing are very sensitive to drive cycles. Fuel-saving benefits of hybrid vehicles are typically derived from brake energy recovery and subsequent use of that recovered energy to boost vehicle acceleration. Additionally, fuel is saved when the engine is shut off during vehicle idle time. It is fairly straightforward then to realize that fuel-saving benefits of hybridization are highest in driving cycles that involve frequent start-stop operation. A complicating factor is, of course, that most vocational vehicles that fall into this category operate on a single shift schedule and thus will not typically accumulate many miles. In these cases the percentage fuel saved may be relatively high, but the absolute amount of fuel saved over a given period may not be very attractive at all. One group of vehicles that falls both in the category of frequent start-stop operation and accumulates many miles are city buses. These vehicles operate up to 20 hours per day. Over the years, many different and vocation specific duty cycles have been developed to evaluate and quantify the benefits of hybridization for a given application. A few specific examples include:

Various utility aerial truck duty cycles: This is a unique cycle that combines both a regular driving portion, typically represented by the Federal Test Procedure (FTP) duty cycle, and a representation of the site operations, when the vehicle is idle and the aerial device is operated.

Refuse truck duty cycles: Again, this vocation-specific cycle combines residential area, house to house, high-frequency start-stop operation with a higher-speed driving portion to go from the residential refuse pickup area to the local landfill. The ratio of these two portions of the cycle significantly affects overall efficacy of the hybrid system.

Various urban and suburban bus duty cycles: As mentioned already, city buses appear to be ideal applications for hybrid systems, with frequent start-stop operation and high mileage accumulation. Many duty cycles for buses have been identified; for instance, inner city cycles such as those for New York and Los Angeles, or Beijing, China. These cycles are characterized by extremely low average vehicle speeds, typically less than 8 mph. There are suburban duty cycles such as the Orange County Cycle that are characterized by higher average speeds in the 10–15 mph range and less frequent stops. Finally, there are Rapid Transit duty cycles to represent buses operating on dedicated lanes with fewer stops and average speeds exceeding 25 mph.

Composite International Local and shortened Commuter Cycle: The CILCC cycle was developed by a team of engineers from the National Renewable Energy Laboratory (NREL), Eaton, and International Truck and Engine Company (ITEC) [5.1] to be representative of a range of class 4 through 6 vocational vehicles that operate in and around cities. The cycle consists of four International Local cycles (a cycle used by ITEC to conduct field tests for heavy-duty trucks) and one shortened Commuter cycle (a highway cycle). It contains 57% city, 28% suburban, and 15% highway driving, which is very close to the average distribution information from various fleets such as UPS, FedEx, Coca Cola, and Snap-On. Figure 5.2 shows the CILCC drive cycle speed trace as a function of time.

The primary objective of the CILCC is to define a reasonable cycle to objectively evaluate Class 4–6 commercial vehicle hybrids. Another often used general cycle for this class of vehicles is the West Virginia University City cycle (WVUCITY), developed by West Virginia University, as shown in Fig. 5.3.

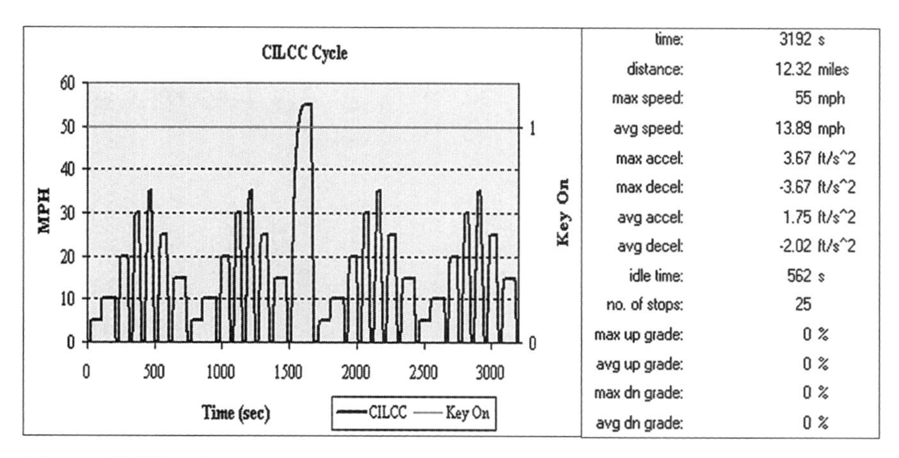

FIG. 5.2 CILCC cycle.

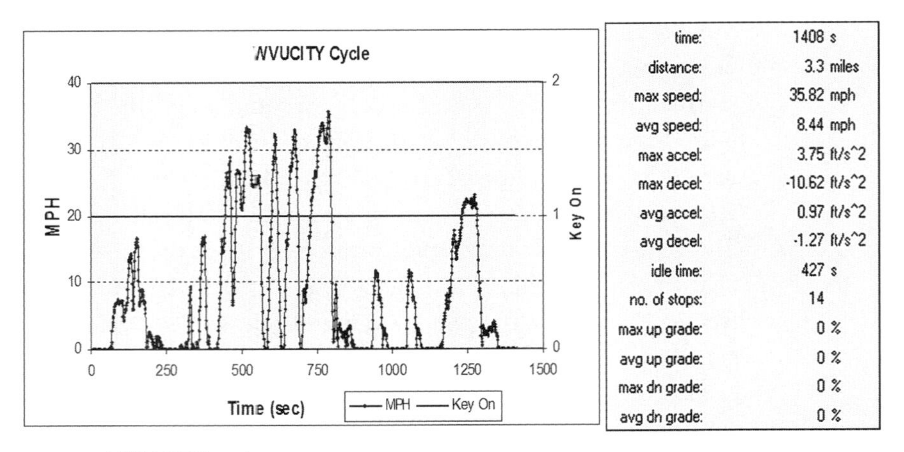

FIG. 5.3 WVUCITY cycle.

5.2 Hybrid System Architectures

The two most common configurations of hybrid vehicles are the series and parallel designs, which are shown schematically in Fig. 5.4. Series and parallel refer to the orientation of the two power plants in the propulsion system.

In the series hybrid, the engine powers a generator that either supplies power to charge the battery pack, or to power the electric drive motor. In the parallel hybrid, the engine supplies mechanical power directly to the propulsion system, while the electric motor is also coupled directly to the propulsion

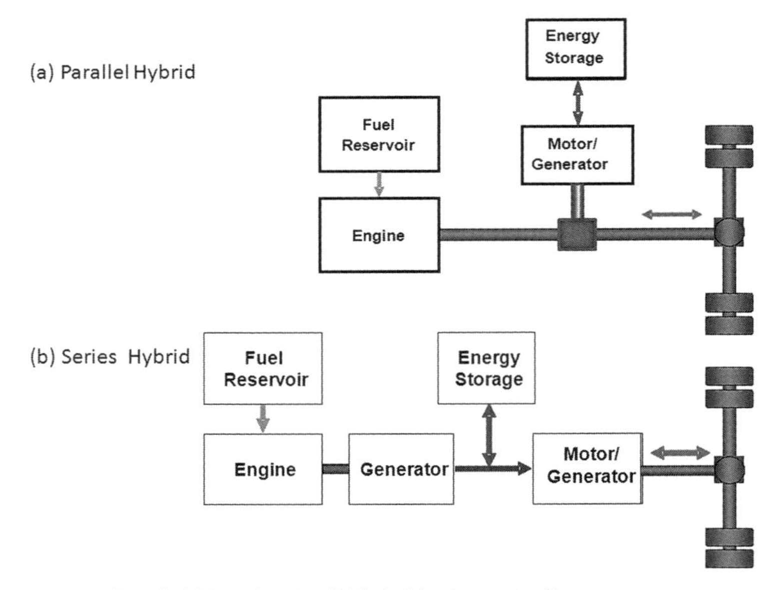

FIG. 5.4 Parallel (a) and series (b) hybrid schematic diagrams.

system. The parallel hybrid vehicle can be run on the engine alone, the electric motor alone, or on both engine and electric motor simultaneously, depending on how control is set up.

A third hybrid architecture is the so-called dual-mode system. Originally patented by General Motors and used in a number of GM passenger cars, it is applied by Allison Transmission in city bus applications [5.3, 5.4]. The hybrid drive system is a combination of transmission, clutches, gear sets, and electric motors (see Fig. 5.5), which are integrated into a single unit, the dual-mode transmission.

In short, a dual-mode is a hybrid vehicle that can operate in two distinct modes. The first mode works much like a regular parallel hybrid. It is the second mode that makes the difference—where the hybrid system can adjust varying amounts of engine and motor function to meet very specific vehicle/task/traffic requirements. Due to the high cost of such a unique and integrated transmission, its successful application has been limited to city bus applications.

The most successful hybrid system to date is no doubt the Toyota Synergy Drive system with more than three million hybrid vehicles on the road globally. Most of these 3 million vehicles are the iconic Prius. The Prius has a complex hybrid system referred to as a series-parallel hybrid (see Fig. 5.6). In the Toyota Prius [5.5, 5.6], a planetary gear set allows for the engine and electric motor to synergistically (in parallel) drive the wheels, or the electric motor to individually drive the wheels (with either the engine on or off). This allows for greater control over the engine and battery usage. Power to the wheels can flow directly from the electric motor (for low speed/load all-electric operation), can come directly from the gasoline engine, or can be a combination of gasoline and electric motor contributions. During decelerations the electric motor is

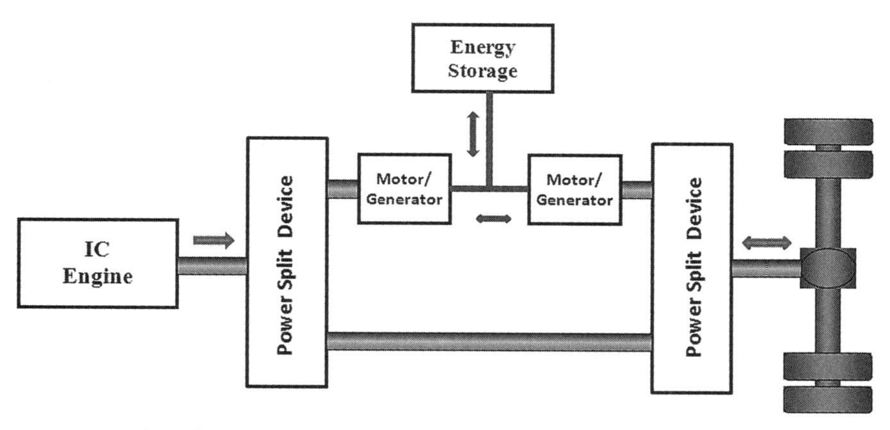

FIG. 5.5 Dual-mode hybrid system diagram.

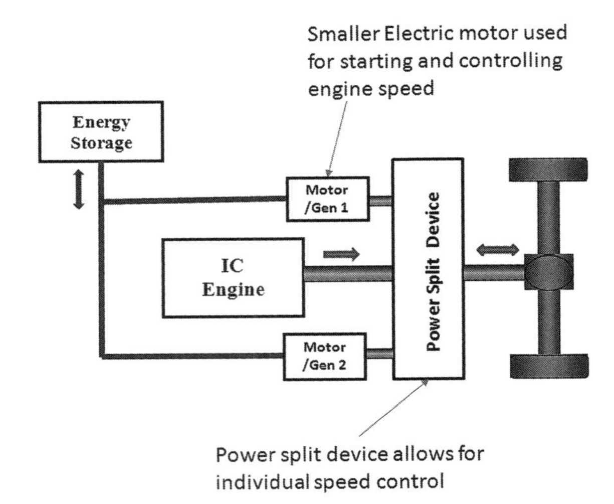

FIG. 5.6 Toyota Prius series-parallel powertrain schematic [5.6].

used to regenerate kinetic energy and charge the battery pack. Energy from the gasoline engine can also be used to power the generator and charge the battery. Relative speeds of the gasoline engine, generator, and electric motor are controlled by the gear ratios of the planetary gear set.

5.2.1 Parallel Hybrid System

The parallel hybrid system configuration is proving the configuration of choice for most commercial vehicle applications. Early pioneers Hino and Eaton Corporation have sold a combined 15,000 units globally. Parallel hybrid systems from companies such as BAE Systems, ZF, and Allison are expected to enter the market by 2012. In the parallel hybrid configuration, both the internal combustion engine and the electric motor are mechanically coupled to the drive wheels. Either the engine or the electric motor can individually supply power to the drive wheels, or they can do so simultaneously. The electric motor can also be used as a generator to charge the batteries. Since the internal combustion engine must be capable of charging the batteries as well as propelling the vehicle at cruise speeds, the engine is larger, and the electric motor is smaller as compared to a series configuration design for a similarly sized vehicle. The internal combustion engine is sized for medium- and high-speed cruise loads and usually provides better highway fuel economy compared to the series configuration. This is due to the fact that in the parallel hybrid configuration under highway cruise conditions the internal combustion motor directly drives the wheels as in a conventional vehicle. With a series hybrid system, however, the mechanical energy from the internal combustion engine is converted to electricity by the generator and then powers an electric motor connected to the wheels. Even with an efficient electric motor and generator, the overall efficiency of the power transfer from the engine to the wheels for a series hybrid system is only around 80%. In the parallel hybrid configuration this efficiency is greater than 97%, assuming a manual or automated manual transmission (AMT) is used. Figure 5.7 shows a parallel hybrid drive unit, as manufactured by Eaton Corporation. It includes an electrically actuated clutch in front, followed by an electric motor in the middle and then the AMT base box. Mounted on top of the transmission box are the X-Y shifters that automatically shift the transmission, as commanded by the transmission controller. Both the Transmission Control Unit (TCU) and Hybrid Control Unit (HCU) are mounted on the side of the transmission base box.

The parallel configuration can also be used to drive the vehicle in Zero Emissions Vehicle (ZEV) mode for a limited period of time if the vehicle is capable of operating fully electrically. The latter means that all accessory system must be able to operate electrically, including steering, braking, HVAC, and other subsystems. This capability is not yet available on mainstream hybrid truck systems. Because the internal combustion engine is required to be on for full vehicle power, full vehicle power is not attainable as a ZEV for a parallel hybrid. The disadvantage of the parallel configuration is that the direct coupling of the internal combustion engine, electric motor, and drive wheels often requires a complex control system. In essence, the controller must continuously optimize the power contributions from the internal combustion engine and electric motor/generator to deliver both optimal performance as well as best possible fuel economy. Also, because the internal combustion engine must

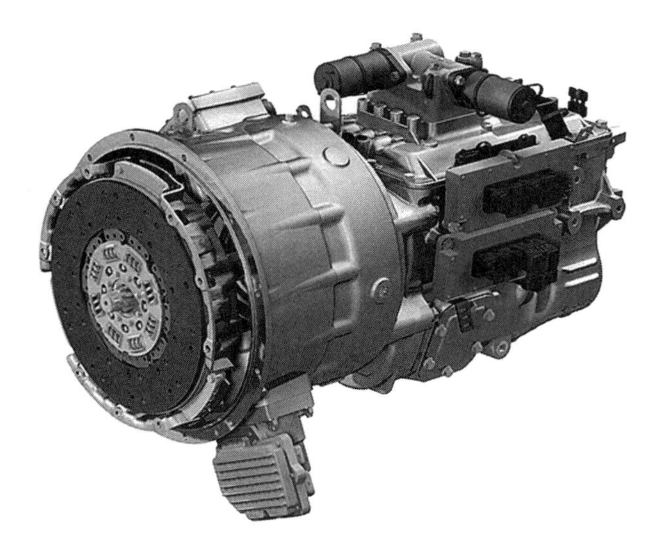

FIG. 5.7 Eaton Hybrid Drive Unit. (Courtesy of Eaton Corporation)

operate over a wide range of speeds and loads, it can't be run at optimum efficiency or emissions points all of the time as in the series configuration.

5.2.1.1 *Parallel Hybrid System Energy Flows*

In a parallel hybrid system, power and energy can flow in a variety of ways to and from the wheels and electric motor/generator, as shown in Fig. 5.8.

1. During mild acceleration from a standstill, the vehicle can be powered entirely by the electric motor, assuming there is sufficient charge in the energy storage system. In this mode of operation, the clutch between the engine and transmission is open. The engine is either idling to continue powering accessory systems such as steering, braking, and HVAC or shut down if those systems can be powered electrically.
2. Because parallel hybrid systems typically use a smaller engine compared to the equivalent conventional vehicle, the internal combustion engine is assisted by the electric motor during hard accelerations. The level of assist from the electric motor and duration will depend greatly on energy storage system capacity and state of charge. For example, an ultracapacitor-based energy storage system can provide a high level of power, but only for a very short duration. However, a reasonably sized Lithium-ion based energy storage system can provide sufficient power for a long acceleration event.
3. Deceleration mode is where the hybrid system truly shows its value. Conventional vehicles use a range of means to decelerate: foundation brakes, engine coasting, Jake-brakes, and retarders. All these methods convert the vehicle's kinetic energy to waste heat. In a hybrid vehicle, however, the clutch is open to disengage the engine from the drivetrain while the electric motor switches to generating mode. The general deceleration control strategy is to have the electric system absorb as much energy as it can, given the constraints of power limits, energy storage system capacity, and thermal management capabilities. If the electric motor cannot slow down the vehicle as fast as commanded by the driver, then the control system will automatically blend in additional deceleration forces via the foundation brakes, Jake-brake, and/or retarder.
4. During cruise mode, the vehicle hybrid powertrain will mostly behave as a conventional vehicle with the internal combustion engine providing motive power. However, even during cruise conditions, especially over articulated terrain, there may be opportunities for fuel efficiency improvements through selective electric motor assist or regeneration. This is achieved by avoiding engine transients when the vehicle is moving up or down a grade.

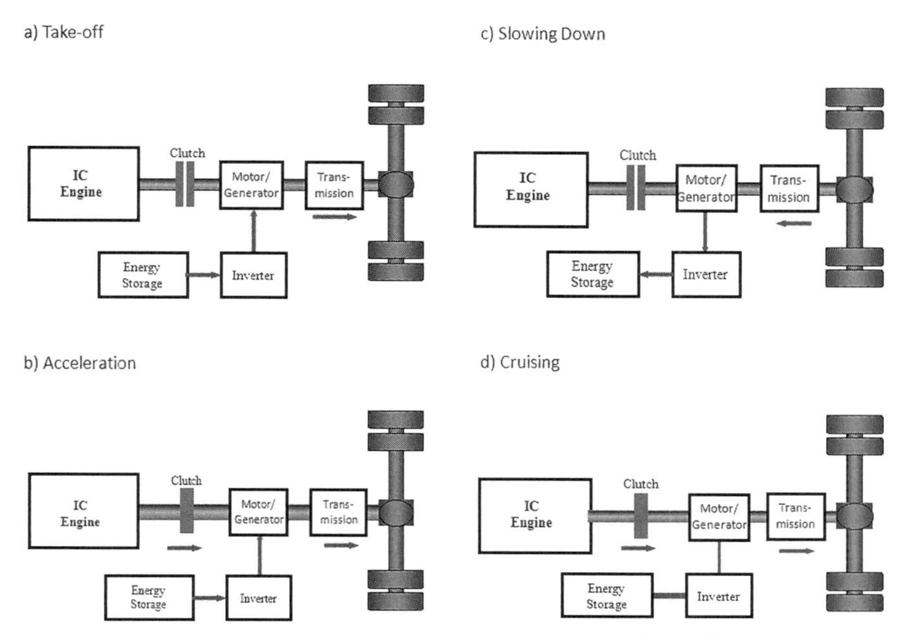

FIG. 5.8 Operating conditions for a parallel hybrid system: a) Take-off, b) Acceleration, c) Deceleration, and d) Cruise mode.

5.2.2 Series Hybrid System

Series hybrid vehicles are similar to purely electric vehicles, except the series hybrid vehicle includes an onboard generator system. The internal combustion engine is used to power a generator to generate electricity, which is then used to power the electric drive motor or charge the batteries. The internal combustion engine is typically sized for the vehicle's high-speed cruise loads. These loads are generally small in comparison to acceleration and hill-climbing loads, so the result is a smaller engine than would be used if the vehicle were conventionally powered. The electric drive motor is then sized to handle the acceleration and hill-climbing loads. The limiting factor is again the capacity of the energy storage system. This must be sized to assure the ability to provide sufficient electrical energy beyond what the generator can provide, to provide acceptable performance when driving up a grade. The series configuration hybrid vehicle results in a relatively simple connection of the electric drive motor to the drive wheels. In most cases, a multiple-speed transmission is not required due to the favorable torque and speed properties of electric motors. The vehicle can also be operated for a finite amount of time as a zero emissions vehicle (ZEV) by running off the batteries only. Full vehicle power is also available while running as a ZEV if the energy storage system is sized for full vehicle power.

However, because the vehicle has an onboard generator, one could opt for a design with fewer batteries, which saves cost and weight as compared to an electric vehicle. In such a case, engine power is required during peak load conditions. The biggest disadvantage to the series configuration is that all of the engine's power must be transmitted through the generator and drive motor. Because of the inefficiencies of these two components, some power is lost that is not lost in vehicle designs where the mechanical power of the engine is directly coupled to the wheels. Another disadvantage to the series configuration is that both an electric motor and generator are required, which usually results in a heavier and more costly vehicle as compared to the parallel configuration. Because of these disadvantages, series hybrid systems have been successful primarily and only in city bus applications. Average speeds are very low in these applications, thus avoiding most of the energy transfer inefficiencies. Buses drive many more miles per unit time than most other commercial vehicles except for class 8 long-haul trucks, making the higher cost of these systems less of an issue. Finally, the inherently smooth acceleration performance of electric drive systems is highly desired in the bus market segment.

5.2.3 Dual-mode Hybrid System

The dual-mode hybrid system was originally patented and developed by General Motors [5.3, 5.4]. When Allison Transmission became an independent company, it obtained license to use the Dual Mode technology for commercial vehicle applications. Distilled down to its most basic components and elements, it is a system in which a conventional automatic transmission with gears and bands and clutches has been replaced with an externally similar shell that houses a pair of electric motors and several sets of planet gears. Figure 5.9 shows a Dual Mode Hybrid Transmission, manufactured by Allison Transmission.

The two modes of operation can be described as a low-speed, low-load mode; and a higher-speed, heavier-load mode that operate as follows:

1. **First mode:** At low speed and light load, the vehicle can move with either the electric motors alone, the internal combustion engine (ICE) alone, or a combination of the two. In this mode, the engine (if running) can be shut down under appropriate conditions, and all accessories as well as vehicle driving continue to operate exclusively on electric power. The hybrid system will restart the internal combustion engine at any time it is deemed necessary. One of the motors, actually better described as motor/generators (M/Gs), acts as a generator to keep the battery charged, and the other works as a motor to propel, or assist in propelling, the vehicle.

FIG. 5.9 Allison Dual Mode Hybrid Transmission [5.3].

2. **Second Mode:** At higher loads and speeds, the internal combustion engine always runs. In this mode, operation is quite complex, as the M/Gs and planet gear sets phase in and out of operation to keep torque and horsepower at a maximum. Basically, it works like this: At the threshold of second mode, both M/Gs act as motors to give full boost to the engine. As the vehicle's speed increases, certain combinations of the four fixed-ratio planet gears engage and/or disengage to continue multiplying engine torque, while allowing one or the other of the M/Gs to switch back to generator mode. This interplay among the two M/Gs and four planet gears continues as vehicle speed and/or load fluctuates across road and traffic conditions. The benefit lies in avoiding engine transients as much as possible and allowing the internal combustion engine to operate at or near its most efficient operating point. Similar to the series hybrid system, the disadvantage lies in efficiency losses of the motor/generator pair.

5.3 Hybrid System Selection

The selection of the right hybrid technology or system for commercial vehicles is dependent on many factors:

- **System cost:** Hybrid system cost is one of the leading factors in selecting the right hybrid technology. Parallel hybrid systems can range from very simple and low cost for a post transmission mild hybrid to a very complex and expensive pretransmission solution with a high level of hybridization. Typical retail cost differential between conventional and parallel hybrid commercial vehicles is in the range of $10,000 to almost $100,000. The parallel hybrid system offers a significant amount of flexibility in terms of selecting the level of hybridization as well as overall hybrid drivetrain architecture. Series hybrid systems tend to be the most expensive option, exceeding $150,000 per vehicle. The reason for this is that the electrical system must be sized for full vehicle performance, and a separate generator and internal combustion engine are also required. System controls are simpler compared to a parallel hybrid because vehicle driving controls are essentially the same as they are for a pure electric vehicle. The internal combustion engine and generator combination are simply controlled to sustain energy storage system capacity. A dual-mode hybrid system falls somewhere between a series and parallel system from a cost perspective, with a cost between $100,000 and $150,000.

- **Annual fuel consumption:** The amount of miles driven per year, as well as conventional vehicle efficiency, determine the amount of fuel that can potentially be saved. Fuel savings are a prime motivator for purchasing a hybrid vehicle, and the incremental cost of the hybrid vehicle should be recoverable within a reasonable amount of time.
- **Average speed:** High fuel consumption alone is not a sufficient indication of suitability for hybridization or the type of hybrid system to select. Vehicle average speed is another good indicator.
- **Application:** Conventional classification of commercial vehicles is generally done by vehicle weight. A more suitable classification for hybrid vehicles, however, is by application or duty cycle. Many applications have unique duty cycles that are more or less suited to hybridization. For example:
  - **Package and Delivery Vehicles:** These vehicles are typically 14,000 to 26,000 lb box trucks, delivering packages or goods from a local distribution center. This application can be further subdivided into urban, suburban, and rural delivery. Urban delivery is generally characterized by low average speed (due to congestion) but also low total miles driven per year, usually less than 8000 mi. While high percentage fuel savings are possible, the absolute fuel savings are small because these vehicles do not consume much fuel to begin with. Rural delivery is characterized by high average speeds and relatively few transients. Annual mileage generally exceeds 25,000. There simply are few opportunities for energy recovery for this type of driving. The suburban delivery subset is typified by average speeds between 10 and 15 mph and between 12,000 and 18,000

miles driven per year. Suburban delivery applications can present an attractive segment for hybridization if the cost of hybridization is below $25,000 per vehicle. The later is based on average conventional vehicle fuel efficiency of approximately 8 mpg. With the given miles driven per year and typical 30% improvement in fuel economy due to hybridization, the annual fuel cost will be around $7500. Therefore, the payback time is close to three years. This coarse analysis is provided to illustrate that a parallel hybrid system is really the only potentially viable solution in this market segment.

- **Refuse Trucks:** These vehicles typically weigh 33,000 lb and up. Their duty cycle is characterized by two distinct portions: residential refuse pickup with frequent hard accelerations and decelerations, and return to the landfill typically in a suburban or rural driving pattern. Total miles driven can be high, but only if the return-to-landfill portion of the duty cycle is large. Generally, most of the fuel saving benefits will derive from the residential pickup portion. A similar analysis shown in the previous section suggests that a parallel hybrid system is the best option. The only caveat is that to capture a significant portion of the deceleration energy, a fairly high-powered electrical system will be required.
- **City Buses:** The third, and again very different, example is the city bus. This market segment is characterized by low to very low average speeds, 10 mph to as little as 3 mph. At the same time, however, annual mileage is quite high, typically greater than 40,000 miles per year. This is because these vehicles often operate 20 hours per day, every day of the year. In Europe and North America, all hybrid buses deployed to date are equipped with either series hybrid or dual-mode hybrid systems. Two additional factors are responsible for this trend; first of all, funding mechanism. In North America, for instance, the federal government pays for approximately 85% of any city bus purchased by a municipality. The buying decisions therefore are highly driven by operational cost, with fuel being a major component of that cost. Second, from a performance perspective, it is highly valued for a bus to exhibit extremely smooth and consistent acceleration; a parallel hybrid with an automated transmission can simply not meet that requirement. In developing countries, however, conventional buses are cheap vehicles, without comfort features such as heating or cooling, and they typically use a manual transmission. Cost is a major focus, and for this reason, most hybrid buses in China, for instance, are equipped with parallel hybrid systems. Even in Europe and North America, more consideration is being given to parallel

hybrid systems due to the inherently lower cost of such systems and the continued improvements in transmission technology.

In the following discussion, the focus is primarily on hybrid component selection and system optimization for the pre-transmission parallel hybrid system, as this is clearly becoming the preferred hybrid architecture for commercial vehicle hybridization.

5.3.1 Electric Motor/Generator Selection

The traction motor is one of the most important components in a hybrid-electric vehicle. To identify and use the most appropriate motor technology for a hybrid development program, careful consideration should be given to the various available motor/generator and corresponding inverter technologies. Some of the motor technologies available are the Induction Motor (IM) and Brushless Permanent Magnet Motor (BLPMM), the latter including Surface Mounted Permanent Magnet Motor (SMPMM), Surface Inset Permanent Magnet Motor (SIPMM), Interior Permanent Magnet Motor (IPMM), and Flux Concentrated Interior Permanent Magnet Motor (FCIPMM), as shown in Fig. 5.10. Another motor technology that more recently is receiving some attention is the Switched Reluctance Motor (SRM).

In the early 2000s, most hybrid systems were still using induction motors due to their long history in industrial applications and proven durability and reliability. Since then, permanent magnet (PM) motors have become more popular, in large part due to their superior efficiency. This is especially important given that the energy recovered during vehicle braking passes through the motor/generator twice, once through the generator to be stored in the energy

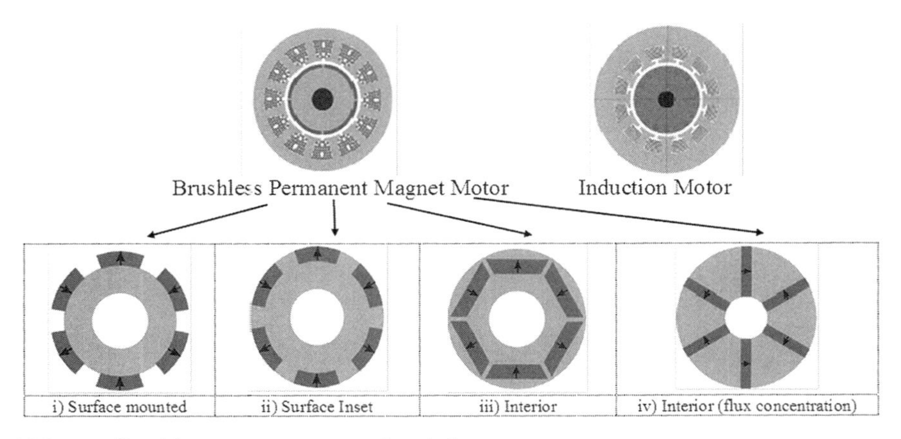

FIG. 5.10 Brushless motor cross-sectional diagrams.

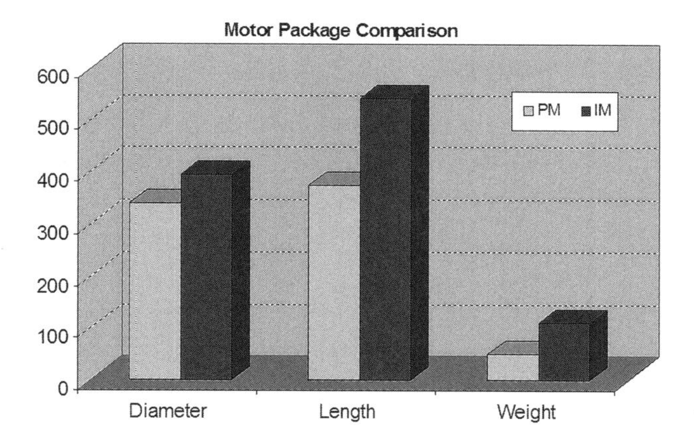

FIG. 5.11 Comparison of motor package.

storage system, and once through the motor when the stored energy is used to assist with vehicle acceleration. An induction motor is typically around 80% efficient, giving a "roundtrip" efficiency of 64%. A PM motor is typically 90% efficient, giving a roundtrip efficiency of 81%, a significant improvement. More recent advances in permanent magnet motor design have resulted in motor/generator efficiencies as high as 95%, for a roundtrip efficiency greater than 90%. Other important motor selection criteria are power density, reliability, fault tolerance, and cost.

Compared with an induction motor, the permanent magnet motor offers the following advantages:

- Higher power/torque density, which will result in a smaller package for the motor system and lower weight for a specific power/torque rating, as illustrated in Fig. 5.11.
- Higher round trip efficiency due to lower rotor loss and effective use of the reluctance torque, which leads to greater vehicle level fuel economy.
- The permanent magnet motor is disadvantaged by its higher cost, but this disadvantage will greatly decline under full production volumes, as shown in Table 5.1.

TABLE 5.1 Relative Comparison of Motor Cost (50 KW)

	IM	PM Motor
Prototype Cost	1.5 X	4.5 X
Production Cost	X	X
Cycle Life Cost	2 X	2 X

Compared with the permanent magnet surface mounted motor, an interior permanent magnet motor offers the following advantages:

- Use of flux weakening control based on salient pole behavior supports a wider range of speed at any given output level, with maximum rotational speed approximately ten times the motor base speed. For a surface mounted permanent magnet motor, maximum speed is closer to three times base speed.
- Mechanically robust construction suitable to high-speed applications.
- Use of salient pole behavior permits sensorless control with benefits of reliability and simple maintenance.

5.3.1.1 *Motor/Generator Sizing*

Motor/generator sizing is a very important step in optimizing a hybrid system for a given vehicle application. According to a study by Eaton Corporation, [5.2] there exists an optimum motor/generator size for a given application and duty cycle. In general, if the motor/generator power rating selected is too small for the application, only a small portion of the vehicle kinetic energy can be captured during deceleration events. As motor/generator power rating is increased, more vehicle kinetic energy can be recovered. At some point, most of the braking energy can be recovered, and any increase in motor/generator power rating beyond this point will not help improve efficiency. However, as motor/generator power rating is chosen to be higher, parasitic losses also increase. Parasitic losses include the thermal management system for the motor/generator as well as any motor/generator internal losses. The combined effect of improving braking energy recovery and increasing parasitic losses is shown in Fig. 5.12.

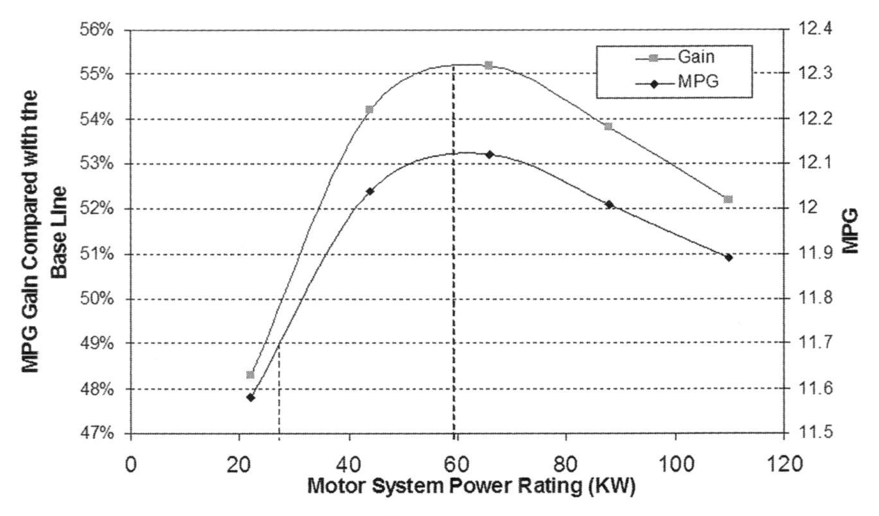

FIG. 5.12 Fuel economy gains with different motor sizes.

The primary reasons for the increased parasitic losses with increasing motor/generator size are:

- The motor/generator power rating cannot be increased without also increasing the ratings for the motor controller/inverter as well as the energy storage system. The resulting disadvantages include:
  - A larger thermal management system is required to cool the motor, inverter, and energy storage system, leading to increased thermal management energy use.
  - The motor/generator may end up being operated at less-efficient torque/speed points. An electric motor/generator has an efficiency map with defined regions of best efficiency and areas of less-efficient operation, analogous to an internal combustion engine. Figure 5.13 shows torque speed maps for two selected motor/generator ratings, 22 kW and 110 kW, respectively. On both maps, the torque-speed points of the CILCC driving cycle (Section 5.1.2) are overlaid (red "x"). Clearly, a much greater percentage of the operating points falls in the more-efficient regions of the 22-kW motor/generator map (top). The opposite is true for the 110-kW rated motor/generator. Most of the CILCC operating points fall in a very inefficient region.
  - The hybrid system cost will be greater than strictly necessary. From a cost-benefit perspective, it may be preferable to have a slightly under-sized hybrid system. The hybrid system designer should be cognizant of the diminishing returns of increased power ratings.

    Aside from the motor/generator power rating, there are several key motor/generator parameters that must be considered carefully: peak torque, maximum speed, and base speed, as shown in Fig. 5.14. In the pre-transmission parallel hybrid architecture, the motor/generator shares a common shaft with the engine and transmission. Therefore, engine, motor/generator, and transmission input shaft speed are all the same. Motor/generator design parameters should therefore be selected carefully to best match operation of the other powertrain systems.
- Maximum speed. Most commercial vehicle diesel engines are rated at 2500 to 3000 RPM. Electric motor/generator maximum speed should therefore be selected to at least match this speed rating.
- Base Speed. Motor/generator peak efficiency typically occurs between one and two times base speed. Figure 5.15 shows an engine speed histogram for a typical medium-duty diesel engine over the CILCC drive cycle. Not including engine idle and vehicle deceleration events, the large majority of engine operation takes place between 1200 and 1800 RPM.

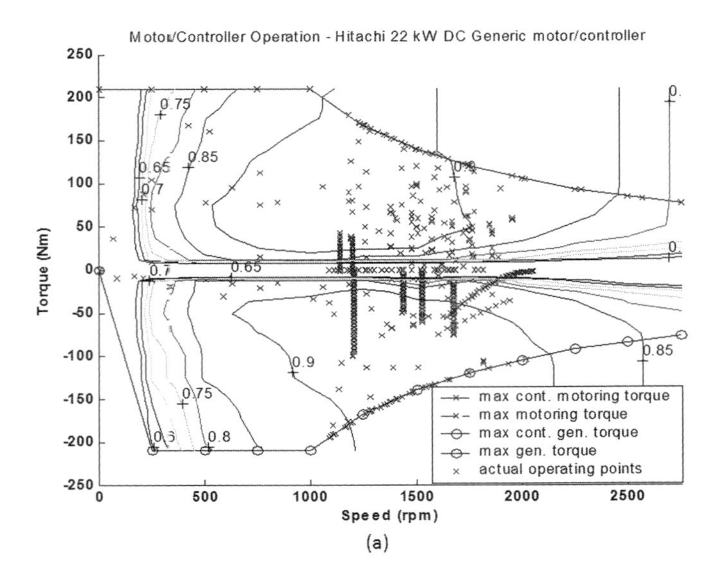

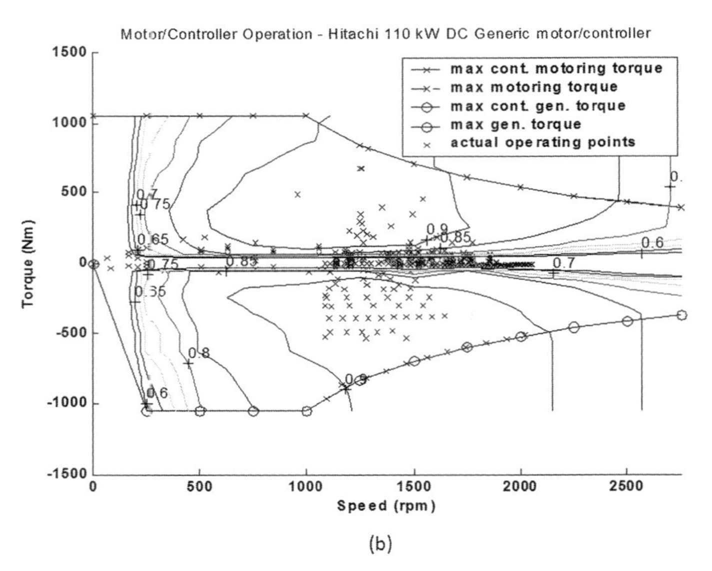

FIG. 5.13 Motor operation points comparison (22 kW vs. 110 kW).

Taking into account that for most permanent magnet motor/generators, maximum speed equals approximately three times base speed, a reasonable base speed selection would be in the 1000 to 1200 RPM range.

- Peak torque: Drivability of a hybrid-electric vehicle is determined by the combined torque from engine and motor. Given that the electric motor can provide torque at speeds well below where an internal combustion

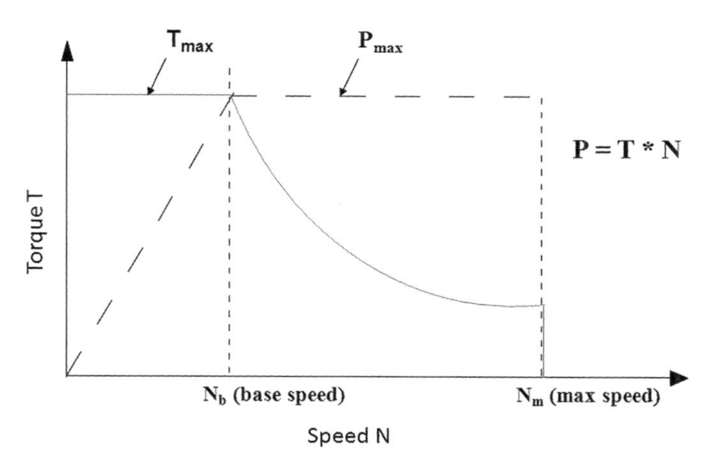

FIG. 5.14 Motor torque T and power P vs. speed N.

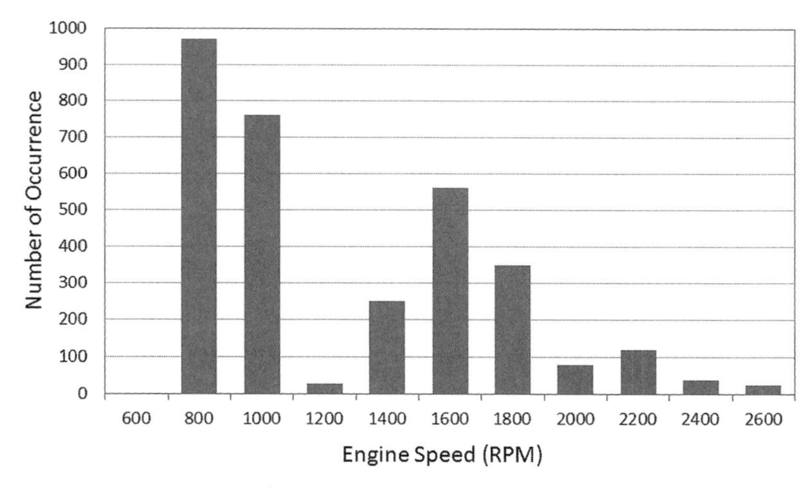

FIG. 5.15 Engine speed distribution over the CILCC cycle.

engine operated, ideally motor peak torque should be close to the diesel engine low-speed torque. This will ensure a smooth transition from low-speed, primarily electric motor-derived torque and higher-speed engine torque.

5.3.2 Energy Storage System Selection

It is well known that energy storage technology has been a significant hurdle to the adoption of electric-drive technology in automotive and commercial vehicle markets. Added to that is the challenge that different types of electrified vehicles have different requirements for the energy storage system. For example, a high-energy-density energy storage system is more suitable for pure electric or series hybrid-electric vehicles, while a high-power-energy storage system is more suitable for parallel hybrid vehicles. Energy storage systems for electrified vehicles can be divided in the electrochemical storage systems, i.e., batteries; or electrostatic storage systems, i.e., capacitor based. Capacitor-based energy storage systems are characterized by a high power density and are capable of delivering a high amount of energy in a short period of time. However, the low energy density of capacitor-based systems makes them less attractive for most hybrid-electric applications. Battery-based systems have been widely deployed in electric and hybrid-electric vehicles, both automotive and commercial vehicles, for more than ten years. The most successful hybrid-electric vehicle in the world, the Toyota Prius, to this day uses a Nickel Metal Hydride (NiMH) technology. Most new hybrid-electric vehicles, including commercial vehicle hybrids, have started using Lithium-ion (Li-ion) technology. Lead-acid batteries were used in some early systems and continue to be used today in some low-cost niche applications. Figure 5.16 shows weight and volume comparisons for Li-ion, NiMH, and advanced lead-acid batteries for a given energy-capacity rating.

From Fig. 5.16, it should be clear why Li-ion technology has received so much attention over the last few years as one of the most promising technologies to solve the energy-capacity hurdle that has so far impeded universal adoption of electrified vehicle technology. Governments around the world have invested in excess of $20 billion to spur research and development as well as manufacturing of advanced Li-ion battery technology.

With early leadership and guidance provided by the U.S. Freedom Car program and U.S. Advanced Battery Consortium (USABC), a range of evaluation criteria and test methods have been established for Li-ion batteries:

- Pulse discharge power (10 s)
- Peak regenerative pulse power (10 s)

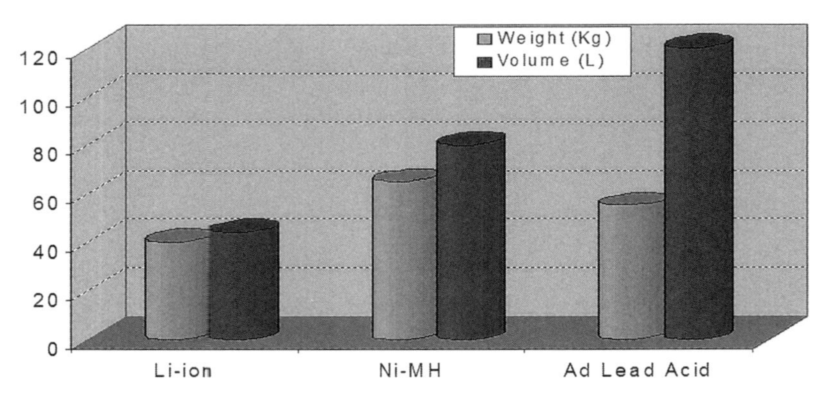

FIG. 5.16 Comparison of battery size.

- Weight and volume
- Total available energy @ C/1 rate
- Cycle life and Calendar life
- Minimum round trip efficiency
- Production price
- Cycle life cost
- Operating voltage limits
- Maximum allowable self-discharge rate

The above is by no means an exhaustive list of evaluation criteria. With thousands of candidate materials for the electrode, cathode, and electrolyte of Li-ion batteries, the research opportunities are almost infinite and thus require a well-defined set of evaluation criteria.

5.3.2.1 *Battery Sizing*

Similarly to the sizing of the electric motor, battery sizing should be done carefully, and with the full vehicle powertrain system in mind. Electrical system power rating and energy storage system capacity are ideally derived from a careful evaluation of the intended application and duty cycle. A study by Eaton Corporation evaluated available brake energy for a 16,000 lb GVW package and delivery vehicle over a range of duty cycles [5.1]. The results are shown in Table 5.2.

The study evaluated four typical drive cycles including the CILCC, West Virginia University (WVU) city cycle, WVU Suburban cycle, WVU Interstate cycle, and the City Suburban Heavy-duty Vehicle Route (CSHVR). For every

TABLE 5. 2 Energy Consumption over Various Drive Cycles

Cycles	Section	Duration(s)	Veh. Decel. Engy. (kJ)	Rolling (kJ)	Aero (kJ)	Aux (kJ)	Friction Brake (kJ)	Full Regen (kJ)	Required Full Regen Pwr (kW)
CILCC	A	22	2850	152	247	137	2143	2648	121
	B	27	1113	121	80	243	380	903	31
WVUCITY	A	15	1225	77	71	85	853	1007	67
	B	18	980	82	56	131	569	916	51
WVUSUB	A	17	1943	90	111	91	1547	1823	107
	B	15	1348	89	89	98	999	1220	81
WVUINTER	A	34	1621	271	399	222	779	1298	38
	B	60	1106	332	424	264	325	736	12
CSHVR	A	21	1645	122	129	124	984	1466	70
	B	28	1066	142	125	176	497	834	30

duty cycle evaluated, two events were selected for evaluation purposes: Section A, representing a maximum severity braking event in terms of total energy availability (power and duration) and Section B, representing an average braking event. Power required for a full regeneration (i.e., braking achieved though full conversion of vehicle kinetic energy to battery stored electrochemical energy) can range from 38 to 121 kW, with available energy ranging from 0.2 kW·h to 0.8 kW·h.

From a rated power perspective, the above data suggest that a battery power absorption capability of between 35 and 75 kW is desired to get significant brake energy recovery for the evaluated example. Similarly, a desired available battery energy capacity should be in the 0.2 to 0.8 kW·h range. For Li-ion batteries, available energy is generally around 30% of maximum capacity for a power cell and up to 90% for an energy cell. These constraints are driven by the effect of energy capacity use on battery life. As a Li-ion battery is cycled deeper, battery life tends to reduce drastically.

Battery sizing for commercial vehicles is faced with one additional challenge borne out of practical manufacturing-related conditions. Li-ion batteries are made up of a series of low-voltage battery cells that in aggregate deliver the system-level voltage. For commercial vehicle and bus applications, nominal system voltages are either in the 350 V dc range or 600–700 V dc range. Li-ion battery cells typically range from 1.6 V dc to 3.7 V dc per cell, depending on selected cell chemistry. Cell energy storage capacity, measured in amp hours (Ah), is generally defined by applications that demand the highest manufacturing volumes. For Li-ion cells currently that means cells for automotive hybrid-electric vehicle applications in the 3–4 Ah range. For automotive electric vehicle applications, cell capacity is typically in the 20–25 Ah range, such as the AMP20 prismatic pouch cell made by A123 Systems [5.9]. Since commercial vehicle hybrid sales to date are very low in volume, cells are not designed or manufactured for their specific needs. This leaves battery system designers for commercial vehicles to chose from available, relatively low cost but suboptimal, cell sizes, or select a custom-designed cell at a much higher cost. As an example, consider a commercial vehicle system operating at a nominal voltage of 360 V dc. Using a 3.7-V dc Li-ion cell technology, 96 cells in series will be required to reach the nominal system voltage. If these cells have a 4 Ah capacity, then overall system capacity would be 1.4 kW·h. Available energy would be around one-third of this total or less than 0.5 kW·h. This is likely not sufficient for a typical commercial vehicle application. The next level up would be to have two parallel strings of 96 cells in series, resulting in a total capacity of 2.8 kW·h and available energy of approximately 1 kW·h. This example illustrates that the battery system designed does not have the luxury of selecting battery capacity from a continuum of available sizes, but rather is limited to what is available.

In summary, battery capacity and charge/discharge capacity should be considered carefully in context of the intended application. In addition to this, practical considerations of available cell size manufactured in high volume, and thus lower cost, must be taken into account.

5.3.3 Electrical System Voltage

Electrical system voltage is the nominal voltage the entire electrical system operates at. Industry standards for commercial vehicle hybrid systems have emerged around 350 V dc and 600–700 V dc. This is a decision that should not be taken lightly. From an efficiency standpoint, it may seem desirable to select the higher voltage. At the higher voltage, less electrical current would be required to reach the same power level. However, greater cost may be incurred for several reasons. First of all, more battery cells would be required to get to the higher system voltage. This could lead to a higher energy capacity than strictly needed, depending on available cell sizes. In addition, greater care must be taken around electrical insulation of the system to avoid leakage or electrical shock hazards. The lower system voltage, however, could lead to exceedingly high currents flowing through the system, resulting in high resistive energy losses. Again, a careful consideration of the tradeoffs is warranted.

5.3.4 Thermal Management

Thermal management can be a significant challenge for hybrid-electric vehicles that include a pure electric vehicle driving mode. Obviously, the internal combustion engine requires cooling when it is running. However, when the vehicle is operating in EV mode, cabin cooling and heating can no longer depend on an engine-driven A/C compressor, or engine exhaust heat-driven heating system. To provide cabin heating and cooling, the hybrid-electric vehicle system incorporates, for example, an electrically driven A/C compressor as well as a Positive Temperature Coefficient (PTC) heater. The electric motor also requires cooling. While the diesel engine coolant temperature normally operates around 90°C, the electric motor and other power electronics devices require a coolant temperature closer to 60°C. To complicate matters further, the energy storage system requires a cooling temperature below 30°C to ensure sufficient battery life. The hybrid system includes separate high-temperature (60°C) and low-temperature (30°C) cooling loops to maintain all subsystems at their required temperature. Figure 5.17 shows a schematic overview of the thermal management system as designed for a 19,000 lb GVW hybrid-electric vehicle [5.8].

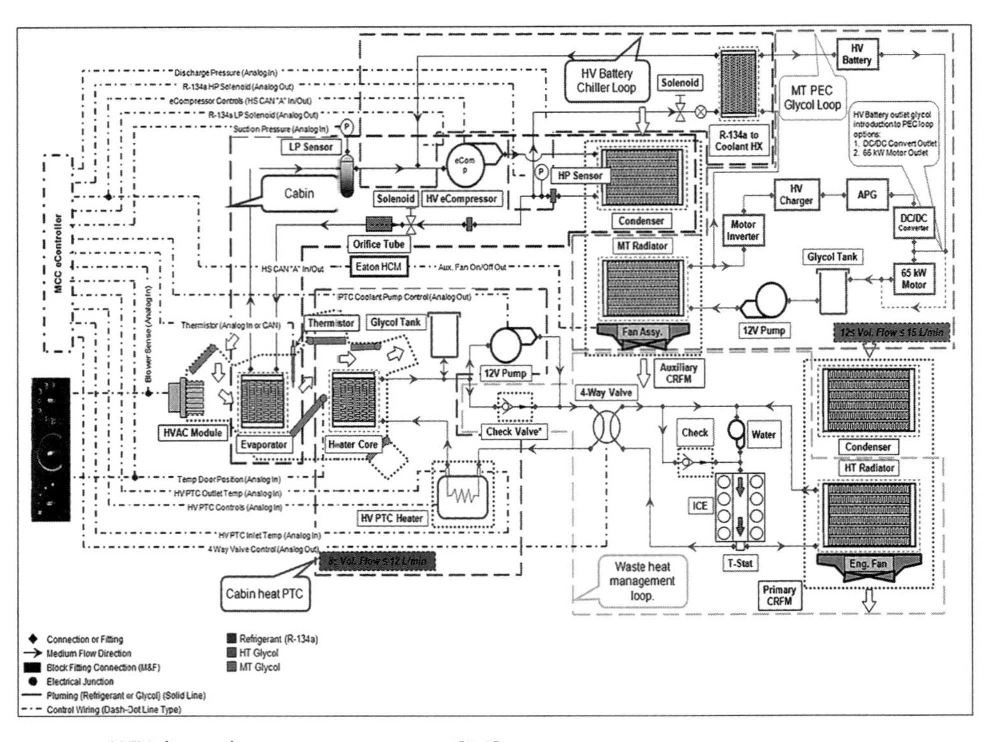

FIG. 5.17 HEV thermal management system [5.8].

5.4 References

5.1 Zou, Z. et al., "A New Composite Drive Cycle for Heavy-Duty Hybrid-ElectricClass 4-6 Vehicles," SAE Paper No. 2004-01-1052, SAE International, Warrendale, PA, 2004.

5.2 Final Report to the National Renewable Energy Laboratory by Eaton Corporation, "Advanced Parallel Hybrid Featuring Fully Integrated Powertrain Systems" Subcontract Number ZCL-2-32060-01.

5.3 Chiang, P. K., "Two-Mode Urban Transit Hybrid Bus In-Use Fuel Economy Results from 20 Million Fleet Miles," SAE Paper No. 2007-01-0272, SAE International, Warrendale, PA, 2007.

5.4 Grewe, T. M. et al., "Defining the General Motors 2-Mode Hybrid Transmission," SAE Paper No. 2007-01-0273, SAE International, Warrendale, PA, 2007.

5.5 Miller, J. M., "Hybrid Electric Vehicle Propulsion System Architectures of the e-CVT Type," IEEE Transaction on Power Electronics, Vol. 21, No. 3, May 2006.

5.6 Muta K. et al., "Development of New-Generation Hybrid System THS II—Drastic Improvement of Power Performance and Fuel Economy," SAE Paper No. 2004-01-0064, SAE International, Warrendale, PA, 2004.

5.7 Burke, A. et al., "Simulated Performance of Alternative Hybrid-Electric Powertrains in Vehicles on Various Driving Cycles," EVS24, Stavager, Norway, May 13-16, 2009.

5.8 Smaling, R. and Cornils, H., "A Plug-in Hybrid Electric Powertrain for Commercial Vehicles," The 25th World Battery, Hybrid and Fuel Cell Electric Vehicle Symposium & Exhibition, EVS-25 Shenzhen, China, Nov. 5-9, 2010.

5.9 AMP20 Prismatic Pouch Cell, http://www.a123systems.com/products-cells-prismatic-pouch-cell.htm, assessed March 22, 2012.

Chapter 6

Hybrid-electric Power Conversion Systems

6.1 Basic Three-phase Motor Theory

Electric motor operation is based on the force exerted on a current-carrying conductor placed in a magnetic field. Lorentz' law of electromagnetic force and Faraday's laws of induction describe the underlying phenomena. The reverse process, producing electrical current from mechanical force, is done by generators such as an alternator or a dynamo; some electric motors can also be used as generators. A key example is the traction motor used on hybrid and electric vehicles, which may perform both tasks. Electric motors and generators are commonly referred to as electric machines. The premise for motor operation is that if you can create a rotating magnetic field in the stator of the motor, it will induce a voltage in the armature that will have magnetic properties causing it to "chase" the field in the stator. This premise applies to AC motors that employ a squirrel cage rotor, and it is probably the most simple and basic of all motor designs. The three-phase motor is widely used in industry because of its low maintenance characteristics. Due to the nature of three-phase power, creating a rotating magnetic field in the stator of this motor is simple and straightforward.

Most traction motors rely on a three-phase system to operate. This section will give a basic overview as to how this three-phase system is produced and how it relates to motor and generator operation for permanent magnet and induction motors. In a three-phase system, three circuit conductors carry three alternating currents (of the same frequency), which reach their instantaneous peak values at different times. Taking one conductor as the reference, the other two currents are delayed in time by one-third and two-thirds of one cycle of the electric current. This delay between phases has the effect of giving constant power transfer over each cycle of the current and also makes it possible to produce a rotating magnetic field in an electric motor. It is not essential that the motor/generator operate on a three-phase system; in fact, any number of phases can be used. However, it is generally agreed that a three-phase system

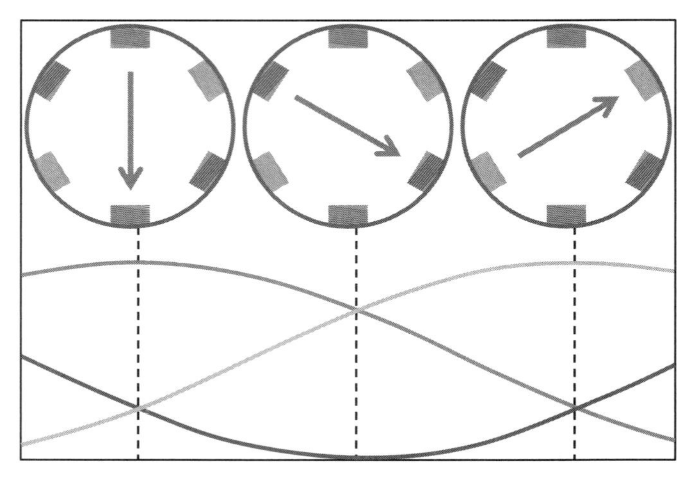

FIG. 6.1 Rotating field in a three-phase motor.

is most cost effective and most efficient [6.1]. Reluctance motors are somewhat different, but even then three phases are generally preferred.

The stator windings are arranged on the stator poles in a way that results in magnetic flux lines that seem to rotate. In this example, the direction of rotation would be counterclockwise.

The strength of the magnetic field changes, as the current flow in the coils of wire around the stator poles changes. Just as the current in the field windings rises and falls 120 electrical degrees apart, so does the resulting magnetic field on the pole face. In other words, when the magnetic polarity on pole "red" reaches its peak, 120 electrical degrees later pole "blue" will reach its peak, and 120 electrical degrees after that, the "yellow" pole will reach its peak. Then the cycle repeats itself, and a rotating magnetic field is developed in the stator of the motor. Swap any two of the three pole face polarities by changing any two leads of the supply voltage, and you will see that the "RBY" order reverses.

The frequency of the applied voltage, and the number of poles, determine the speed of the rotating magnetic field in the stator of a motor. In the simplified illustration in Fig. 6.1 of a six-pole motor, the stator windings on the "red" poles are experiencing maximum polarity in the left-most diagram; 120° later the "blue" poles are at maximum, and in another 120°, the "yellow" poles are the strongest.

Lines of magnetic flux protrude into the area occupied by the armature. As the flux lines move around the inside of the stator, they cut the conductive bars of the rotor, inducing a voltage in the armature. The conductive bars run the length of the armature, and are connected at each end. This provides numerous paths for current to flow, due to the induced voltage potential. The armature is constructed in such a manner that the current flow in the armature produces magnetic fields that follow the rotating magnetic field in the stator, resulting in torque. Induction of voltage in the armature only takes place when there is relative motion between the flux and the conductive bars of the rotor. Consequently, the rotor speed can never reach the synchronous speed of the rotating magnetic field in the stator. If it did, there would be no relative motion, no induced voltage, no magnetic field on the armature, and no torque.

When voltage is applied to the motor leads, current flows in the stator windings. As the rotor starts to turn, the magnetic fields of the armature induce a counter or back Electro-Motive Force (EMF) into the stator windings. The back EMF magnitude grows with rotor speed. Because this back EMF opposes the applied EMF, it reduces the current flow in the stator. This is why the starting current of an induction motor is typically around six times greater than the running current (starting current is sometimes referred to as locked rotor current).

The peak output voltage (back EMF) for a single coil in a three-phase system is given by

$$E = \frac{1}{2}\varphi \cdot \omega \cdot P \tag{6.1}$$

The instantaneous voltage for each of the three coils is given by

$$\begin{aligned} ea &= E \cdot \sin \omega t \\ ea &= E \cdot \sin(\omega t + 120°) \\ ea &= E \cdot \sin(\omega t + 240°) \end{aligned} \tag{6.2}$$

Normally the coils are joined in a "Y" configuration, as shown in Fig. 6.2 (also called a "star" configuration). Another connection, called "delta," is widely used in industrial motors but typically is not used in traction motors. The net result is effectively the same, although the detailed design considerations are more complicated and beyond the scope of this book.

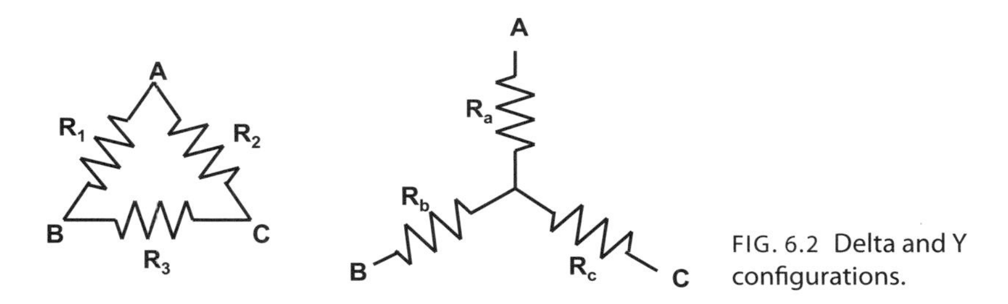

FIG. 6.2 Delta and Y configurations.

The output from each individual coil is called the phase voltage. When connected in a Y configuration, the voltages between the three remaining ends are called the "line" voltages. The line voltage is given by

$$V_1 = V_{ph} \cdot \frac{\sqrt{3}}{2} = 1.732 \cdot V_{ph} \tag{6.3}$$

Assuming a resistive load, the power output from the machine is given by

$$P = 3 \cdot V_{ph} \cdot I = 1.732 \cdot V_1 \cdot I \tag{6.4}$$

Given that the battery providing power to the motor is a DC source or sink, the three-phase AC input or output must be converted to DC. This can be done with a simple six-diode bridge rectifier, as shown in Fig. 6.3.

The average value of the rectified voltage is

$$V_{dc} = 1.654 \cdot V_{phase\ peak} \tag{6.5}$$

This is not a satisfactory method for the following reasons:

- The phase voltage varies, hence the output voltage could be anywhere between zero and the design maximum.
- The battery voltage varies with load, state of charge, and a number of other factors.
- Control of current is required based on state of charge and the level of power the generator is required to deliver.

For a permanent magnet motor, it is impossible to control the magnetic field excitation, so it will be necessary to use active rectification. The method of controlling both voltage and current is by using active devices instead of the six passive diodes. The active devices are typically Insulated Gate Bi-polar Transistor (IGBT) components for high-voltage systems and Field Effect Transistor (FET) devices for lower-voltage systems. Six diodes are still required

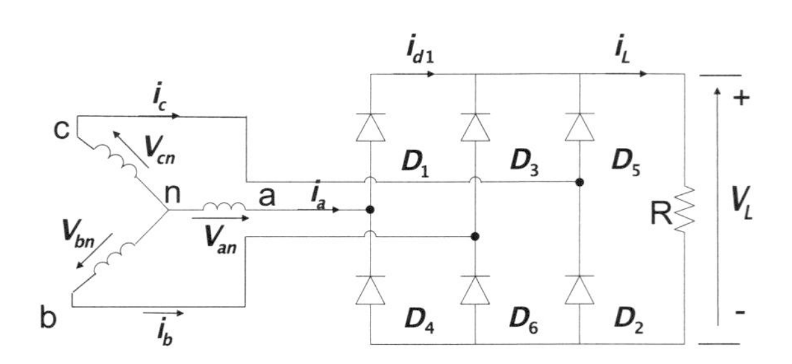

FIG. 6.3 Six-diode bridge voltage rectifier.

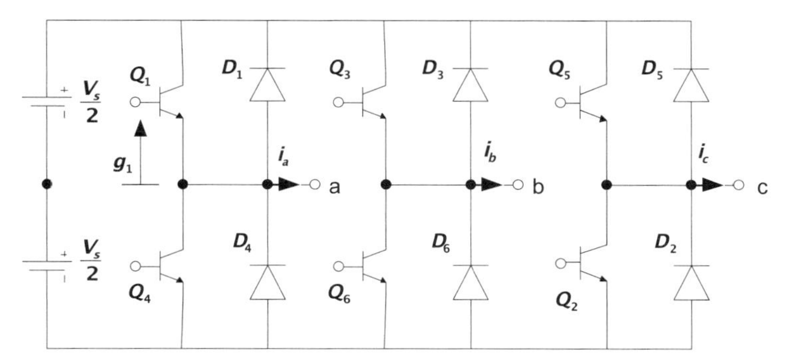

FIG. 6.4 Active rectifier basic circuit.

to take care of inductive and other effects. The basic circuit now becomes as shown in Fig. 6.4.

6.2 Basic Power Inverter Design and Operation

Power inverters for hybrid-electric cars and trucks combine both DC to AC inverters and AC to DC converters (rectifier) to manage the power and recharging. An inverter is an electrical device that converts electricity derived from a DC source (e.g. battery) to AC that can be used to drive an AC appliance (e.g., electric motor/generator). The theory of operation is relatively simple. DC power from the hybrid battery is fed to the primary winding in a transformer within the inverter housing. Through an electronic switch (generally a set of semiconductor transistors), the direction of the flow of current is continuously and regularly broken (the electrical charge travels into the primary winding, then abruptly reverses and flows back out). The in/out flow of electricity produces AC current in the transformer's secondary winding circuit. Ultimately, this induced alternating current electricity flows into, and produces power in, an AC load (for example an electric vehicle's traction motor). A rectifier is a similar device to an inverter except that it does the opposite, converting AC power to DC power to recharge the hybrid battery while the traction motor operates in generator mode [6.2, 6.3].

There are two basic types of inverters, i.e., voltage source inverter and current source inverter. As shown in Fig. 6.5(a), the voltage source inverter (VSI) requires a very-high-performance dc bus capacitor to maintain a near ideal voltage source and absorb the ripple current generated by the switching actions of the inverter. The root mean square (rms) value of the ripple current can range from 50 to 80% of the motor current. Concerns about the reliability of electrolytic capacitors have forced hybrid-electric vehicle (HEV) makers to use film capacitors, and currently available film capacitors that can meet the demanding requirements of this environment are costly and bulky, taking up one-third of the inverter volume and making up one-fifth of the cost. The reliability of the inverter is also limited by the capacitors and further hampered by the possible shoot-throughs of the phase legs in the VSI. In addition, steep rising and falling edges of the pulse width modulated (PWM) output voltage generate high *dv/dt* related electromagnetic-interference noises, cause motor insulation degradation due to the voltage surges resulting from these rapid voltage transitions, and produce high frequency losses in the windings and cores of the motor as well as generate bearing-leakage currents that erode the bearings over time. Furthermore, for the VSI to operate from a low-voltage battery, a bidirectional boost converter is needed.

Another type of inverter circuit is the current source input (CSI) inverter, shown in Fig. 6.5 (b). With a current source inverter, the DC power supply is configured as a current source rather than a voltage source. The CSI does not require any dc bus capacitors but instead uses only three AC filter capacitors of a much smaller capacitance. The switches of the current source inverter have to be reverse blocking. If IGBTs are used for the current source inverter,

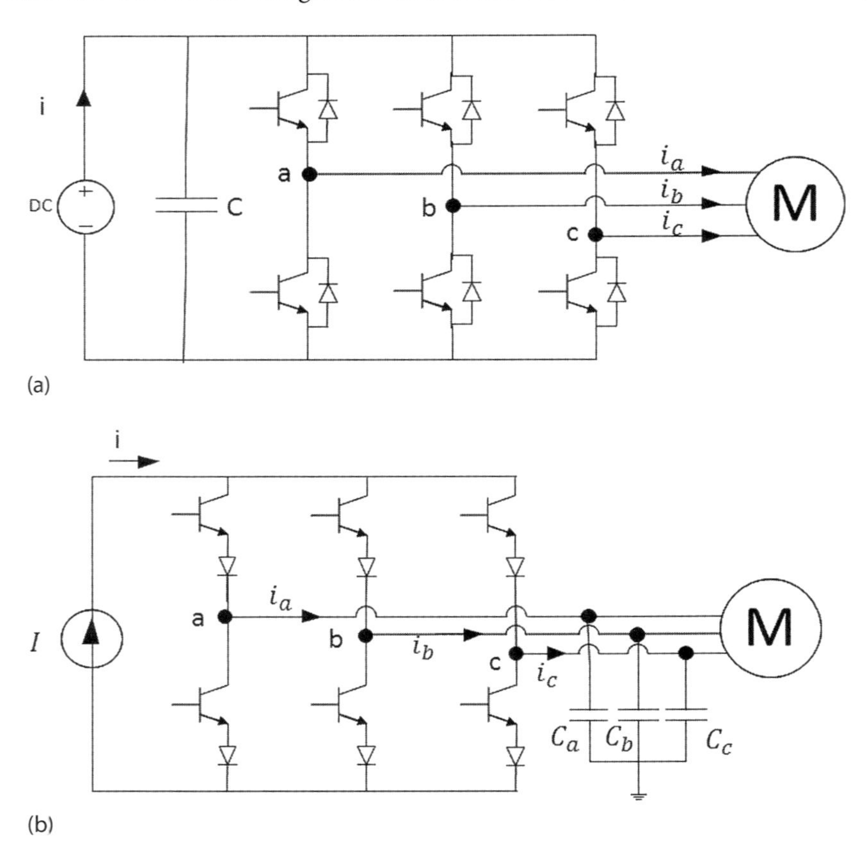

FIG. 6.5 (a) Voltage source inverter, (b) Current source inverter.

the blocking capability can be achieved with diodes connected in series to the IGBTs. This yields to relatively high semiconductor conduction losses.

The CSI offers many advantages important for electric vehicle (EV)/ HEV applications, including that (1) it does not need antiparallel diodes in the switches, (2) it provides natural shoot-through short-circuit protection capability, (3) it produces sinusoid-shaped voltages to the motor due to the effect of the output AC filter capacitors, and (4) it can boost the output voltage to a higher level than the source voltage to enable the motor to operate at higher speeds. These advantages could translate into a substantial reduction in inverter cost and volume, increased reliability, a much higher constant-power speed range, and improved motor efficiency and lifetime. All these features make the CSI-based motor drive attractive for EV/HEV applications [6.2].

Virtually all hybrid-electric vehicles presently use silicon-based insulated gate bipolar transistors (IGBTs), combined with suitable diodes, as the main driving switch [6.4]. Unfortunately, silicon-based switches have a thermal limit below 85°C. Along with typical hybrid battery thermal constraints, this provides a significant and costly thermal management challenge for hybrid vehicles with three different thermal limits; one for the hybrid battery, one for the inverter, and yet another for the internal combustion engine. In addition to higher temperature capability, materials with higher breakdown voltages are desired to enable the use of higher-power motors and to protect against surge voltages. Silicon carbide (SiC) semiconductor material is a promising candidate for a next generation of IGBTs. The material has the potential to operate at temperatures up to 600°C. This means that an inverter based on this material can be cooled by the standard IC engine cooling circuit, allowing one of the cooling circuits on current hybrid vehicles to be eliminated. Also, the breakdown voltage for silicon carbide is ten times higher than that of silicon. Since SiC devices can be made smaller, the capacitance will be lower and switching can be faster. Losses can be lower during both turn-on and turn-off. SiC also has much better thermal conductivity, allowing fast heat transfer out of the device. However, SiC is a difficult material to grow and work. Developments are still needed in terms of wafer quality and size, gate isolation, and packaging to handle the higher-temperature environment. Once mature, this technology will enable significant reductions in weight and size, and increases in efficiency.

6.3 Basic Motor/Generator Designs

Electric motors can be divided into alternating current (AC) and direct current (DC) types. Direct current motors use direct-unidirectional current. DC motors are used in special applications in which high torque starting or smooth acceleration over a broad speed range is required. Speed control of DC motors does not affect the quality of the power supply and is relatively easy to implement in two ways:

- Armature voltage: Increasing the armature voltage will increase speed.
- Field current: Reducing field current will increase speed.

DC motors are available in a wide range of sizes, but their use is generally limited to a few low-speed and low- to medium-power applications like machine tools and rolling mills because of problems with mechanical commutation at large sizes. Also, the risk of arcing at the brushes restricts their use to clean, non-hazardous locations. DC motors are also more expensive relative to AC motors.

AC motors use an electrical current which reverses its direction at regular intervals. An AC motor has two basic electrical parts: a stator and rotor. The primary advantage DC motors have over AC motors is the ease of speed control. To compensate for this, AC motors can be equipped with variable-frequency drives. However, the improved speed control comes at the cost of reduced power quality. Induction motors are the most popular in industry due to their ruggedness and low maintenance requirements. AC induction motors are inexpensive at less than half the cost of DC motors and also provide almost double the power density, or power-to-weight ratio, compared to a DC motor.

A number of classifications are in use to distinguish different AC motor types, but in the context of this book, three types of motor/generator designs will be discussed:

- Induction motors
- Permanent magnet motors
- Switched reluctance motors

6.3.1 Induction Motor/Generator

The three-phase induction motor has a three-phase stator winding assembly. When it is supplied with a three-phase sinusoidal voltage and frequency, a rotating magnetic flux is created in the air gap of the machine. The rotor is an iron structure generally with a die-cast aluminum cage, as shown in Fig. 6.6.

When the rotor is stationary, the rotating magnetic flux created by the stator will interact with the rotor, and currents will flow in the conductors. Assuming the rotor is purely resistive, these currents will set up another magnetic flux that is 90° displaced from the stator flux. The two magnetic fluxes will interact, and torque will be created on the rotor, causing it to rotate in the direction of the stator flux. The rotor does, however, have inductance, and the currents that flow will lag the voltage. We can write

$$e = k \cdot \omega s \tag{6.6}$$

FIG. 6.6 Three-phase single-pole induction motor sketch.

and

$$I = \frac{e}{\sqrt{R^2 + (\omega s \cdot L)^2}} \tag{6.7}$$

where

ωs is the stator flux speed

k is a constant for a given machine design

In actual operation, the rotor speed always lags the magnetic field's speed, allowing the rotor to produce useful torque. This speed difference is called slip. Slip increases with load and is necessary for torque production. Slip speed is equal to the difference between rotor speed and synchronous speed. When the rotor is not turning, the slip is *100%*:

$$\omega s = 2 \cdot \pi \cdot \frac{f}{(P/2)} \tag{6.8}$$

Slip is defined as

$$s = \frac{\omega s - \omega r}{\omega s} \tag{6.9}$$

where

f is the applied stator frequency

P is the number of magnetic poles for the given machine

ωs is the stator flux speed

ωr is the rotor speed

With the rotor stationary

$$s = 1$$

If the rotor is rotating at synchronous speed

$$s = 0$$

In normal operation, "s" lies between 0.005 and 0.01 (depending on details of the motor design).

The rotor current can now be written as

$$I = s \cdot \frac{e}{\sqrt{R^2 + (s \cdot \omega s \cdot L)^2}} \tag{6.10}$$

If the rotor speed is zero, motor stalled, $s = 1$. Then

$$I = \frac{e}{\sqrt{R^2 + (\omega s \cdot L)^2}} \tag{6.11}$$

The calculation of torque for any given speed is quite involved and beyond the scope of this book. It is sufficient to show that the net torque speed curve can be as shown in Fig. 6.7.

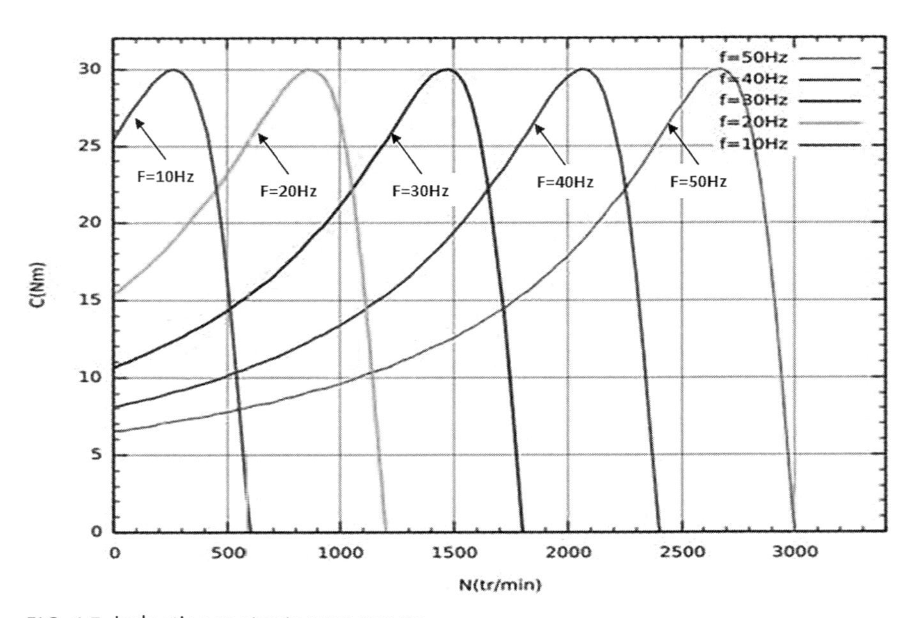

FIG. 6.7 Induction motor torque curves.

The induction motor can be made to generate if it is driven by an external source such as the vehicle inertia or a downgrade by selecting an operating frequency that is lower than the speed at which the rotor is rotating. This can only occur when the motor stator windings are connected to a suitable three-phase supply that can back feed the energy to a suitable sink such as a battery. This is easily achieved in the same main power drive circuit that is required for motor operation from a DC supply.

Assuming that regeneration is required for braking or energy recovery on a downhill slope, the frequency (and the voltage to keep V/f constant) is continuously adjusted downward with speed. Maximum regeneration and high efficiency occurs when the "slip" is between 0.005 and 0.01. It follows that a control system must continuously monitor actual speed to adjust frequency and voltage to appropriate levels.

It is important in the induction motor that the stator flux be sinusoidal in distribution. Consider an extreme condition with the flux vector as a square wave (not actually possible). Using Fourier, we can describe the square wave using normal mathematics. The key issue lies with the harmonic frequencies that accompany the primary frequency. These are traveling at different speeds and directions from the main flux. These harmonics can create torque that may counteract the intended torque delivered by the primary magnetic flux. The net result is a large heating loss in the rotor and reduction in operating efficiency. Even minor divergence from a sinusoidal flux can have a significant and negative effect.

When a distributed winding is used, the motor designer has a number of tools available to minimize these losses, and it is also important to obtain a supply voltage that is close to, but not an absolute, sine wave. An important part of this is to have at least three slots per pole, and preferably more. High-power-density traction motors usually run with a pole number greater than 12, and often more. The high number of slots will significantly increase stator winding complexity. There are other types of winding that can achieve the same effect, but currently distributed winding is the most widely used for all induction motors.

The ratio of slots in the stator and rotor is an important number for the reduction of "magnetic noise." This is a sound created by the action of the slot edges passing each other on the rotor and stator. Skewing of either the stator or rotor slots is widely used. It is much easier to do this on the rotor, as the "winding" is in reality a die-cast aluminum cage.

The stator and rotor laminations are generally punched from the same sheet of material in a progression die. On very large motors in low volume, this can sometimes be impractical. In these cases two simple round blanks are produced. Each blank is then placed in a machine called a "notching" press.

Each slot is then individually punched and the blank indexed for the next slot. This is slower but has a much lower tool cost and in lower volume is a practical method of production. The rotor winding is almost always a die-cast aluminum cage, which includes all of the rotor conductors and the end ring. This is a very cost-effective technique. It should be noted, however, that the use of copper, which can also be die-cast for this application, will give better efficiency. Many large industrial motors use copper bars that are brazed to the end rings. Both of these techniques are applicable to automotive and truck applications but are used only in low-volume production at this time.

6.3.2 Permanent Magnet Motor/Generator

A permanent-magnet motor does not have a field winding on the rotor frame, instead relying on permanent magnets to provide the magnetic field against which the stator field interacts to produce torque [6.5]. The Permanent Magnet (PM) motor stator design is similar to that of the stator of an induction motor. There are typically two types of stator design. First is the distributed winding type, which is effectively the same as the type used in an induction motor. Some details are different, but the basic construction and method of obtaining a rotating flux pattern is exactly the same. The second type is the concentrated pole design.

Figure 6.8 shows a typical construction for a brushless dc motor with six (magnetic) poles. These projecting poles are also called salient poles. It is a three-phase machine; hence, three pole pieces for a pair of North South magnets results in six pole pieces that are wound with individual coils, as shown. The coils for each phase maybe connected in series or parallel at the designer's choice. The resultant three coils are typically connected in a Y configuration.

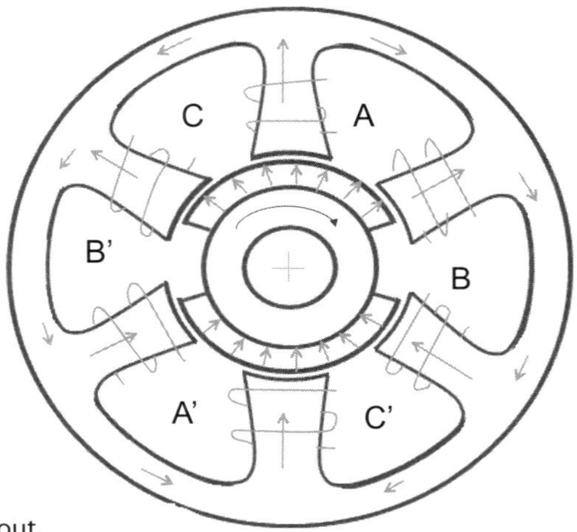

FIG. 6.8 Two-pole magnetic layout.

To understand how a PM motor with a concentrated pole design operates, refer to Fig. 6.8. Power is applied to winding on A and A', and current flow sets up a "north" and "south" pole, respectively, which the permanent magnets will react to, and begin movement. This movement will cease when the appropriate poles are in alignment. However, if, at the appropriate time, current is shut off in windings A and A', and turned on in windings B and B', then the rotor continues to move. Again at the appropriate time, B and B' are shut off, and C and C' are turned on. By continuation of this timing sequence, complete rotation occurs. In effect, rotation of the rotor is created by continuously switching the magnetic field set up in the stator, forcing the rotor to try and catch up. In this example, the explanation was simplified by exciting only one winding at a time. In reality, the stator consists of a three-phase Y-connected winding, and two or three windings are actually energized. This makes efficient use of windings and development of higher motor torques.

Current is being switched from winding to winding (which is identical in function to the mechanical commutator in a DC motor), and this action of switching current in the PM motor has been termed "electronic commutation." The next logical question then is, "How does the motor know when to switch current, or commutate?" Actually the motor itself does not perform the commutation function. Commutation is accomplished by the control system, which is running the motor. A rotor assembly includes a small "sensor" magnet. This sensor magnet will turn Hall Effect devices "on" and "off." The Hall Effect device feedback can then be used to determine shaft position and provide information about location of the rotor magnet.

The control system consists of logic circuitry and a power stage to drive the motor. The control's logic circuitry is designed to switch current at the optimum timing point. It receives information about the shaft/magnet location via signals from the Hall Effect sensors, and outputs a signal to turn on a specific power device, to apply power from the power supply (e.g., the vehicle high-voltage battery) to specific windings of the PM motor.

6.3.3 Switched Reluctance Motor/Generator

Switched reluctance motors are relatively new, having been developed in the 1980s. Switched reluctance motors differ from reluctance motors in that they have salient poles both on the rotor and stator. Reluctance motors are arguably some of the simplest motors around. The stator design is similar to that of an induction motor. The rotor is made of a stack of laminations shaped simply to automatically align itself with the magnetic field generated by the stator. A reluctance type rotor resembles the cage construction of an induction motor, but with parts of the periphery cut away to force the flux from the stator to enter the rotor in the remaining regions where the air gap between rotor and stator is small. The rotor will tend to align itself with the field and hence is able to remain synchronized with the traveling field generated by the three-phase windings on the stator in much the same fashion as a PM motor.

The switched reluctance motor differs from the reluctance motor described above in that it has salient poles on both the rotor and stator. This double salient pole arrangement has proven to be very effective as far as electromagnetic energy conversion is concerned.

6.4 Stator Design

Two stator designs are currently in widespread use: distributed winding and concentrated pole design. A more detailed description of the two is given in this section.

6.4.1 Distributed Winding

In distributed windings, all the winding turns are arranged in several full-pitch or fractional-pitch coils, as shown in Fig. 6.9. These coils are then housed in the slots spread around the air-gap periphery to form the phase windings. This stator configuration is used in most AC and DC motors. Well-established mass volume manufacturing equipment has been available for many years, as this design is used in most high-volume industrial motors. Assembly time is fast. In essence the coils are pre-wound on a custom-tooled former. The coils are then picked up and pushed into the stator slots with appropriate tooling. For smaller volumes, the coils are pre-wound and hand inserted.

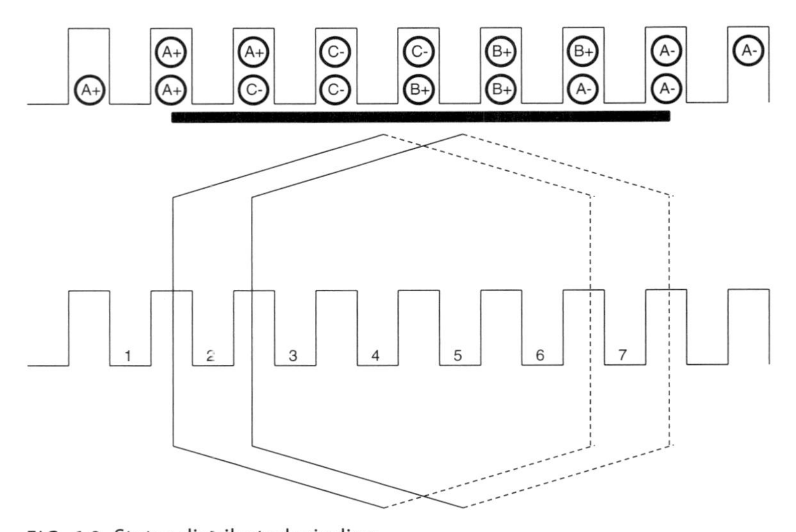

FIG. 6.9 Stator distributed winding.

There are some disadvantages to the distributed winding design. First, is the "fill factor." The actual copper cumulative cross section compared to available slot area is typically in the 40 to 50% range. Second, the length of the end winding can be quite long to facilitate easy insertion of the coils and a satisfactory mechanical arrangement at the end windings. Especially for electric motors subject to axial space constraints such as those used as traction motors in hybrid vehicles, these long end turns can lead to significant packaging constraints or affect the power and torque that a motor can provide given available space in the vehicle drivetrain. Both slot fill and long end turns add to cumulative winding resistance, which for a given size of machine will limit maximum continuous power output and efficiency. For industrial motors this is not a major issue. Mass is not critical and making the motor physically larger will provide the extra space to minimize winding resistance. For an automotive traction motor, the increased mass and size are significant issues. With new manufacturing and winding techniques, it is possible to achieve upwards of 80% slot fill and very short end windings. There are a number of variations, but generally precut rectangular cross-section material is used. Using rectangular cross-section materials poses significant manufacturing challenges that typically require a high level of automation to wind continuous "square" wires. Another method applied is the high-voltage hairpin design used by Remy Motors, leading to both a high slot fill ratio as well as very short end windings. The techniques are not new and date back to the 1920 era, but the attempts to use them in high-volume manufacturing are. The increase in power density is significant, and total material use can also be lower.

6.4.2 Concentrated Pole Construction

As opposed to distributed windings, concentrated windings use a one-slot coil pitch. Because of the very short end connections, the winding losses are reduced compared to distributed winding. The number of coils N_c and the number of poles $2\,p$ in a 3-phase electric machine must meet the following condition:

$$\frac{N_c}{GCD(N_c, 2p)} = 3K \tag{6.12}$$

where

GCD = Greatest Common Divider of N_c and $2\,p$

K = Integer value

For example, the following number of coils and poles can be designed: $N_c = 9$ and $2p = 6$, $N_c = 12$ and $2p = 8$, $N_c = 18$ and $2p = 12$, and so on. Generally a higher slot fill factor for the copper is achieved, 60 to 70% being typical, and the end windings are very short.

A further variation on the concentrated pole is to use a segmented construction. If the individual pole pieces are made as separate parts (to be laminated in a subsequent manufacturing step), a high-speed coil winder can be used to wind each individual pole piece. This cuts winding time significantly. More importantly, it allows a slot fill factor exceeding 75% with copper round wire. Even higher slot fill can be achieved using rectangular-cross-section copper.

6.5 Rotor Design

6.5.1 Squirrel Cage Rotor

Most common AC motors use the squirrel cage rotor, which will be found in many industrial AC motors. The squirrel cage takes its name from its shape—a ring at either end of the rotor, with bars connecting the rings running the length of the rotor. It is typically cast aluminum or copper poured between the iron laminates of the rotor, and usually only the end rings will be visible. The vast majority of the rotor currents will flow through the bars rather than the higher-resistance and usually varnished laminates. Very low voltages at very high currents are typical in the bars and end rings; high-efficiency motors will often use cast copper to reduce the resistance in the rotor.

In operation, the squirrel cage motor may be viewed as a transformer with a rotating secondary—when the rotor is not rotating in sync with the magnetic field, large rotor currents are induced; the large rotor currents magnetize the rotor and interact with the stator's magnetic fields to bring the rotor into synchronization with the stator's field. An unloaded squirrel cage motor at synchronous speed will consume electrical power only to maintain rotor speed against friction and resistance losses; as the mechanical load increases, so will the electrical load—the electrical load is inherently related to the mechanical load. This is similar to a transformer, where the primary's electrical load is related to the secondary's electrical load.

6.5.2 Wound Rotor

An alternate design, called the wound rotor, is used when variable speed is required. In this case, the rotor has the same number of poles as the stator and the windings are made of wire, connected to slip rings on the shaft. Carbon brushes connect the slip rings to an external controller, such as a variable resistor, that allows changing of the motor's slip rate. In certain high-power, variable-speed wound-rotor drives, the slip-frequency energy is captured, rectified, and returned to the power supply through an inverter. Compared to squirrel cage rotors, wound rotor motors are expensive and require maintenance of the slip rings and brushes, but they were the standard form for variable speed control before the advent of compact power electronic devices. Transistorized inverters with variable-frequency drive can now be used for speed control, and wound rotor motors are becoming less common.

6.5.3 Permanent Magnet Rotor

A permanent-magnet motor does not have a field winding on the rotor frame, instead relying on permanent magnets to provide the magnetic field against which the stator field interacts to produce torque. Permanent-magnet fields are convenient in electric motors to eliminate the power consumption of the field winding. The elimination of copper losses in the rotor is the primary source of efficiency gains of the permanent magnet motor over an induction motor. Historically, permanent magnets could not be made to retain high flux, especially when exposed to higher temperatures; field windings were more practical to consistently obtain the needed amount of flux. However, large permanent magnets are costly and must be handled with care during manufacture and motor assembly, as their strong magnetic flux can pose a danger when handling many of them. To minimize overall weight and size, permanent magnet motors may use high-energy magnets made with neodymium or other strategic elements; most of these are neodymium-iron-boron alloys. Permanent magnets are mounted on the rotor body in one of two ways:

- Surface mounted
- Interior mounted

The differences between the surface-mounted permanent magnet (SPM), as shown in Fig. 6.10, and the interior permanent magnet (IPM) go well beyond manufacturing considerations. This apparently subtle variation causes major changes to the operating characteristics of the motor.

IPM designs embed the magnets inside the rotor. This allows for greater strength, which permits higher running speeds. It also creates magnetic saliency with variations of inductance that are measurable at the terminals according to the rotor position. While the specifics of this concept deserve more extensive discussion, the practical effect is that the motor also develops reluctance torque in addition to permanent magnet torque.

SPM designs usually fix the magnets on the surface of the rotor with some sort of adhesive, so the strength of that bond is a practical determinant for maximum speed and overall robustness. Moreover, surface mounting does not create saliency, so there is no reluctance torque.

6.5.3.1 *Surface-mounted Permanent Magnets*

Many SPM motors used in industry (most of them servo motors) use this construction. It maximizes magnetic flux in the air gap and is generally considered the best magnetic configuration.

The primary challenge is retention of the magnets at high rotational speeds. The magnets are typically bonded to the rotor surface with an epoxy. In addition, the magnets can be banded with carbon fiber filaments and epoxy; however, this process is time consuming and costly. As mentioned before, high temperature exposure for these high-energy magnetic materials is also an issue. The highest-temperature-rated Neodymium-Iron-Boron magnet alloys are rated at 180°C. In SPM motors, the permanent magnet is directly exposed to the magnetic field of the air gap, which makes them more susceptible to demagnetization than in the IPM configuration. In an IPM, the magnets are shielded by the rotor steel, which provides a leakage path for the armature reaction flux.

6.5.3.2 *Interior-mounted Permanent Magnet Rotor*

This type of construction is used widely in traction motors or hybrid and electric vehicle application. Figure 6.11 shows a simple embedded magnet structure. Ideally there is no iron bridge around the surface of the magnet.

Mechanical considerations force the use of a magnetic bridge. The compromise is to make the bridge dimension as small as possible for acceptable stress levels and hence minimize the magnetic flux that will be diverted around this path. There are a large variety of designs of IPM rotors. Another major type of IPM rotor (not shown) employs magnets placed in a "spoke" arrangement along the rotor radius and magnetized tangentially. The "spoke" design has the intrinsic advantage of magnetic flux concentration, so that in

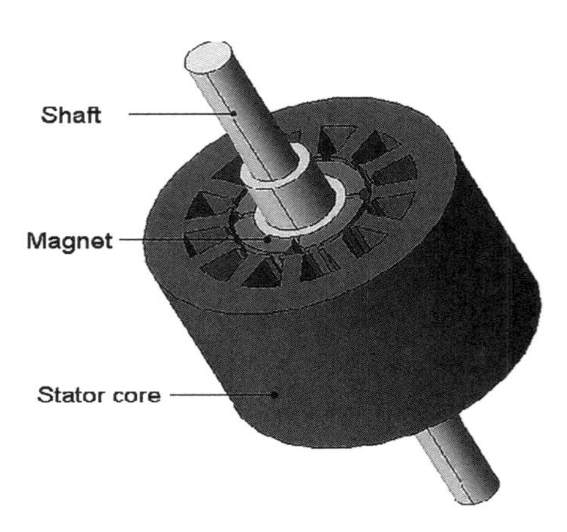

FIG. 6.10 Surface mounted permanent magnets.

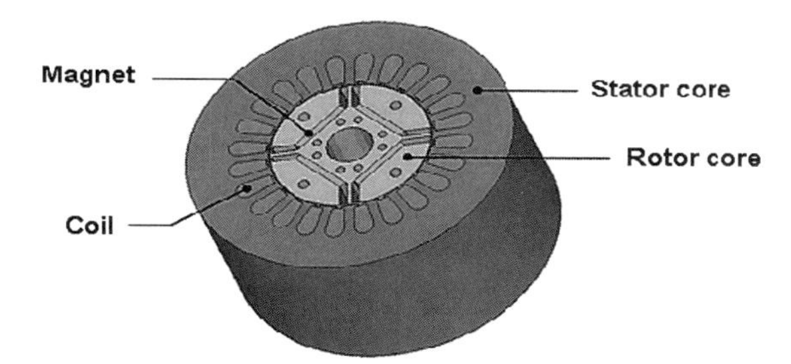

FIG. 6.11 Interior mounted permanent magnets.

high-polarity motors the flux density in the motor air-gap is increased. This leads the way to further performance improvement and/or size reduction. Many combinations of the magnet shape, position, and number of magnets per pole is possible for IPM motors.

6.6 Electric Motor Power Density

A number of factors affect the power density of electric machines. First and foremost, of course, is the design of the machine. Generally, permanent magnet motors exhibit greater power density compared to induction machines. For any given electric motor design, a number of factors affect power density; they are described in the following sections.

6.6.1 Temperature

The continuous power output of an electric motor is primarily determined by the operating temperature. In general it is acceptable to run the copper windings up to 200°C. With permanent magnet motors, the magnet peak temperature limitations must also be considered. For neodymium magnets there are many grades with temperature limits being a significant cost parameter. In general, Neodymium grades used in traction motors will have a temperature limit of approximately 180°C. It is essential that this limit is not exceeded at any time, otherwise partial demagnetization will occur, resulting in permanent degradation of motor performance.

6.6.2 Copper Losses

Copper losses are basically the resistive losses in the copper windings. These are equal to the product of current and winding resistance. Resistance is given by:

$$R = K \cdot \frac{L}{A} \cdot N \tag{6.13}$$

where:

K is the constant for copper

L is the length of one turn of wire

A is the cross section of the wire

The length of turn for a given machine is fixed, and N is also generally fixed based on the required speed and voltage. Also, K varies with temperature of 0.4%/°C from a 20°C base. It follows that for a 160°C rise, the winding resistance will rise by a factor of 1.62, a not insignificant number.

6.6.3 Iron Losses

Iron losses are created by the magnetic flux reversals that occur in the iron. For a given machine, it is desired to keep the magnetic flux as high as possible, but there is a limitation due to magnetic saturation. A more important limitation is at a somewhat lower limit due to heat generation. There are two sources of iron losses. First is the hysteresis loss, and second is the eddy current loss.

Hysteresis loss is given by

$$P_h = K_h \cdot f \cdot B^2 \tag{6.14}$$

and eddy current loss is given by

$$P_e = K_e \cdot t^2 \cdot f^2 \cdot B^2 \tag{6.15}$$

where:

K_e is the excess eddy current loss constant

t is the thickness of the lamination

f is frequency

B is the magnetic flex density

For industrial and low-speed motors, typically 50/60 Hz P_h is significant. For higher speeds and frequency, as in high-power-density traction motors, P_e is the important component. K_e varies with the grade of material used, generally high-Si steel.

The compromise between magnetic flux density and winding area is as follows:

Copper loss is minimized by making the slot larger.

Iron loss is minimized by making the iron section larger.

A number of factors are important to the selection of the right compromise:

1. Copper loss is generally at a maximum near low speed when maximum torque is required, and at low speed, iron loss is negligible.

2. It is a good concept to have maximum efficiency at the load point where the machine spends most of its time operating.
3. High speed normally requires low torque when the vehicle is in cruise mode; hence, the copper losses are relatively low.

It follows that the motor with the best published efficiency is not necessarily the best motor for all applications.

6.6.4 Motor Speed

Consider power output for a machine of a given size:

$$P = T \cdot w \tag{6.16}$$

With w being the rotational speed of the motor, it is immediately obvious that power is proportional to speed.

6.6.5 Cooling

Electric motors can be air or liquid cooled. High-power-density motors are generally liquid cooled. Practical design considerations for traction motors typically lead to an external water jacket cooling design. Water (usually a water/glycol mixture to assure proper cooling in cold climates) passages are located at the periphery of the stator. and heat is removed from the stator core and windings primarily through conduction. Power density with this cooling design is limited by the temperatures of the end windings, which have a poor thermal conductive path to the water jacket, and the rotor magnets (if any). On a hybrid vehicle, a single liquid cooling loop is used to cool the associated power electronics and the motor/generator.

Alternatively, internal cooling can be used, especially when the electric motor is an integral part of the gearbox. In this case, the end windings can be directly cooled by dripping the oil onto them. Some care must be taken to minimize oil entering the motor air gap, as this will generate significant viscous loss.

6.7 Electric Motor Characteristics

A number of factors should be considered in the design or selection of an electric traction motor for hybrid or electric vehicles. They are listed in Table 6.1.

TABLE 6.1 Motor Evaluation Criteria

Motor Selection Criteria	
Peak/Continuous power	System Voltage
Peak/Continuous Torque	Motor BackEMF
Base Speed	Torque Ripple
Rated Speed	Cogging Torque
Peak efficiency	Rotor Design

6.7.1 Power and Torque Characteristics

When selecting a suitable electric motor for an HEV application, the hybrid system topology is a critical parameter. Whether the system entails a parallel or series hybrid configuration has a significant impact on motor selection and, in fact, on the entire electrical system topology. In a series HEV (or pure electric vehicle), for example, the vehicle is entirely propelled by the electric motor. Power and torque ratings of the electric motor must therefore be sufficient to propel the vehicle with acceptable acceleration and top speed. Another consideration is the speed range of the motor. Electric motors have the design flexibility to be more efficient in certain operating regimes, but typically not over the entire speed and load range. For this reason, it may prove advantageous from an energy efficiency perspective to consider the use of a two- or three-speed transmission to limit the motor range of operation to a most efficient regime.

Parallel HEV configurations are widely used in commercial vehicle hybrid systems. In these systems, the electric motor can either be installed before or after the transmission. There are benefits to both approaches. A post transmission motor installation is relatively simple to do from a mechanical and vehicle integration perspective. Similarly, torque input and recovery control can be fairly straightforward to implement. However, at low vehicle speeds the electric motor torque is very small in comparison to engine torque, which is multiplied through the vehicle transmission. A pre-transmission location for the electric motor allows for torque multiplication through the transmission. This means that a relatively small motor can accelerate a heavy commercial vehicle with acceptable performance up to a 15 to 25 mph speed before the IC engine must start providing additional torque. The pre-transmission parallel HEV is fast becoming the standard for commercial vehicle hybrid systems with companies such as Eaton, Hino, BAE Systems, Allison, ZF, and others now or soon offering these systems in a range of power and torque specifications. There are several reasons for this:

- Commercial vehicle operators require a very high level of reliability from their vehicles, and vehicle uptime is a critical operating parameter. With a parallel configuration, even if the electric system (motor, battery, or any other high-voltage system) fails, the vehicle essentially reverts back to standard IC engine operation. Even with diminished performance, the vehicle can complete its mission in most cases.
- The electric motor and IC engine complement each other very well in a parallel hybrid system. A typical commercial vehicle diesel engine has peak torque between 1000 and 2000 RPM. An electric motor has its peak torque between 0 RPM and its base speed. Torque is highest and relatively constant up to the base speed; power is relatively constant

from the base speed up to the rated speed of the motor. Base speed is indicated by the breakpoints in the respective torque and power curves in Fig. 6.12. Base speed results from the design of the electric motor and can thus be selected to best match and complement the torque characteristics of the IC engine. As shown in Fig. 6.12, the red and blue curves reveal very similar peak torque characteristics, but very different peak power levels due to the much higher base speed for the blue curve. In this case the motor represented by the blue curve appears to be overdesigned.

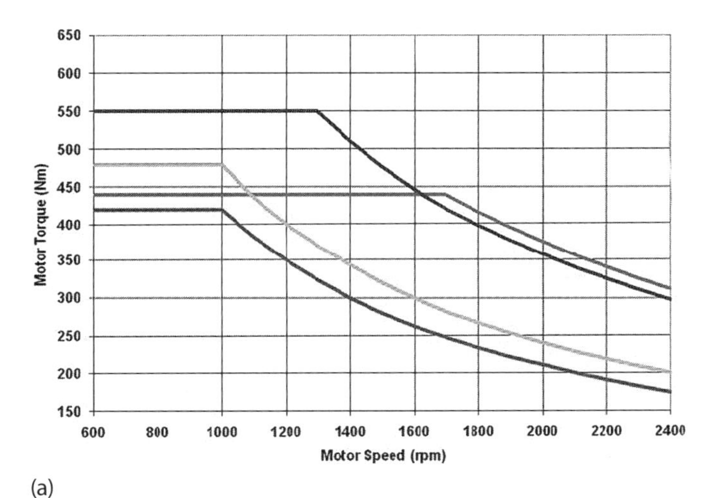

(a)

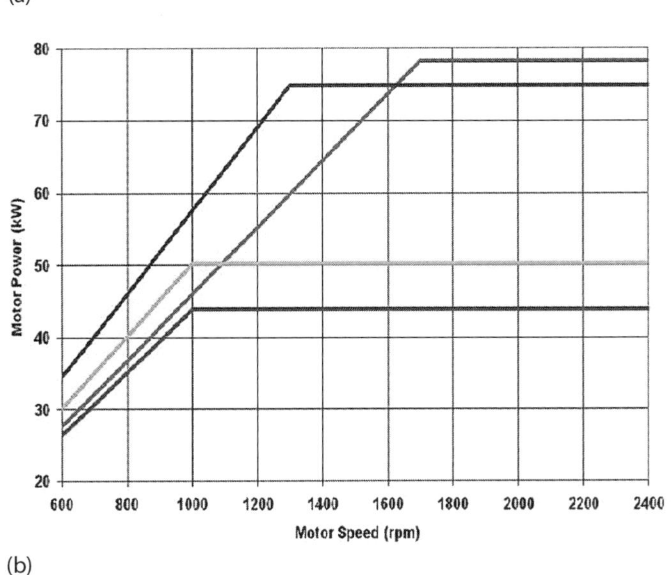

(b)

FIG. 6.12 Electric motor torque and power curves.

- While currently cost for any of the available commercial vehicle hybrid systems remains relatively high, a parallel configuration is more cost effective for most commercial vehicle applications compared to other hybrid system topologies.

Another consideration for a pre-transmission hybrid configuration is the desired level of hybridization. The level of hybridization is typically indicated by the ratio of peak torque provided by the electric motor divided by the cumulative peak torque of the IC engine and electric motor. A hybridization ratio of 25 to 35% is considered ideal from a cost and performance tradeoff perspective. With continued advances in technology and cost reductions in high-voltage electric systems, this ratio will no doubt be subject to change. It should be noted that electric motor power and torque selection have profound effects on hybrid system cost beyond that of the electric motor itself. All the major components of the high-voltage electrical system: traction motor, high-voltage battery, and traction inverter and their thermal management systems must be matched.

6.7.2 System Level Voltage

As mentioned in the previous section, all the high-voltage subsystems and components making up the electrical system in a hybrid vehicle must be matched. Another constraint in the design of the electrical system is the amount of current carried by any conductor in the high-voltage electrical system, which is generally limited at 250 to 300 A. While this is not a hard limit, currents much higher than this level require conductors with large cross sections and can lead to significant resistive losses in the system. Automotive hybrid systems typically have system nominal voltages in the 260–300 V_{dc} range, while commercial vehicles and buses typically have either a 300–350 V_{dc} or 600–700 V_{dc} system voltage. Combining the current limits with typical system voltage levels leads to a power limitation of around 100 kW for a traction drive motor for the lower voltage range, which is widely used in commercial vehicles. Because bus hybrid systems tend to be series drive systems, the 100-kW limit is too low. Hence almost all hybrid bus systems tend to have a 600–700 V_{dc} system voltage with traction motor power significantly higher than 100 kW. Since the cost for a higher-voltage system is significantly higher compared to the lower voltage range, one should consider carefully whether the benefits of a greater than approximately 100-kW motor outweighs the higher cost.

6.7.3 Back EMF

A notable disadvantage associated with PM motors is the generated back EMF, whether a current is applied to the stator windings or not. The latter is due to the fact that the permanent magnets always provide a magnetic flux, as opposed to the induced magnetic field in an induction motor. According to Faraday's law of electromagnetic induction, when a current-carrying conductor is placed in a magnetic field (that is, if the conductor cuts the magnetic field), an EMF is induced or produced in the conductor, and if a closed path is provided, current flows through it. When the same thing happens in an electric motor as a result of motor torque, the EMF produced is known as "back EMF." It is so called because this EMF that is induced in the motor opposes the EMF of the generator. This back EMF that is induced in the electric motor is directly proportional to the speed of the armature (rotor) and field strength of the electric motor, which means that if the speed of the electric motor or the field strength is increased, the back EMF will be increased; and if the speed of the motor or field strength is decreased, the back EMF is decreased.

This back EMF acts as a resistance, and any resistance in a line reduces and opposes the current flow. If the speed of the motor or field strength increases, the back EMF increases, which in turn increases the resistance to the current flow in windings, and hence less current is delivered to the armature of the electric motor. Also, if the speed of the electric motor armature or field strength decreases, the back EMF decreases, which in turn reduces the resistance and hence results in more current flow to the armature of the motor.

When the motor is first started, there is no back EMF induced, and as discussed above there is maximum current flow to the motor armature; as a result, the motor toque will be at a maximum. In this case there is no resistance offered by back EMF. The only resistance available is the motor winding resistance.

6.7.4 Torque Ripple

Torque ripple in three-phase PM motors is caused by several factors. The primary cause is any deviation of the back EMF voltage or current waveforms from a perfect sinusoidal signal. If the current waveform for each of the phases is perfectly sinusoidal, then the resulting sum of the torque generated in each of the three phases is a constant torque. However, practical design considerations for the PM motor lead to higher-order harmonics being introduced into the phase current waveforms. These harmonics lead to a "ripple" superimposed on the average torque generated by the motor. Therefore, providing a sinusoidal current and voltage waveform is the best way to minimize torque ripple on three-phase brushless motors.

Other sources of torque ripple can be found in features such as mechanical imbalance, magnetic saturation in the stator and rotor cores, cogging torque, and so on. The effect of torque ripple is the generation of noise and vibration in the driveline of a vehicle. It is, therefore, important to limit torque ripple as much as practically possible.

6.7.5 Cogging Torque

Cogging torque of electric motors comes from variations in magnetic field density around a rotor's permanent magnets as they pass the nonuniform geometry of the slot openings in the stator. It is also known as detent or "no-current" torque. Major factors affecting cogging torque include magnetic wave shapes, air-gap length, slot opening, number of stator slots and rotor poles, skewing, copper fill, pole pitch, flux distribution or density, magnet volume, and material weight.

Cogging torque is an undesirable component for the operation of permanent magnet motors. It is especially prominent at lower speeds, with the symptom of jerkiness. Cogging torque results in torque as well as speed ripple; however, at high speed the motor moment of inertia filters out the effect of cogging torque. A number of analytical tools are available to evaluate the tradeoffs of the various parameters affecting cogging torque.

6.8 References

6.1 Miller, J. M., *Propulsion Systems for Hybrid Vehicles*, The Institution of Electric Engineers, 2004.

6.2 Su, G-J. et al., "A Current Source Inverter Based Motor Drive for EV/HEV Applications," SAE Paper No. 2011-01-0346, SAE International, Warrendale, PA, 2011.

6.3 Yoshimoto, K. et al., "A Novel Multiple DC-Inputs Direct Electric-Power Converter," SAE Paper No. 2009-01-0203, SAE International, Warrendale, PA, 2009.

6.4 Giesselmann, M. et al., "Design of an Ultra High Power IGBT Inverter for Rapid Capacitor Charging," SAE Paper No. 2000-01-3652, SAE International, Warrendale, PA, 2000.

6.5 Sato, Y. et al., "Development of High Response Motor and Inverter System for the Nissan LEAF Electric Vehicle," SAE Paper No. 2011-01-0350, SAE International, Warrendale, PA, 2011.

Chapter 7

Hydraulic Hybrid Powertrain System Design

7.1 Introduction

There are several examples of hydraulic hybrid commercial vehicle trials dating back to 1978 [7.1]. Many hydrostatic drive systems (series hydraulic drives without regeneration) are deployed in off-highway vehicles ranging in size from lawn and garden equipment to large earth-moving machines. As a result, there is a substantial industry infrastructure manufacturing hydraulic equipment for mobile off-highway applications as well as for industrial uses. The reason for this is the rugged reliability, high power density, and controllability of hydraulic systems. For on-highway vehicles, the hydrostatic drive system was not as efficient as a mechanical transmission and was therefore not widely adopted.

Hydraulic hybrid vehicle systems feature hydrostatic pumps and motors for power transmission, and hydraulic accumulators for energy storage. The energy storage capability of the hydraulic accumulator is orders of magnitude less than that of a battery, but charge and discharge is more rapid than that of a typical battery (high power density). Therefore, systems are designed to maximize regenerative braking and power buffering rather than sustained propulsion from stored energy. This may enable some engine downsizing and significantly reduce fuel consumption and emissions.

In general, hydraulic components are made using readily available materials and conventional manufacturing processes, and, therefore, the additional costs associated with large batteries and high-power electronic components are avoided.

7.1.1 The Hydrostatic Transmission

A hydrostatic transmission is simply a pump and a motor connected in a circuit to perform the desired functions. The pump is driven by the prime mover (e.g., engine or electric motor), and the hydraulic motor drives the load (e.g., driveshaft or machine). If machines with variable capacity are used, the transmission is infinitely variable (IVT), and mechanical gearshifts are eliminated.

A hydrostatic transmission provides improved maneuverability, but at a cost. The efficiency of a hydraulic transmission is generally lower than a mechanical transmission. A mechanical transmission will typically have an efficiency of 92% or greater, meaning that 92% of the input energy is delivered to the load (wheels). A hydrostatic transmission has an efficiency of approximately 80 to 85% with traditional pumps and motors.

Application of such a drive system to road vehicles has traditionally been uneconomical due to the relatively low efficiency of conventional hydraulic machines used in the hydrostatic systems relative to a mechanical transmission. The development of hydraulic machines with much improved efficiency and enhanced controllability has allowed hydrostatic drives to be reconsidered for road vehicles, particularly in combination with hydrostatic regenerative braking.

7.1.1.1 Configurations

Four basic configurations of hydrostatic transmission are shown in Fig. 7.1. They are in-line, U-shape, S-shape, and split [7.2]. In a hydrostatic transmission, the prime mover (e.g., engine) drives a hydraulic pump and generates oil flow at pressure, which powers a hydraulic motor connected to the road wheels or sprockets of a track drive. Generally, one or both of the pump and motor have variable capacity, and a control system interprets operator pedal input to adjust flows, pressure, and engine power to give required vehicle movement.

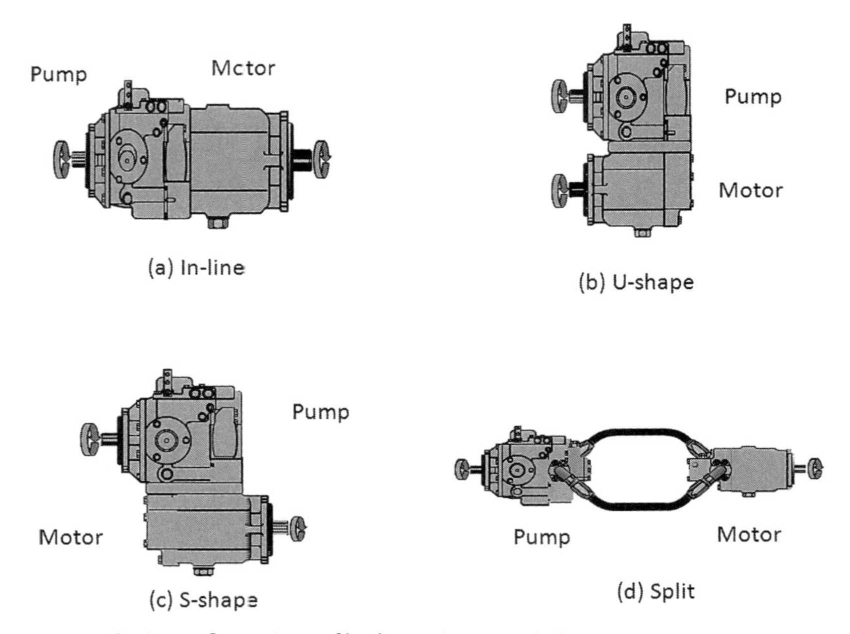

FIG. 7.1 Basic configurations of hydrostatic transmissions.

Often with off-highway vehicles, additional implement functions (e.g., bucket lift) are powered hydraulically using pressure and flow from the engine-driven pump or separate engine-driven pumps.

7.1.1.2 *Classification of Hydrostatic Transmissions*

Hydrostatic transmissions can be classified as shown in Fig. 7.2. An open circuit is one in which oil is delivered from the reservoir to the motor by the pump, and flow from the motor returns to the reservoir. In a closed-circuit hydrostatic transmission, fluid flows from the pump to the motor and directly back to the pump.

In an open-circuit system, the low pressure level is determined by the reservoir pressure (usually atmospheric), but this is not the case with a closed system. The closed system requires an additional facility for maintaining fluid volume (typically a small pump fed by an atmospheric reservoir).

7.1.1.3 *Operating Characteristics of Hydrostatic Transmissions*

If the pump and the motor have the same displacement, the theoretical performance of both is shown as the heavy dark line in the middle of the family of curves shown in Fig. 7.3. The pump performance curves are shown below the theoretical line, and the motor curves above the line [7.2]. It also shows that the motor speed varies with flow rate. If precise control of a motor speed is required, some means must be provided to sense the actual motor speed and adjust pump output until the target speed is achieved.

7.1.2 Hydrostatic Regenerative Braking

For hydrostatic regenerative braking, vehicle inertial energy is captured by generating high-pressure oil flow and storing this oil in a hydraulic accumulator. This is done using a hydraulic machine that can act as both a pump and

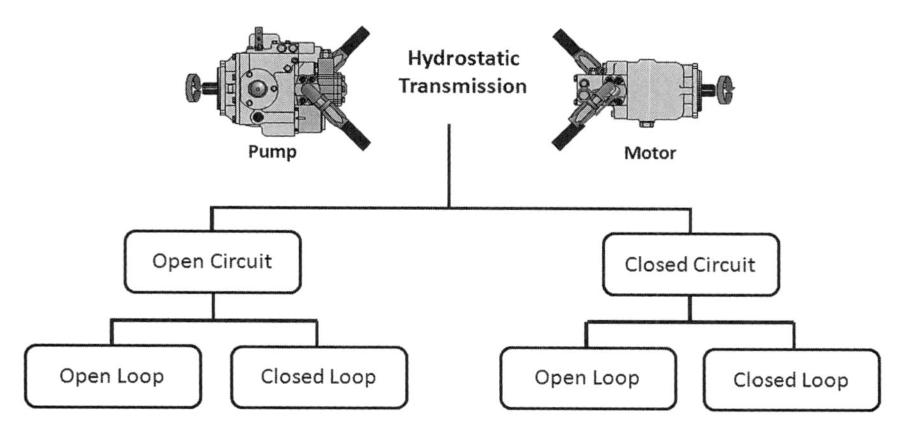

FIG. 7.2 Classification of hydrostatic transmissions.

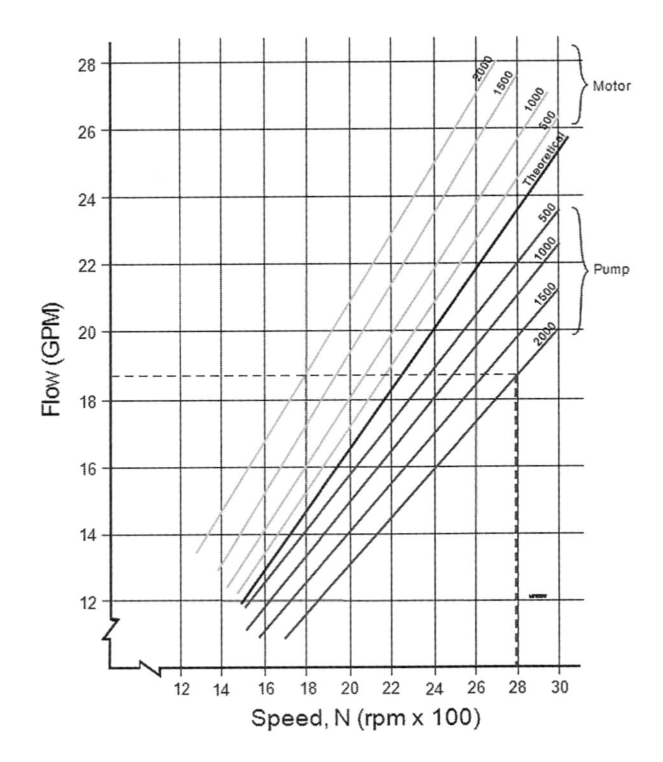

FIG. 7.3 Typical performance curves for a pump and motor in a hydrostatic transmission.

a motor and which is connected to at least one vehicle axle. In operation, the hydraulic machine acts as a pump during vehicle deceleration and charges the accumulator. The stored energy is used during subsequent acceleration, when the hydraulic machine acts as a motor powered by high pressure flow from the accumulator. The engine is not required to provide as much power, and fuel is therefore saved. In a conventional vehicle, inertial energy is lost as heat generated in brake components. This means that there is an additional advantage from incorporating hydrostatic regenerative braking in a vehicle, which is a significant reduction in brake wear, and hence brake maintenance cost.

Hydrostatic regenerative braking may be applied to a vehicle as a standalone system. This is called a parallel hydraulic hybrid system.

Hydrostatic regenerative braking can also be incorporated with a hydrostatic drive system if the drive motor is configured to also act as a pump. This system is called a series hydraulic hybrid drive.

Other potential hybrid architectures include split powerflow systems where some power is transmitted mechanically and some hydrostatically, and two-mode systems in which the transmission is mechanical for part of the drive cycle and hydrostatic for the remainder. Both of these configurations will include regenerative braking using accumulators for energy storage. The hybrid systems discussed here are described in more detail in Sections 7.2 and 7.3.

7.1.3 User Requirements

There is an extensive range of requirements for commercial vehicle hydraulic hybrid systems from both the customer (vehicle builder) and end user (the vehicle builder's customer). The vehicle must be able to confidently perform specific tasks under specified road conditions with low cost of ownership. For commercial vehicles, cost of ownership includes initial cost, operating costs, and lifetime maintenance costs. There is also an absolute requirement for the vehicle builder to achieve and certify regulatory requirements for safety and environmental compliance.

There are many standardized drive cycles for commercial vehicles. Some of these are produced by regulatory bodies for fuel consumption assessment and others by industry bodies for comparison between similar truck models. Most relate to the type of vehicle and its intended duty. Other cycles are issued by vehicle manufacturers for evaluation of competing powertrain systems. A developer of powertrain systems will need to model the vehicle analytically with the hybrid powertrain to design an optimum system and quantify the benefits.

Hydrostatic regenerative braking is of greatest value in reducing fuel consumption for situations in which the vehicle duty profile encounters frequent brake applications (stop/start duty cycle). This type of duty cycle can typically be found with refuse collection or package delivery vehicles. Off-highway vehicle operation typically involves high natural resistance to motion and relatively low speeds, so the potential for energy recovery is limited, although there are exceptions to this, e.g., some warehouse and port material handling vehicles.

The energy storage capability of the accumulator can also be used as a "buffer" to enable the engine to operate at high efficiency for a larger proportion of the drive profile than a nonhybrid vehicle. This feature enables additional fuel savings for almost all vehicle classes. An example could be the situation in which the engine operates at higher speed and torque (in its maximum efficiency zone) than required to power the vehicle during light deceleration or hill descent and uses the excess power to charge the accumulator. The stored energy would be used later to power the vehicle with the engine at idle or stopped, or to augment engine power.

Engine emissions are obviously directly related to fuel consumption, but it is worth noting that more-efficient engine operation will reduce emissions further. Research has shown that engine starts and fast transients contribute significantly to emissions, particularly of particulates from diesels [7.3]. Hybrid systems can reduce transient-induced emissions by using stored energy to reduce the rate of change of engine power and speed during vehicle operation. Credit for this depends on the legislative environment, however. When a commercial vehicle engine is certified for emissions on a test stand using a standardized test schedule, the benefits of hybrid operation are irrelevant. When the vehicle is certified based on total tailpipe emissions over a drive cycle, the benefits of a hybrid powertrain can be substantial and may even enable reductions in aftertreatment.

7.2 Hydraulic Hybrid System Architecture

7.2.1 Parallel Hydraulic Hybrid Systems

In most vehicle applications, the conventional mechanical drivetrain is retained and a hydraulic system is provided to add to or subtract torque from the final drive. Figure 7.4 shows a schematic of such a system for a two-wheel-drive vehicle (a) and also for a four-wheel-drive commercial vehicle (b) with the hydraulic system interacting with the driveshaft between the transfer case and the rear axle.

Additional components of the system include a clutch at the driveshaft to disengage the hydraulic system when not in use, a valve to isolate the accumulator, and a low-pressure oil storage tank which may be slightly pressurized [7.4, 7.5].

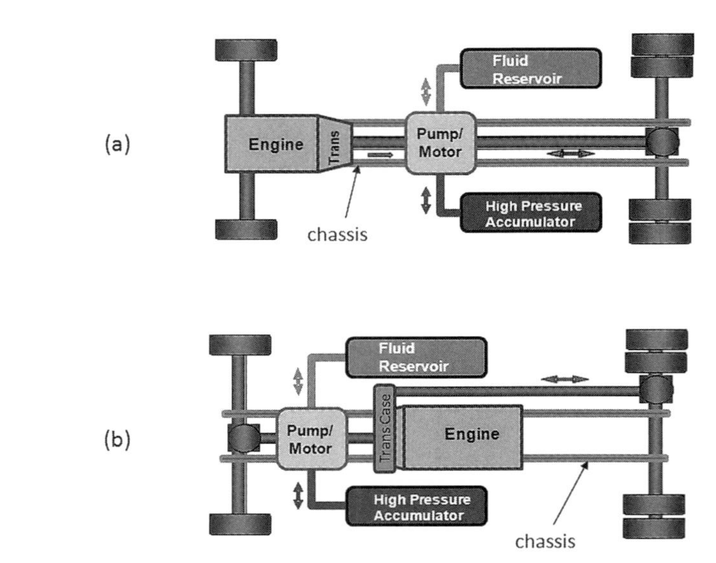

FIG. 7.4 Parallel hydraulic hybrid system

The hydraulic pump/motor is mounted on a transfer gearbox, which interfaces with the vehicle driveshaft between the transmission and the drive axle. The gearbox function is to match the speed range of the hydraulic machine to that of the driveshaft over the regenerative portion of the drive cycle. It is usual to fit a clutch in the transfer gearbox at the driveshaft interface so that the pump motor and gearbox parasitic loss is eliminated when regenerative braking is not in use. For a typical refuse vehicle application, regenerative braking would be employed during the collection cycle between say 3 to 15 and 20 mph. At higher vehicle speed, the system would be clutched out but would still be available over the same speed range during normal driving.

Regenerative braking operates as follows [7.5, 7.6]:

- During vehicle deceleration, torque is absorbed by the hydraulic machine acting as a pump. This torque generates high-pressure fluid flow from the pump, which is stored in a hydraulic accumulator. This represents a significant proportion of the kinetic energy of the vehicle. Foundation brakes would only be applied if deceleration provided by the hydraulic system is insufficient to meet driver demand.
- The energy stored in the accumulator is reused for vehicle acceleration and propulsion when the fluid at pressure is passed back to the hydraulic machine, now acting as a motor, to provide torque to the driveshaft. This reduces the engine torque demand, which results in fuel savings.

For commercial vehicles, single or two-pedal operation is used during braking. With single pedal control, the driver operates the brake pedal to slow the vehicle, and the hybrid control system smoothly blends the regenerative braking with the foundation brake, giving priority to regeneration.

With two-pedal operation, regeneration is activated as the driver lifts his foot from the accelerator pedal, and typically regenerative braking occurs over the last 10% or so of accelerator travel. This is often called "retarder mode," as the commercial vehicle retarder is activated this way (although here no energy is stored for reuse). The foundation brake system is used when additional stopping power is required.

It is important for vehicle safety that the system control give priority to the antilock brake system in the event of wheel slip.

With the parallel systems this is done by deactivating the hydraulic system by activating the clutch to disengage the hydraulics. The need to do this in about 200 ms is one of the system design parameters. It is worth noting that the regenerative braking only operates on the driven wheels, whereas the foundation brakes operate on all road wheels, so the probability of wheel slip is increased somewhat during regeneration.

It is also feasible to specify a pre-transmission parallel hybrid system. Here the hydraulic pump motor is connected to the engine output shaft through the accessory drive or a "power take-off" (PTO). Advantages are that the limited speed range of the engine (particularly a diesel engine) enables smaller pump/motors to be specified for a given power, and that regeneration can be selected over the entire vehicle speed range. However more energy is lost in the transmission with this arrangement, particularly if a torque converter is operational.

7.2.2 Series Hydraulic Hybrid Systems

Figure 7.5 shows a schematic of a series hydraulic hybrid in which the engine directly rotates a hydraulic pump, and there is generally no mechanical transmission fitted. The pump provides sufficient flow at pressure to drive a hydraulic motor/pump which generates torque that is transmitted to the drive wheels of the vehicle. A hydraulic accumulator is connected to the high-pressure fluid line to store braking energy during deceleration. During deceleration the drive motor/pump operates as a pump to absorb energy from the road wheels and pump up the accumulator. The energy stored is used during subsequent acceleration and cruise to provide pressure and flow to the motor/pump and therefore reduce the engine power required.

The elimination of the mechanical transmission gives a weight and cost offset. However, the pump and motor/pump require the highest efficiencies, particularly at part load, to maximize vehicle fuel economy.

The pump and pump/motor are of variable capacity type (the flow can be varied independently of pressure and speed). This enables the system to operate the engine at the highest efficiency for a given power requirement, and torque to the road wheels is independent of road speed.

Leaving aside the accumulator, the system is effectively an infinitely variable hydrostatic transmission. A clutch or torque converter is usually not required.

Also, the engine-driven pump can easily be configured as a pump/motor. This enables it to be used to restart the engine using energy stored in the

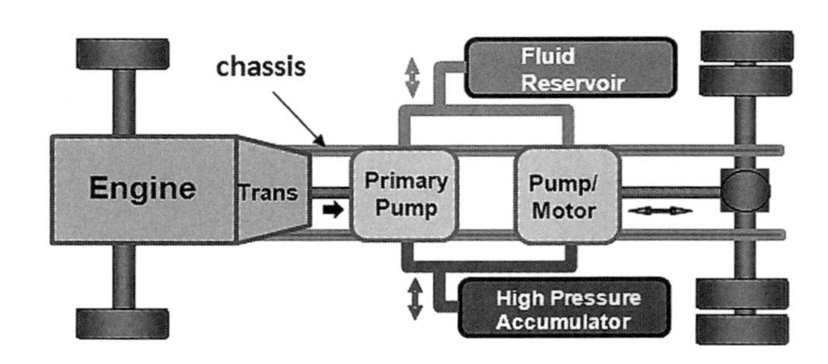

FIG. 7.5 Series hydraulic hybrid system.

accumulator, and the engine can be turned off when not required to further reduce fuel use, provided that means are provided to power key accessory functions (brakes, steering, A/C, etc.).

In any particular commercial vehicle application, the fuel savings achieved depend very much on the duty cycle (speed range, frequency of stops, idle time, etc.). For the example of the package delivery vehicle, the target of 50% fuel consumption reduction is achievable using a combination of efficient engine operation, regenerative braking, engine-off operation, and modest engine downsizing [7.9, 7.10].

The basic system just described can have some variations. Hydraulic pumps/motors suitable for commercial vehicles are most efficient between 750 and 2000 RPM and have maximum speeds of 2500–3000 RPM. Therefore, they are ideally suited to direct drive from a diesel engine. With a gasoline engine, a reduction gear is probably required. The drive motor/pump has to provide launch torque (zero speed) and operate effectively at vehicle maximum road speed. For some applications, a two-speed "shift on the fly" gearbox or transfer case is used to give good low-speed torque at one ratio and enable maximum vehicle speed within the motor speed range at the second ratio. More than two ratios are possible but at the price of increased cost and complexity.

An example of the use of a two-speed transfer case is the tactical military transport vehicle, which is required to launch fully laden on a 60% grade and achieve a maximum sustained speed of 65 mph [7.7, 7.8].

7.2.3 Multimode and Split Powerflow Hydraulic Hybrid Systems

These systems are proposed to combine the high efficiency of a mechanical transmission with continuous ratio change and capability for regenerative braking provided by a hydrostatic drive. An example of the multimode hydraulic hybrid system is shown in Fig. 7.6. There are a very large number of possible configurations, just as there are for electric hybrid systems.

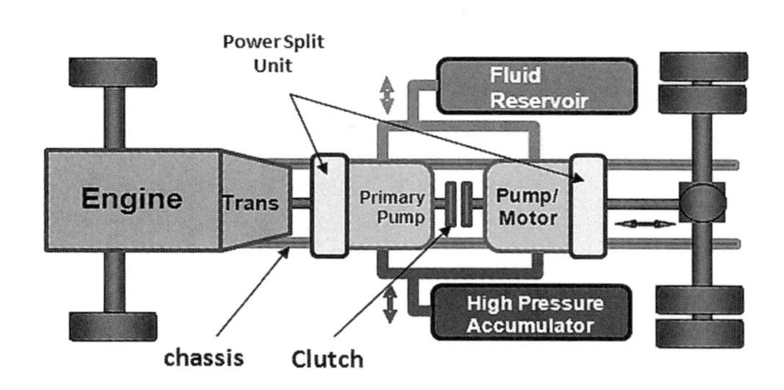

FIG. 7.6 Schematic of multimode hydraulic hybrid system.

However, because hydraulic machines and electric motors have different torque/speed characteristics and operating speed ranges, optimum configurations for a given vehicle application will likely be different for electric and hydraulic hybrid systems.

The advantages of these configurations include increased transmission efficiency relative to series and parallel hydraulic hybrid systems and reduced size of the hydraulic machines used (because not all power is transferred hydraulically and pump/motor speed range can be optimized). However, the complete transmission is more complex, and extensive testing must be carried out to validate the combined unit for vehicle use.

The simplest example of a two-mode system is provided by the Parker "Runwise" transmission for refuse collection vehicles [7.6, 7.11]. This features a hydrostatic drive system with hydraulic regenerative braking (series hydraulic hybrid system) for low-speed operation, including refuse collection. Above approximately 35 mph, the hydraulic system is disconnected mechanically and direct mechanical drive is used.

A potential development for a vehicle with a less specific drive cycle profile, such as would be used for package delivery or a city bus, features three modes. Mode 1 is a hydrostatic transmission used for start and low speed. Mode 2 for intermediate speeds uses a blend of mechanical and hydrostatic power flow (by varying the capacity of the hydraulic machines), and Mode 3 at higher speeds uses mechanical drive only.

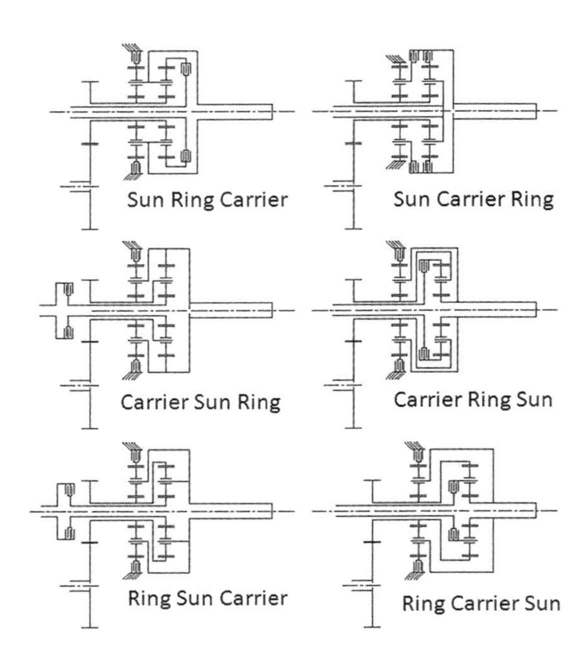

FIG. 7.7 Six configurations for possible split powerflow hydraulic transmissions [7.16].

Fully variable split power flow systems generally utilize planetary gear sets to split and sum torques [7.16, 7.17, 7.18]. With two planetary gear sets and a pair of pump/motors, there are over a thousand possible configurations. Some examples are shown in Fig. 7.7. Clutches are employed to enable several operational ranges for different road speed and torque requirements. Energy storage using a hydraulic accumulator is incorporated within the hydraulic circuit.

Optimum configurations are designed by setting up and analyzing a transmission model and evaluating its performance in a vehicle simulation against a list of vehicle functional and performance specifications. This process can be economically automated using advanced dynamic programming techniques [7.9]. In general, one needs to ensure that transmission and engine efficiency is high for high-use parts of the drive cycle, and that vehicle specification points are met with good drivability.

These systems have great potential for optimum solutions for higher-volume applications such as class 8 line-haul trucks and large cars/light trucks, where high transmission efficiency is essential and development costs can be amortized.

7.3 Design and Specification of Hydraulic Hybrid Systems

The modeling of a commercial vehicle is described in Chapter 11. This section provides information necessary to specify and to design hydraulic components for drivetrain systems.

7.3.1 Hydraulic Pumps and Motors

Construction and operation of the machines used in hybrid vehicles are discussed in Section 8.2.

7.3.1.1 Sizing of Pumps and Motors

Parallel hydraulic hybrid systems are designed to capture braking energy and reuse it for propulsion. The operational vehicle requirements are met by the specification of the mechanical driveline. The pump/motor is sized to maximize fuel savings, and this is determined by careful analysis of vehicle performance over intended drive cycles. The pump speed range is matched to the vehicle driveshaft by choice of gear ratio to operate during frequent deceleration events. The hydraulic system is disengaged with a clutch when not in use or outside the applicable speed range.

For a series system, the pump and motor are sized to transmit power to the vehicle wheels under all operating conditions. The components are rated to a maximum operating pressure (say 350 bar), and the pump should be large enough to absorb maximum rated engine power when operating at maximum capacity at this pressure with a small margin. A commercial vehicle diesel engine operates to a maximum speed of 2200 to 2500 RPM typically, and the pump would then be driven directly. With a gasoline engine or high-speed diesel, a reduction gear would be specified.

Pumps sized this way should provide required power for acceleration and grade requirements.

Hydraulic machines operate most efficiently over a limited speed range—typically below 2000 rpm. Efficiency losses increase up to maximum speeds of 3000–4000 RPM, depending on size and design. Above these speeds, cavitation damage may occur.

The motor will operate at constant torque, proportional to pressure. The motor is sized to provide launch torque to the drive wheels at the required load and grade. The differential ratio to the drive wheels should permit the motor to operate at about 2000 RPM at economic cruise speeds and not exceed the maximum pump speed at maximum sustained vehicle speed. For very large vehicles, two or more motors may be required to meet these criteria, or a multispeed gearbox added between the motor and the differential.

7.3.1.2 Performance of Pumps and Motors

Ideal torque is proportional to the difference between inlet and outlet pressure, and ideal flow is proportional to machine speed and capacity (volume per revolution). However, there are some torque and flow losses in real machines due to friction and leakage, and these losses are a function of operating temperature and pressure, and oil used as well as mechanical construction, and it is usual to characterize these with a "mechanical" and "volumetric" efficiency. Further discussion of this subject is provided in Section 8.2.6.

Manufacturers of pumps and motors provide performance data in the form of graphs and tables. These are usually applicable for a limited range of oils and temperatures and may omit some real effects such as control actuator leakage. For detailed models, it is important to obtain data for the actual operating conditions of the hybrid system under investigation. This may require specific tests conducted by the manufacturer or an accredited test house.

Data can be entered as tables, equations with calibrated constants, or a combination.

The following equations apply to displacement-type pumps and motors usually specified for vehicle hybrid systems.

Ideal leak free flow Q_i (L/min) of a hydraulic machine is:

$$Q_i = \frac{xND}{1000} \tag{7.1}$$

where:

- x is displacement fraction
- N is RPM
- D is maximum pump displacement (cc/rev)

Actual pump delivery flow Q_a is

$$Q_a = \eta_{\text{v,pump}} Q_i \tag{7.2}$$

where:

- η_v is volumetric efficiency

Efficiency may be entered as a series of tables, but this requires a lot of test data. The following equation may be used with constants derived from more limited test data for the pump type under consideration:

$$\eta_{v,pump} = 1 - \frac{C_s}{xS} - \frac{\Delta P}{\beta} - \frac{C_{st}}{x\sigma} \tag{7.3}$$

where:

- C_s is a constant for laminar (viscous) leakage
- C_{st} is a constant for turbulent leakage
- S is a nondimensional number $\mu\omega/\Delta p$
- μ is oil dynamic viscosity (Ns/m^2)
- ω is rotational speed (rad/s)
- Δp is pressure difference (Bar)
- β ρ is oil bulk modulus (Pa)
- σ is a nondimensional number $\dfrac{\omega D^{1/3}}{\left(2\dfrac{\Delta P}{\rho}\right)^{1/2}}$

Equivalent equations for a hydraulic motor are as follows:

Actual flow to the motor is:

$$Q_a = \frac{Q_i}{\eta_{v,motor}} \tag{7.4}$$

$$\eta_{v,motor} = \frac{1}{1 + \dfrac{C_s}{xS} + \dfrac{\Delta P}{\beta} + \dfrac{C_{st}}{x\sigma}} \tag{7.5}$$

Ideal shaft torque T_i (N·m) of a hydraulic machine is

$$T_i = \frac{xD\Delta P}{2\pi} \tag{7.6}$$

Actual pump torque T_a is

$$T_a = \frac{T_i}{\eta_{t,pump}} \tag{7.7}$$

where:

η_t is torque (or mechanical) efficiency

Again the following equation may be used with constants derived from limited test data for the pump type under consideration:

$$\eta_{t,pump} = \frac{1}{1 + \frac{C_v S}{x} + \frac{C_f}{x} + C_h x^2 \sigma^2} \tag{7.8}$$

where:

C_v is a constant for viscous friction loss

C_f is a constant for mechanical friction loss

C_h is a constant for hydrodynamic torque loss

Equivalent equations for a hydraulic motor are as follows:

$$T_a = \frac{T_i}{\eta_{t,pump}} \tag{7.9}$$

$$\eta_{t,motor} = 1 - \frac{C_v S}{x} - \frac{C_f}{x} - C_h x^2 \sigma^2 \tag{7.10}$$

7.3.2 Hydraulic Accumulators

7.3.2.1 *Sizing of Accumulators*

The hydro-pneumatic (or gas) accumulator with a fixed charge of nitrogen gas is currently specified for energy storage in hydraulic hybrid systems [7.12].

Accumulators are manufactured in a range of sizes (capacities) and certified maximum pressures.

They are catalogued by "gas volume," which is the volume of gas inside the accumulator when there is no fluid present. The accumulator is precharged with gas (usually nitrogen) to a pressure typically about half of the rated maximum pressure.

The optimum size of the accumulator for a particular vehicle is determined from extensive model evaluation using expected operational drive cycles. Too small an accumulator will not capture enough vehicle kinetic energy (proportional to mass times the square of speed) to give optimum fuel savings, whereas too large an accumulator is more costly and heavier than is justified by the fuel saved.

For the parallel system, maximum benefit is obtained where the greater part of vehicle operation consists of frequent stops and starts and relatively low speed between stops. Such applications are found with refuse collection vehicles and city buses.

For series systems, choice of accumulator optimum size is more complicated and interacts strongly with the power management control strategy adopted. An approach for addressing this optimization is described in Ref. [7.13].

It is important to recognize that the accumulator stores energy over a range of pressures between charge pressure and maximum pressure, and this "instantaneous" pressure also defines the hydraulic system pressure unless the accumulator is isolated by the shutoff valve.

Precharge pressure is a system selective variable. However, too low a precharge pressure will result in inadequate power at the motor for vehicle drive, and too high a pre-charge pressure will reduce the energy storage capacity and therefore require a larger accumulator.

7.3.2.2 Performance of Gas Accumulators

For a real gas, the Benedict Web Rubin (BWR) equation is used to describe the relationship between gas pressure (p_g), temperature (T), and specific volume (v) [7.12].

$$p_g = \frac{RT}{v} + \frac{\left(B_0RT - A_0 - \frac{C_0}{T^2}\right)}{v^2} + \frac{(bRT - a)}{v^3} + \frac{a\alpha}{v^6} + \left(c\left(1 + \frac{\gamma}{v^2}\right)e^{-\gamma/v^2}\right)/v^3T^2 \tag{7.11}$$

where:

R is the gas constant ($Nm/kg \cdot K$)

A_0, B_0, C_0, a, b, c, α, γ are constants

For an accumulator without additional mass within the gas bladder, the compression and expansion of the gas is usually considered isothermal due to conduction of heat from the gas to the much greater mass of the casing and oil. The following equation applies to displacement-type pumps and motors usually specified for vehicle hybrid systems [7.12].

$$\frac{dT}{dt} = \frac{1}{m_g C_v}\left(Q_w - m_g T\left(\frac{\delta p_g}{\delta T}\right)_v \frac{dV}{dt}\right) \tag{7.12}$$

where:

- m_g is the mass of the gas (kg)
- C_v is the specific heat of the gas (0.214 J/g-mol·K for nitrogen)
- Q_w is the heat transfer to the oil and wall
- $\frac{dV}{dt}$ is the rate of change of gas volume = oil flow into the accumulator
- T is gas temperature (K)

7.3.3 Hydraulic Lines and Other Components

Pipes and hoses used for connecting pump, motor, accumulator, and reservoir should be designed for low-pressure loss by minimizing length and number of bends and by avoiding constriction. Flow is turbulent, and the basic equation is as follows:

$$\Delta P = KL\left(Q_p / A_p\right)^2 \tag{7.13}$$

where:

- ΔP is pressure drop (bar)
- L is length of pipe (m)
- Q_p is pipe flow (L/min)
- A_p is internal flow area of the pipe (m^2)
- K may be considered as a constant

K may be derived from manufacturers' data sheets. Note that internal diameter of pipes may be slightly different from nominal catalogue size. For instance, a nominal 1-in pipe ID may vary between 0.815 in and 1.185 in, depending on wall thickness.

It is customary to estimate an "equivalent length" of a pipe run by adding a factor for bends. Again, manufacturers publish tables for their fittings. For example, a 1-in pipe with a 45° bend with mean radius of two times pipe radius would add 1.1 ft to equivalent pipe length. A 90° bend could add 2 ft, and a sharper bend could add 5 ft. It is important to maintain desired cross section through the bend.

Other components in the fluid lines, such as shutoff valves, etc., should have the same or larger "open area" as the pipe/hose.

For vehicle hybrid system pipe/hose assemblies, the following maximum fluid velocities are recommended:

High-pressure lines:	6–9 m/s (20–30 ft/s)
Pressurized return line	3 m/s (10 ft/s)
Suction line (nonpressurized)	1.2 m/s (4 ft/s)

7.3.4 Fluid Conditioning

Fluid conditioning systems for hydraulic hybrid systems should be designed to match the specific needs of the application. Primary design consideration should be given to maintaining required fluid cleanliness levels and removing excess heat present in the system (see Section 8.5.2).

It is recommended to use an external filter system during commissioning of the vehicle when it is new, or after major new hydraulic system components are fitted. This will remove a large percentage of contaminant that is built in or generated by initial running.

7.3.5 Noise and Vibration

Hydraulic transmission elements generate noise and vibration characteristics when installed in vehicle systems, just as mechanical or electrical components do.

One characteristic of particular concern is pressure fluctuation in high-pressure pipes as a result of compressibility effects in pumps and motors that feature seal plates. The origin of these pressure pulses is discussed in Section 8.2.6, but the net result for the system is the presence of exciting frequencies at harmonics (multiples) of the pumping frequency (number of pistons × the rotational speed). This produces a characteristic whine when emitted as air-borne noise by vibrating system components. Methods to alleviate this include isolation of components using energy-absorbing mountings and inclusion of pressure pulse damping elements into the pipe run. Dynamic analysis of system response to pressure perturbations in both time and frequency domain is useful during design of fluid transmission lines and placing of components to avoid amplification of pressure pulses.

It is worth noting that the radial piston pump described in Section 8.2.5.2 does not have a seal plate and therefore exhibits very low pressure fluctuations and hence very low levels of system noise.

Pumps and motors also generate torque fluctuations as a result of geometric construction, and pumps can transmit engine torque fluctuations as additional pressure pulses. Torque fluctuation can give vibration and roughness in the system but can be minimized by the use of energy-absorbing shaft couplings tuned for frequencies of concern, and by dynamic analysis of the shaft system to design elements of the system to avoid amplification of torque pulses.

It is important to note that the rotational inertias of pumps and motors are much less than engines or mechanical transmissions.

There are some cyclic out-of-balance forces generated by operational pumps and motors due to piston friction, torque, and pressure pulses. Therefore, again it is important to effectively isolate the pump/motor unit from the chassis with anti-vibration mounts to reduce transmitted noise and vibration in the hybrid system [7.19].

7.4 Examples of Systems at or near Production Application

Some applications are described below with more detailed descriptions of the systems described in Section 7.2.

7.4.1 Refuse Collection Vehicles

Refuse collection vehicles on many routes have the potential for significant kinetic energy recovery and reuse (regenerative braking). Current refuse body systems use hydraulics for loading and compaction of waste, so there is familiarity with operation and maintenance of hydraulic systems. Parallel hybrid systems (regenerative braking only) are now being sold for refuse vehicles in the United States and Europe, and are being considered for some bus applications. Maximum benefits are seen for cases in which vehicles are required to accelerate to a significant speed between frequent stops for refuse pick-up.

Figure 7.8 shows a parallel hydraulic hybrid power unit and an installation on a refuse truck. The system configuration calls for the hydraulic pump/motor unit to be inserted into the drive shaft between the transmission and axle (Fig. 7.9). This installation is relatively low cost and noninvasive. In the event of most hydraulic system failures, the system can be deactivated and the vehicle will still operate. Retrofit of existing vehicles is feasible.

These vehicles can show significant fuel savings in service operation, and brake life increases between two and four times have been experienced relative to vehicles without hydraulic regenerative braking.

Figure 7.10 illustrates the distance between stops for collecting trash, and shows the regenerative energy recovery during braking for a specific hydraulic

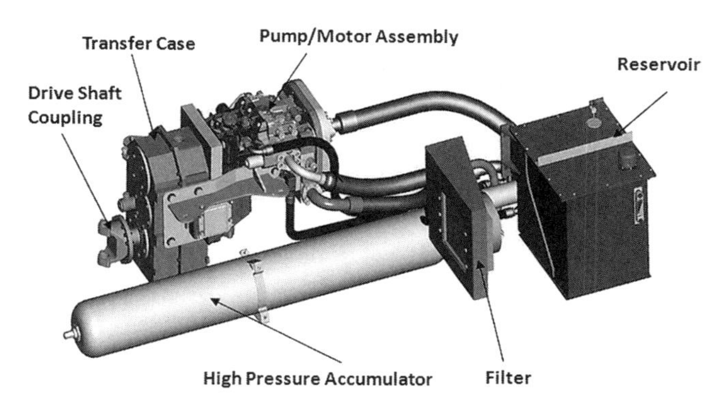

FIG. 7.8 Typical refuse truck system and installation—hydraulic hybrid power unit. (Courtesy of Eaton Corporation)

hybrid power unit design. At point 1 when the vehicle starts to brake, part of the kinetic energy is recovered and stored in the accumulator. When the accumulator is full at point 2, the kinetic energy can no longer be recovered, and the vehicle will continue to slow down to point 3. At point 4, the stored hydraulic energy can be used to drive a hydraulic motor to propel the vehicle, which will reduce the engine fuel consumption during the acceleration of the vehicle until the accumulator is fully discharged at point 5. Fuel savings as a function of collection distance with and without hydraulic regenerative braking for this specific design of hydraulic power unit is also shown.

Test data for the refuse truck shows a reduction of fuel consumption (gpm) from 15 to 30%. In a particular study, the parallel hydraulic hybrid system also provides 26% higher acceleration than the standard truck, which increases the truck productivity about 11% [7.5, 7.15].

FIG. 7.9 Typical vehicle installation. (Courtesy of Eaton Corporation)

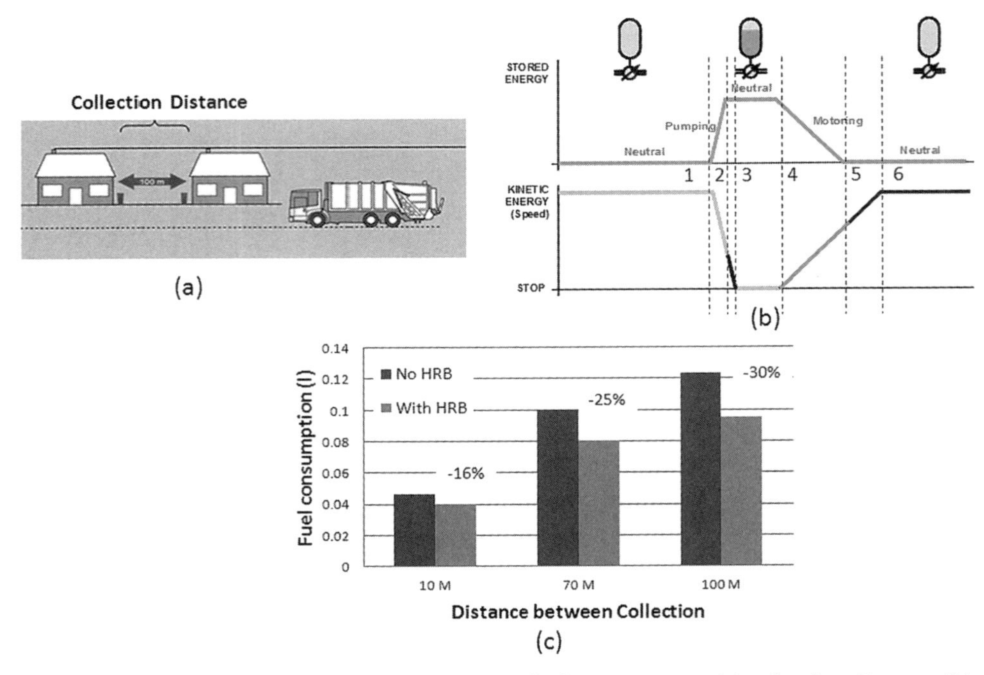

FIG. 7.10 Effect of distance between collections on fuel consumption: (a) collection distance; (b) braking kinetic energy recovery; (c) fuel saving with hydraulic regenerative braking.

One manufacturer is producing a two-mode system in which the vehicle gearbox is replaced by a new transmission that works as a hydrostatic unit at low vehicle speeds and transmits power mechanically at higher speeds. Regenerative braking is included with the hydrostatic operation.

Similar systems can be cost effective in some bus applications with appropriate drive cycles, particularly when government subsidies for electric systems are not available [7.6].

7.4.2 Package Delivery Vehicles

There are several examples of package delivery vehicles with series hydraulic hybrid drive systems on the road. Figure 7.11 shows the series hydraulic hybrid vehicle for UPS. The urban drive cycle is appropriate for energy recovery, and a combination of series hydraulic infinitely variable transmission, engine downsizing, regenerative braking, and engine idle reduction has been shown to have the capability to reduce fuel use by up to 50% [7.10]. Limited engine-off mobility in the warehouse space is available to limit exhaust pollution in the enclosed space. Engine-off operation requires operation of several accessories independently of the engine. This is also the case for electric systems, but with the hydraulic systems the possibility exists to drive some of these

FIG. 7.11 Series hydraulic hybrid delivery vehicle. (Photo taken at 2011 HTUF, Baltimore, Maryland)

with a hydraulic motor. This can give efficiency advantages and can reduce electrical load.

In addition to the ability of the system to capture kinetic energy that is normally lost to heat generated in the brake system during vehicle braking events, another important characteristic of the series hydraulic hybrid system is the ability to manage the power output of the engine independently of the driver power demand. Figure 7.12 illustrates the difference in the areas of primary operation for an engine in a conventional vehicle as compared to the optimized operation zone for a vehicle fitted with the series hybrid drive system [7.9]. In a traditional vehicle, the engine must operate over a wide operating range to meet the power demand of the driver while simultaneously reacting to the large step changes in effective gear ratio of the transmission. However, the series hydraulic hybrid system is able to operate the engine at an optimized operation zone to minimize fuel consumption due to the buffering effect of the hydraulic system. The hybrid energy control strategy simultaneously manages the power flow out of the engine and into the hydraulic system while monitoring the instantaneous amount of stored energy and the instantaneous amount of stored energy being released to meet the driver power demand. At the optimized operation zone, the vehicle fuel economy is maximized, and harmful emissions are minimized [7.10].

The parcel delivery application is well suited for the hydraulic hybrid system. A typical urban delivery route includes hundreds of stop/start cycles.

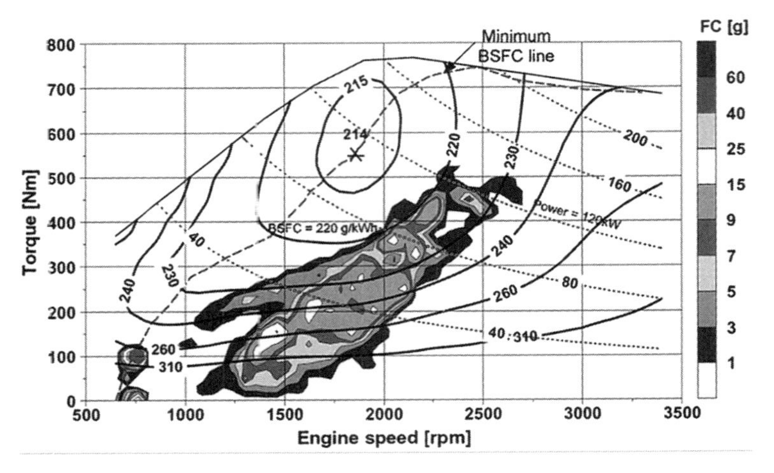

FIG. 7.12 Engine visitation points on the BSFC map for a conventional vehicle. Shading indicates the relative amount of fuel consumed at a given zone during FUDS.

Each stop/start event is typically quite short in duration and involves a large amount of energy flow. Furthermore, the distance traveled between each successive stop/start cycle is typically short. This property allows the vehicle to operate for a significant proportion of the time using only stored hydraulic energy captured during braking. A similar configuration in a shuttle bus can be expected to give similar savings.

7.4.3 Multimode Powersplit System in a Warehouse Vehicle

The multimode system features a split-power-flow hydromechanical multimode gearbox with added hydraulic power generation, as shown in Fig. 7.13 [7.8]. The power split system has one hydraulic path and one mechanical path in parallel. In the low vehicle speed range (0–65 km/h) the power is transferred hydraulically. From 65 to 100 km/h, direct mechanical drive is used with disconnected hydraulics. These two gear ratios ensure high efficiency at all speeds. The hydraulic recovering system can handle brake energy from 65–0 km/h.

The power split hydraulic hybrid has the benefits of improving fuel economy, enhancing driver handling, and improving productivity. This is achieved via a continuously variable transmission with three modes for different vehicle speed conditions. Ample hydraulic power is provided for load lifting, and the engine is always operating optimally.

Figure 7.14 shows the fuel savings of a power split hydraulic hybrid system. Warehouse vehicles are designed to carry pallets or other loads over a paved

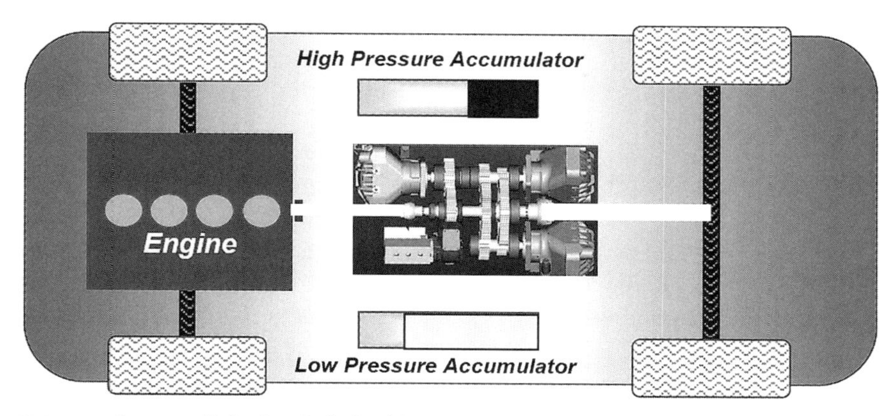

FIG. 7.13 Power split hydraulic hybrid [7.8].

floor, and to lift and retrieve the loads from elevated shelving or to stack loads. They generally use implement hydraulics for lifting and may have a mechanical or hydrostatic transmission for vehicle motion.

Because the floor is generally concrete, the rolling resistance is low, and therefore fuel economy can be improved by including regenerative instead of friction braking. The energy available from lowering of loads can also be recovered and reused to save fuel. With hydraulic regenerative braking, there is a 13% fuel savings. Additional 20% fuel savings was achieved when power split technology was applied.

Other applications are possible, such as for port vehicles used to transfer shipping containers from ships to trains or trucks for onward shipping. These vehicles also travel over hard surfaces at speeds that make regenerative braking worthwhile.

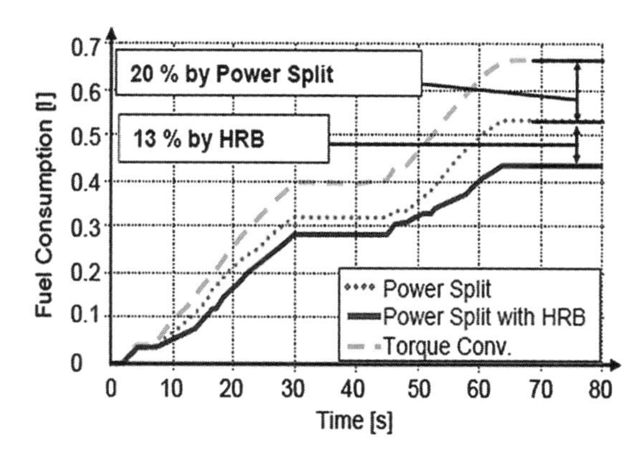

FIG. 7.14 Warehouse cycle—fuel use with and without regenerative braking [7.5].

7.5 References

7.1 Wendel, G. R., "Hydraulic Hybrid Vehicle System," Michigan Clean Fleet Conference, May 17, 2007.

7.2 "Mobile Hydraulics Manual—Your Comprehensive Guild to Mobile Hydraulics," Eaton Fluid Power Training, pp. 321–357, Eaton Corporation, 2006.

7.3 Technologies and Approaches to Reducing the Fuel Consumption of Medium- and Heavy-Duty Vehicles, The National Academies Press, 2010.

7.4 Conrad, M., "Hydraulic Hybrid Vehicle Technologies," Clean Technologies Forum, Sacramento, CA, September 09, 2008.

7.5 Baseley, S. et al., "Hydraulic Hybrid System for Commercial Vehicles," SAE Paper No. 2007-01-4150, SAE International, Warrendale, PA, 2007.

7.6 Rydberg, K-E, "Energy Efficient Hydraulic Hybrid Drives," The 11th Scandinavian International Conference on Fluid Power, SICFP'09, Linköping, Sweden, June 2–4, 2009.

7.7 Patil, C. et al., "Model-Based Approach to Estimate Fuel Savings from Series Hydraulic Hybrid Vehicle: Model Development and Validation," SAE Paper No. 2011-01-2274, SAE International, Warrendale, PA, 2011.

7.8 Surampudi, B. et al., "Design and Control Considerations for a Series Heavy Duty Hybrid Hydraulic Vehicle," SAE Paper No. 2009-01-2717, SAE International, Warrendale, PA, 2009.

7.9 Kim, Y. J. et al., "Simulation Study of a Series Hydraulic Hybrid Propulsion System for a Light Truck," SAE Paper No. 2007-01-4151, SAE International, Warrendale, PA, 2007.

7.10 Van Batavia, B., "Hydraulic Hybrid Vehicle Energy Management System," SAE Paper No. 2009-01-1772, SAE International, Warrendale, PA, 2009.

7.11 Schärlund, L., Hydraulic Hybrids, Parker PMDE Trollhättan. Presentation at IFS meeting in Eskilstuna, November 5, 2008.

7.12 Pourmovahed, A. et. al. "Modeling of a hydraulic energy regeneration system. Part I. Analytical treatment," Journal of Dynamic Systems, Measurement and Control, Transactions of the ASME, vol. 114, no. 1, pp. 155–159, 1992, ASME.

7.13 O'Brien II, J. et al., "Optimal Accumulator Sizing for Mobile Applications," SAE Paper No. 2011-01-0715, SAE International, Warrendale, PA, 2011.

7.14 Wu, B., et al., "Optimal Power Management for a Hydraulic Hybrid Delivery Truck," Vehicle System Dynamics, 2004, Vol. 42, Nos. 1–2, pp. 23–40.

7.15 "Eaton's HLA® system delivers 15–30% better fuel economy in many applications while reducing emissions," Hydraulic Launch Assist the Eaton HLA® System, Eaton Corporation, 2008.

7.16 Fussner, D. et al., "Analysis of a Hybrid Multi-Mode Hydro-mechanical Transmission," SAE Paper No. 2007-01-1455, SAE International, Warrendale, PA, 2007.

7.17 Li., P. Y., "Optimization and Control of a Hydro-Mechanical Transmission based Hybrid Hydraulic Passenger Vehicle," 7th Intentional Fluid Power Conference, Aachen, 2010.

7.18 Carl, B. et al., "Comparison of Operational Characteristics in Power Split Continuously Variable Transmissions," SAE Paper No. 2006-01-3468, SAE International, Warrendale, PA, 2006.
7.19 Elahinia, M. H. et al, "Noise and Vibration Control in Hydraulic Hybrid Vehicles," SAE Paper No. 2006-01-1970, SAE International, Warrendale, PA, 2006.
7.20 Holland, M.A. et al., "Electrically Controlled Fixed-Displacement Pump, Variable-Displacement Motor Hydrostatic Transmission," SAE Paper No. 2006-01-3469, SAE International, Warrendale, PA, 2006.

Chapter 8

Hydraulic Hybrid Components and Controls

8.1 Introduction

Components used in hydraulic hybrid vehicles are generally derived from equipment used in off-highway and industrial fluid power systems. Vehicle drive systems demand high efficiency and robust, lightweight components, and equipment can be specified to provide these attributes. As applications become more numerous, we can expect investment in even lighter, more compact, and more efficient component designs.

8.2 Hydraulic Pumps and Motors

Pumps convert shaft power to hydraulic flow at pressure, and motors convert hydraulic flow at pressure back into shaft power. (Hydraulic power = pressure × flow). The operating principles of commonly used pump types have a long history; some have been seen in Greek and Roman times [8.1]. Early machines were conceived for use with water for irrigation and mine drainage.

There are two main classes of pumps and motors—rotodynamic and positive displacement. Rotodynamic machines convert velocity to pressure within a rotor and stator assembly (and vice versa). Examples are engine coolant pumps and water turbines. In general, pressure ratio is low and efficiency is mediocre, and application to hydraulic drive systems is limited to possible use for charge systems to pressurize positive-displacement pump inlets. This type of machine will not be discussed further.

Positive-displacement machines admit fluid into chambers at one pressure and then eject the fluid at a different pressure as the mechanical shaft rotates. The inlet and outlet pressures are sealed from each other by features of the mechanism. If internal leakage is ignored, then hydraulic flow is independent of pressure and proportional to shaft speed × volumetric displacement (number of chambers × volume change of each chamber between inlet and outlet).

There are several types of positive-displacement machines, and in general they are compact, and some can be efficient enough for use in hydraulic drive systems for vehicles. The availability of hydraulic oils starting in the 19th Century spurred considerable development of positive-displacement pumps and motors and led to the development of the hydraulic power industry as we know it today.

In general, operation is reversible, so pumps can operate as motors and sometimes in both directions of rotation. There are some minor changes for optimization of noise and performance, but a given design can be configured by the manufacturer as a pump, motor, or pump/motor. For a hybrid vehicle, the drive motor must be configured as a pump/motor to facilitate regenerative braking and must also be capable of rotation in both directions for vehicle reverse [8.2, 8.3].

8.2.1 Gear Pumps and Motors

There are two basic types of gear units with either external or internal meshing gears. In both types, the meshing gears are contained in a housing; the pumping chambers are between the gear teeth and are sealed by the gear meshing and the fit of the gear in the housing.

A static port plate (or seal plate) at one or both ends of the rotor with kidney-shaped passages is used to admit or reject oil from the chambers and also to seal the chambers axially. The port plates are designed to maintain close clearances with the rotating gears under all operating conditions, including transients. For high-pressure operation, this involves using floating plates with a pressure balance method using static seals in the housing ends. In some designs of external gear unit, there are also additional oil passages in the housing to assist filling with low-pressure fluid. Gear tooth forms are designed to minimize dead volume when meshed, as this minimizes hydraulic noise. Accuracy of manufacture is essential to good performance and long life. In general, gear pumps and motors have a fixed capacity and thus are not suitable for vehicle propulsion systems. Gear motors are very compact and low cost and can be used to drive vehicle accessories.

8.2.1.1 External Gear Units

The two meshing gears rotate in overlapping circular cavities in the housing. Each gear is supported by shaft bearings. The operation of a typical external gear pump is shown in Fig. 8.1. The external radius of each gear seals against the housing. The gear teeth are generally of involute form. One of the two gears, called the drive gear, will be connected to the drive shaft; the other, called the idler gear, is driven by the drive gear. A gear pump carries oil from the inlet

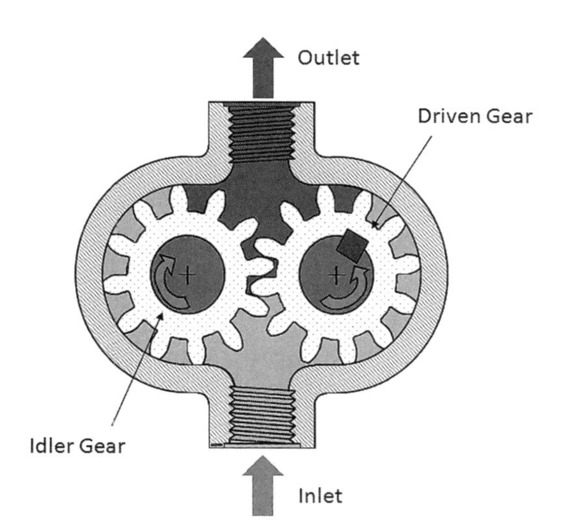

FIG. 8.1 Cross sections of an external gear pump/motor.

to the outlet in the space between gear teeth. As the gear teeth un-mesh at the bottom (Fig. 8.1), a partial vacuum is created, allowing fluid into the spaces between the teeth. As the gears rotate, the fluid is carried around the outlet of the top of Fig. 8.1. The fluid is expelled to the outlet from the spaces between the teeth as the gears mesh.

The constant volume of the pumping chambers between inlet and outlet gives rise to additional hydraulic noise due to compressibility of the oil. Capacity is defined by the following equation where the two gears have the same number of teeth:

$$V = m \times z \times b \times h \times \pi \tag{8.1}$$

where: V is volume flow per revolution
m is a modulus of tooth form (typically 0.5)
z is the number of teeth per gear
b is width of the gears
h is the height of the gear tooth

External gear pumps are inherently unbalanced. The outlet pressure and the force generated from the transmission of the power from the drive gear to the idler gear are unbalanced and must be borne by the shafts and bearings. Despite these drawbacks, gear pumps are very popular due to their simplicity and robustness [8.3].

8.2.1.2 Internal Gear Units

With this type of machine, the two gears are of different sizes, and the smaller gear sits inside the larger gear, as shown in Fig. 8.2. There is only one circular cavity in the housing, which is eccentric to the bearing supporting the inner gear. The outer (larger) gear rotates in the housing and does not require a separate shaft or bearing. A crescent-shaped sealing member is fitted in the area where the gears no longer mesh to seal the pumping chambers, and this component may feature pressure compensation features to ensure good sealing under all operating conditions.

The size and number of spaces between teeth determine the displacement. The internal gear pump is similar to the external gear pump in terms of wear compensation and unbalanced loading. However, these designs tend to be quieter. Due to the crescent and the internal gear, the manufacture is slightly more difficult than for an external gear pump.

Figure 8.3 shows a special case of an internal gear unit for which the smaller gear has one less tooth than the outer. These are generally known as "gerotor" or "ring gear" units, and here the crescent seal is not required as the tooth forms are designed to remain in contact and therefore seal the chambers as they rotate. Usually the outer gear teeth have a semicircular form, and this enables the inner teeth to be formed on a gear-cutting machine (although many are now produced in powder metal).

8.2.2 Vane Pumps and Motors

Figure 8.4 illustrates a simple vane pump, which features vanes fitting tightly in slots in the pump rotor. The vane is pushed outward against the static housing or stator by centrifugal force assisted by pressure under the vane and/or

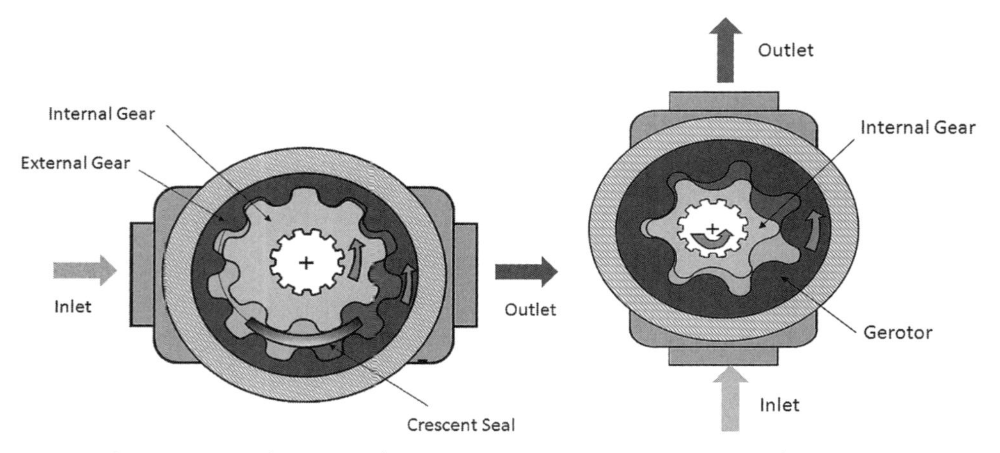

FIG. 8.2 Cross sections of an internal gear pump/motor.

FIG. 8.3 Diagram of a gerotor pumping element.

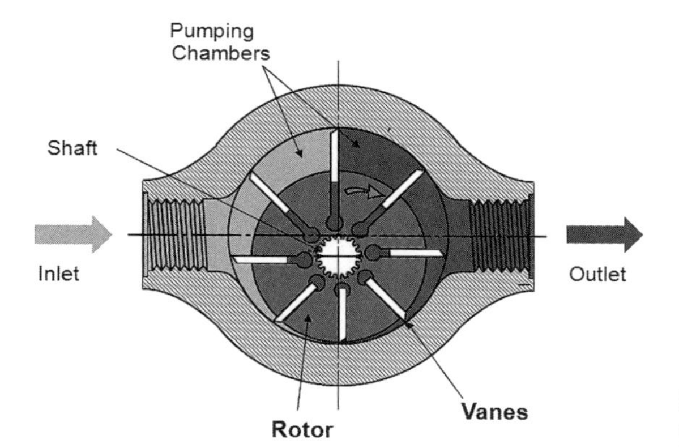

FIG. 8.4 Diagram of a vane pumping element.

spring force. Pumping chambers are formed between two adjacent vanes, with the rotor, cam ring, and two side plates forming the other walls of the chambers. Because the cam ring is eccentric, the chamber changes size with rotor rotation to produce pumping action. Axial sealing and transfer of fluid takes place though port plates/seal plates at the ends of the rotor, which are similar in operation to those used in internal gear pumps.

Another configuration features a noncircular form in the housing, giving two pumping actions each revolution. This has the advantage of reduced side loads on the rotor bearings but requires more vanes.

The vanes are held against the cam ring surface by centrifugal force. The spinning of the rotor causes a centrifugal force, which pushes the vanes out of the rotor slots. Effective sealing at the vane-cam ring surface requires that the rotor be turning at a certain minimum speed, depending upon the pump design and the operating conditions. However, since centrifugal force is usually insufficient to overcome friction in the vane slots and high pressure at the outer tip of the vane on the outlet side, pressurized oil is also supplied under the vanes. Figure 8.5 shows a high-performance pump in which intra-vanes, or small inserts, are used in the vanes. Full outlet pressure is fed into the intravane area continuously to generate enough force to hold the vane tip against the cam ring at all times. Meanwhile, the surface pressure is applied to the underside of the vane.

Vane pumps and motors have low pressure fluctuation but are not efficient enough for use in hydraulic drive systems for vehicles. They can be used for accessory drives and auxiliary functions.

8.2.3 Piston Pumps and Motors

Piston units function in the same way as an internal combustion engine, with pistons oscillating within cylinders and valves operating to admit and discharge fluid, usually hydraulic oil. Pistons and cylinders can be designed to

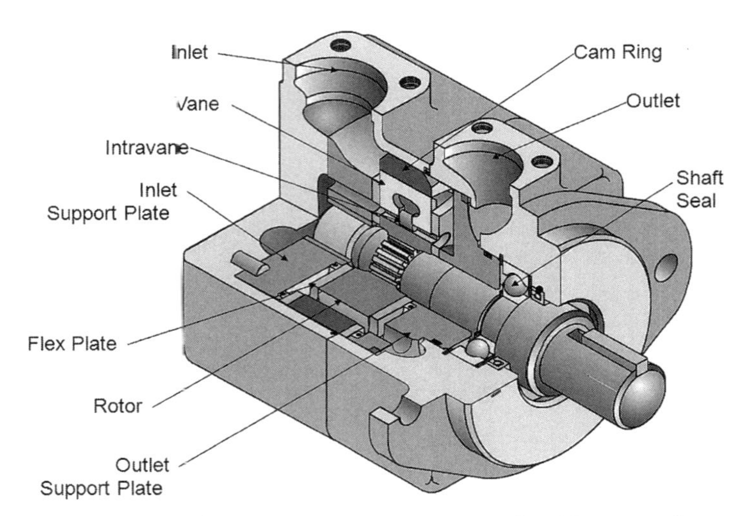

FIG. 8.5 High-performance vane pump construction. (Courtesy of Eaton Corporation)

operate with small clearances and can be manufactured with high precision so that leakage of high-pressure fluid is very low. As a result, for pressures of 200 bar and above, these pump types are the mainstays of hydraulic power systems for industrial and mobile applications.

There are two principal types, with piston movement primarily parallel to the axis of shaft rotation (Axial Piston Pump/Motor) or at right angles to the axis (Radial Piston Pump/Motor). Both types are being considered for use in hydraulic drive systems for vehicles.

The pumping elements are generally surrounded by low-pressure oil, which provides lubrication and temperature stability. Leakage from high pressure through small clearances around the pistons and seals flows into this space. It is usual practice to drain this oil through a separate small pipe to the reservoir which is called a "case drain." The space may also be connected to the pump inlet, but the space is then subject to pressure fluctuations at the inlet. Note that the shaft seal usually sees this case pressure.

8.2.4 Axial Piston Pumps/Motors

There are two principal types of axial piston machines in general use: the "swash plate" type and the "bent axis" design. Both are produced in large quantity for industrial, marine, and mobile applications in a range of sizes giving maximum outputs from about 20 cc/rev to more than 1000 cc/rev for use in high-pressure hydraulic systems. Both types can be produced as pumps, motors. or pump/motors with fixed or variable capacity. For current hydraulic hybrid vehicle applications, variable-capacity pump/motor units which can operate in both directions of rotation are specified.

8.2.4.1 *Swash Plate Pumps/Motors*

The in-line piston pump is the most common design. Figure 8.6 shows one design of such a pump. The cylinder assembly rotates coaxially with the driveshaft. Pumping action takes place because the pistons are constrained to move against a plate held at an angle to the axis of rotation (the swash plate). Construction of a typical unit is illustrated in Fig. 8.6.

The cylinder block, sometimes called the cylinder barrel, is driven by the shaft via splines and is pushed axially against a valve plate (sometimes called a port plate), where flow transfer takes place and which provides sealing of high pressure. The valve plate contains ports connected to high- and low-pressure passages within the pump. The edges of these ports are constrained by the structure to coincide with the extremes of piston motion, to allow smooth inlet and egress of fluid.

The port plate and cylinder block may have a mating spherical surface. This enhances stability of location and is often accompanied with a slight inward angling of the cylinder, which reduces the leakage area and friction loss at the cylinder block/port plate interface. In this arrangement, the valve plate is often called a "lens plate" because of its shape.

The swash plate has static and rotating components. The rotating part retains the individual piston ends and slides against the static swash plate as the pistons rotate (driven by the cylinder block). The static part constrains the motion of the rotating parts to drive the pistons to carry out the pumping action. For variable capacity, the angle of the static part with respect to the axis of rotation is varied by a mechanism. The swash plate actuation mechanism is usually one or more small hydraulic cylinder actuators. An angle of 90° to the shaft axis results in no pumping action. With this type of pump, an angular change up to 20° (plus or minus) gives maximum output with acceptable

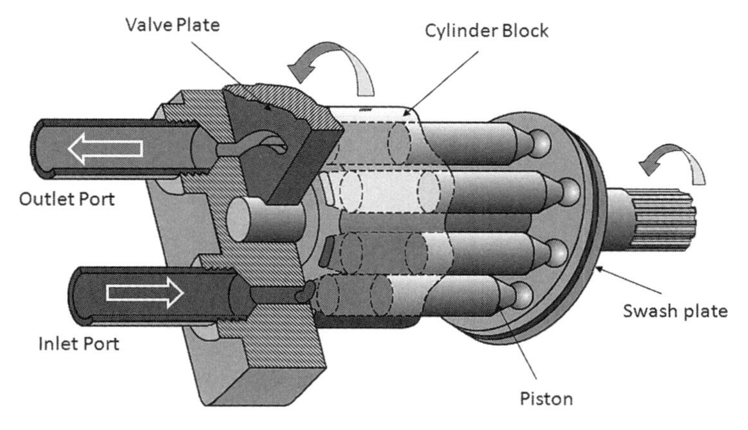

FIG. 8.6 Sectional view of a swash plate pump/motor.

geometry. Angular movement in both directions permits pumping and motoring and reverse direction of rotation without the need for a reversing gear.

The pump shaft can be driven from either end, or indeed several pump units can be connected together in tandem or other accessory components driven from the non-drive end. In these cases, an additional shaft seal is required.

These units are often used in hydrostatic drive systems and are currently deployed in several parallel hydraulic hybrid vehicle applications. They are available in a large range of sizes and are affordable and robust, and have good but not exceptional efficiency.

8.2.4.2 *Bent-axis Pumps/Motors*

With this type of machine, the cylinder block is constrained to rotate at an angle with respect to the driveshaft axis, as shown in Fig. 8.7. The piston rod ends are spherical and are attached to the driveshaft flange sockets by a retention plate, and are forced in and out of their bores as the distance between the driveshaft flange and cylinder block changes. As the shaft rotates, the piston at the bottom starts to retract to intake the fluid, while the one at the top is fully back in the bore ready to discharge the fluid. Side load from the pistons actually drives the cylinder block rotation. Alignment is ensured by the central dummy cylinder or pintle.

Many bent-axis machines are used today, with a fixed capacity determined by a fixed angle built into the structure. However these are not generally suitable for vehicle hybrid systems because of the range of speed and flow required, so mechanisms have been designed to change the cylinder block axis angle. The complication of this is that the port plate and the inlet and outlet flow passages must also move, and that results in additional leakage paths and high actuation loads.

The geometry of motion during rotation in this type of pump does not allow single-piece cylindrical pistons to be used, and therefore either the piston

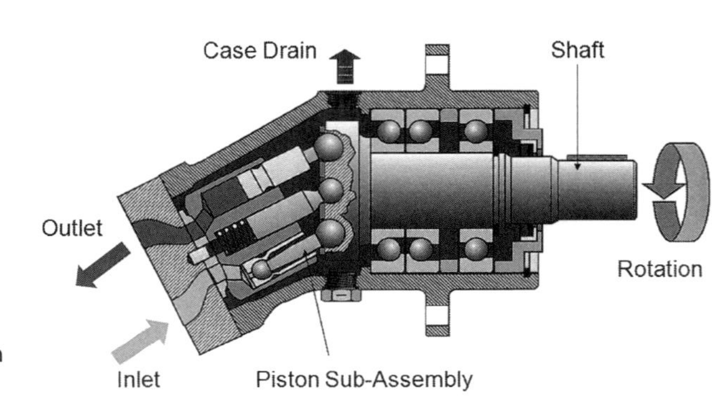

FIG. 8.7 Cross section of a bent-axis unit.

has a spherical end and line contact with the cylinder bore, or a separate cylindrical piston with a spherical joint is used. However, detailed analysis of geometry and loads shows that much larger axis angles than swash angles can be achieved, which results in very compact rotating elements with this type of unit compared with the swash plate pump type. Axis angles in excess of 45° have been demonstrated [8.6]. However, for variable-capacity units, a large part of the component envelope is taken up by enclosed space for angular movement and the necessary actuation mechanism.

Most variable-capacity bent-axis machines vary axis angle in one direction only. This means that the inlet and outlet passages have to be switched by an external valve to enable pumping, motoring, and reversing in a single unit, and this has pressure loss and structural implications. One manufacturer has a design with bi-directional axis movement to eliminate this issue (at the expense of additional envelope volume).

With careful attention to design detail for low friction and leakage, bent-axis machines have been shown to give exceptional efficiency, particularly at high and moderate axis angles. Figure 8.8 shows a variable-capacity bent-axle unit used for a hydraulic hybrid. The ability to operate at high axis angle enables the unit operating as a motor to give high torque for vehicle launch while keeping size small and efficient at higher speeds and flows [8.4].

It can be seen that the pumping elements are very compact, but the size of the unit is mostly made up of space for axis movement, and by the valve and actuation system. It should also be mentioned that the fluid flow passages to

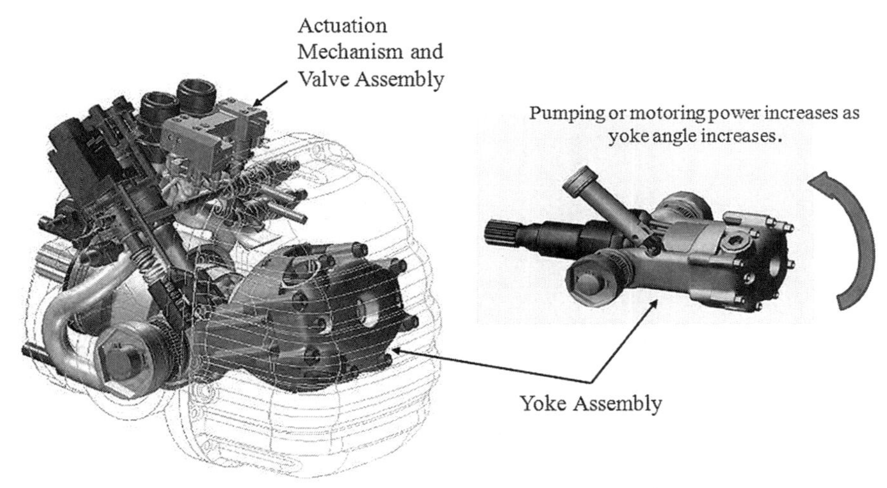

FIG. 8.8 View of a variable-capacity bent-axis unit used for a hydraulic hybrid. (Courtesy of U.S. EPA)

the valve plate that connect to the structure through the axis yolk are somewhat constricted, and this means that a low-pressure reservoir should have a pressure of at least 10 bar to ensure adequate filling of the pumping elements without cavitation (see Section 8.2.6.2).

8.2.5 Radial Piston Pumps/Motors

Radial piston machines have been produced with static cylinders and pistons reciprocated by a cam or eccentric mounted on the driveshaft, or with a rotating cylinder block with piston motion produced by a static cam located in the housing.

8.2.5.1 Fixed-capacity Radial Piston Unit with Static Pistons

Figure 8.9 shows a cross section of a static cylinder design [8.11]. An eccentric cam is mounted on the shaft and moves the pistons in and out to move fluid during rotation of the shaft. A spring inside each piston ensures good sliding contact with the eccentric. The pistons move within the non-rotating cylinders, which tilt slightly against a spherical seat in the housing to enable conforming contact by the piston end with the eccentric. Passages through the housing lead to a rotating valve plate for transfer of high- and low-pressure flow. With a fixed eccentric, the flow capacity of this pump is fixed. These units are robust, axially compact, and have excellent efficiency; they are used to drive wheels and drums at moderate speeds.

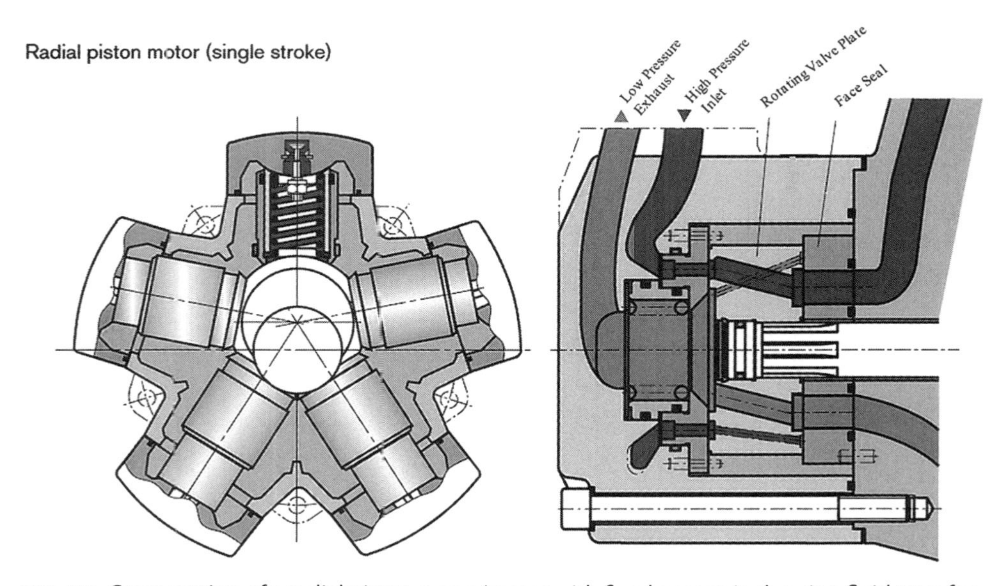

FIG. 8.9 Cross section of a radial piston pump/motor with fixed eccentric showing fluid transfer components. (Courtesy of Bosch Rexroth, The Hydraulic Trainer Vol. 1)

8.2.5.2 *Variable-displacement Radial Piston Machines*

A variant of the above design has been proposed for which each individual cylinder is served by two solenoid-operated valves, one for high pressure and one for low pressure, as shown in Fig. 8.10. With each valve actuation electronically controlled, it is possible to operate with any number of cylinders open to low pressure and not pumping or motoring. It is possible with averaging over several revolutions to achieve fully variable output in this way by choosing the number of "active cylinders" electronically. This concept is also described as a "digital displacement" machine [8.5].

With the seal plate no longer required and with careful design and precise manufacture, this arrangement gives exceptional efficiency, particularly at low flow fractions as sources of friction and leakage can be minimized. Precise timing of high-pressure valve operation enables pressure peaks to be virtually eliminated, which results in very low levels of hydraulic noise.

A digital encoder is fitted to the driveshaft within the unit to provide timing information to the pump controller. Speed of response to flow changes is very rapid, and independent valve action enables the unit to operate as a pump or motor, and to operate at idle (zero flow) and to reverse without additional valves or features. Several radial units may be mounted along the driveshaft in tandem to increase flow capacity required for a particular application [8.5, 8.7].

Specification of the required control elements presents challenges, as large numbers of solenoid drivers are required which must be precisely timed, and this may limit achievable maximum rotational speed. However, the ability to time valve opening for a given speed, pressure, and even temperature is the

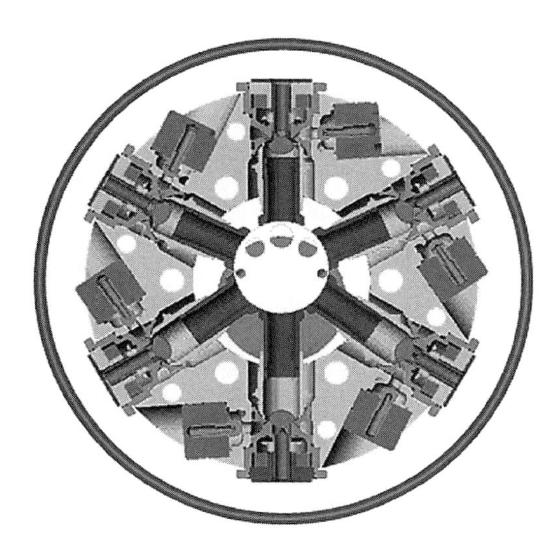

FIG. 8.10 Diagram of a static cylinder radial piston machine with valves at each cylinder [8.5].

key to achieving exceptionally high efficiency and very low pressure fluctuation (and hence low noise).

Traditional test methods for performance measurement do not take into account electrical load, and this is a source of additional loss for this pump/motor in a system.

8.2.5.3 Radial Piston Motors with External Cam

Radial piston motors with an external cam are similar in design to other radial piston motors, and as with radial piston motors, they are generally used in low-speed, high-torque applications. Figure 8.11 shows a cutaway of an external cam type of hydraulic motor. The main components of this type of motor include a cam plate, pistons, piston rollers, a rotor, valves, and an output shaft.

The static cam has multiple lobes, so there are many piston strokes per revolution. Pistons are reciprocated by rollers retained in the rotating housing, which are pushed by the cam form. A static seal plate around the driveshaft enables fluid transfer to and from the rotating cylinder block (and hence pistons) for motoring action. Again, this is a fixed-capacity machine, although there is a variant with a special seal plate with passages that permit half the pistons to be de-activated and hence permit operation with half flow. The output torque of the motor is created by pressure acting against the piston, which rolls along the profile of the cam plate. These machines deliver very high torque at low speeds and have applications as wheel motors in earth moving equipment and in rolling mills.

8.2.6 Considerations of Pump/Motor Performance and Noise

Component performance equations have been developed in Section 7.3.1.2 for use in vehicle system design. The following sections deal with pump/motor performance from a component perspective.

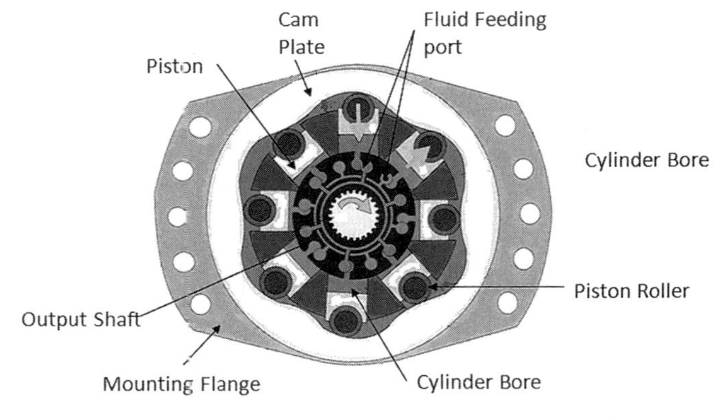

FIG. 8.11 Section showing rotating group of an external cam type of hydraulic motor.

8.2.6.1 *Pump and Motor Efficiency*

The power required to drive a pump of specific delivery and pressure rise, and the heat rejected into the fluid during the pumping process, are functions of pump efficiency, which can be defined as a ratio between the recoverable energy added to the working fluid by the pump and the mechanical energy supplied to the driving shaft. Usually, the pump efficiency is considered to be the product of Volumetric Efficiency (leakage from high pressure) and Mechanical Efficiency (torque losses, friction, etc.).

The volumetric efficiency η_v of a positive-displacement pump is defined as the ratio of the actual delivery to the displacement or the theoretical delivery:

$$\eta_v = \frac{Q_A}{Q_T} = \frac{(Q_T - Q_L)}{Q_T} = 1 - \frac{Q_L}{Q_T} \tag{8.2}$$

where: Q_A is the actual delivery

Q_T is the theoretical delivery

Q_L is the internal leakage

Leakage of pressurized oil occurs internally from the pumping mechanism and includes

- Leakage of oil through fixed clearances
  - Moving surfaces such as piston/cylinder
  - Joints
- Leakage of oil through hydrodynamic sliding surfaces
  - Valve plates

Leakage is dependent on pressure, local viscosity, and geometry. Flow loss from actuation of the variable capacity mechanism is generally not included in manufacturer's data and can be significant in practice.

The mechanical efficiency η_m (also called torque efficiency) is defined as the theoretical torque T_p required to drive a pump at a particular speed N, pressure rise ΔP, and delivery Q_A divided by the actual torque T_A as:

$$\eta_m = \frac{T_P}{T_A} = \frac{T_P}{(T_P + T_F)} \tag{8.3}$$

where T_F is the friction torque.

The torque losses due to friction include:

Piston/cylinder movement (this is increased by side loading)

Rotary movement:

Face seals and cylindrical containment

Bearings (also increased by side loading)

Torque losses depend on speed, oil viscosity, and pressure. Other losses include:

Pressure losses:

Excess pressure peaks above delivery pressure occur due to operation of valves and valve plates (energy not recovered).

Churning loss:

Torque consumed in motion of fluid within the pump mechanism.

All of the losses described above apply to both pumps and motors. The efficiency ratios are reversed for motors, and for high-efficiency units, this results in efficiency numbers being slightly higher for pumps than for motors of the same design (see Section 7.3.1.2).

Figure 8.12 shows typical efficiency curves for a high-performance, variable-capacity axial piston pump at full flow and 50% flow. Partial flow performance is very important for hybrid system operation, and pump development is focused on ensuring high efficiency at reduced flow fractions.

8.2.6.2 *Cavitation*

Cavitation damage to pump components occurs when local static pressure in the oil drops below about 0.5 bar absolute. Under these conditions, vapor bubbles form adjacent to component surfaces and collapse violently, causing damage due to high surface stresses in the material, which allow small particles to be lost from the surface.

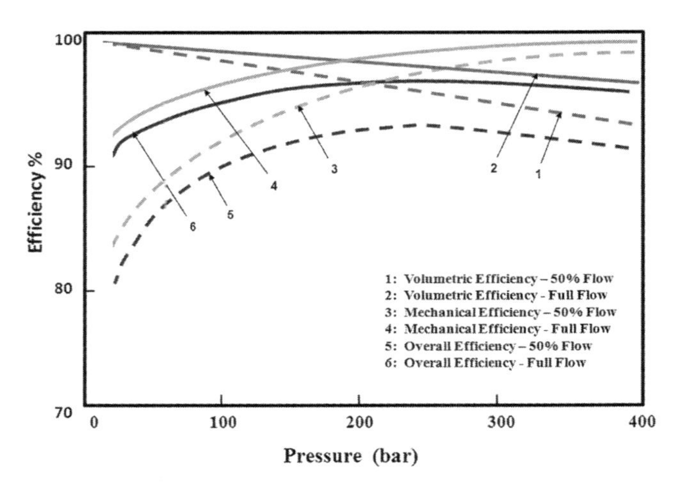

FIG. 8.12 Pump efficiency vs. pressure for an axial piston pump.

Components affected include port plates, cylinders, and some valve surfaces where a combination of high velocity and low inlet pressure occurs. Low absolute pressures can occur due to pressure drops in the pipe from the reservoir to the pump (worse when cold due to high fluid viscosity) and due to transient flow increases, such as when a pumping chamber is expanding. These effects are speed dependent, as both pressure drop and fluid velocity are related to fluid flow, and cavitation concerns generally determine the maximum robust operating speed for a pump in a particular application.

Closed systems are less susceptible as the low-pressure vessel is pressurized. Cavitation concerns are also less for motors, as the motor inlet is at high pressure, and this may permit higher safe operational speeds for motors.

8.2.6.3 Pump Outlet Flow

A positive-displacement pump sucks in a volume of fluid each revolution, determined by the pumping chamber active volume. The output volume is essentially the same minus the leakage.

Theoretical flow is therefore the product of pump capacity (cc/rev or liters/100 RPM) and speed (RPM) divided by 100 (liters/min).

Practically, slightly less fluid is admitted due to pressure losses and vapor content. This is not significant until the static pressure begins to approach the vapor pressure of the oil. At this point cavitation begins, and much less oil is admitted. The speed at which this occurs sets the maximum safe operating RPM for the pump. The cavitation speed is lower at higher oil viscosity (cold) and higher if the inlet oil is "charged" to a higher pressure.

8.2.6.4 Compressibility Considerations

Hydraulic oil is generally considered to be incompressible, but there are some compressibility effects at 300 to 400 bar. The effect is small and due to adiabatic compression/expansion, which causes efficiency levels defined in this section to be overstated by about 1% [8.6].

There is a compressibility loss due to high-pressure oil remaining in "dead volume" after the pressure stroke expanding to low pressure as the valve plate opens the cylinder to inlet. This loss is really a volumetric loss but is included in "mechanical efficiency" as a result of test measurement protocols. Measures to reduce this loss include minimizing "dead volume" and adjusting valve timing (opening after "top dead center") so that compression energy is recovered.

These methods are not very effective in variable-capacity seal plate machines, particularly if they are used as both pump and motor and reversed, as are most hybrid drive motors. With the variable radial piston machine, compressibility effects can be almost eliminated with small dead volume and precise valve timing.

8.2.6.5 Noise Considerations

Hydraulic drive system noise is caused mainly by pressure and flow fluctuations in the working fluid. These fluctuations are generated mostly in the pumps and motors. Unbalanced mechanical force fluctuations can also contribute, as can driveshaft torque fluctuation. Noise is generated by vibration of system components and structures in response to loads from the fluid pressure, force, and torque fluctuations and has been discussed in Section 7.3.4 in a system context.

Flow fluctuation results from the geometry of the rotating elements in pumps and motors and generally occurs at a frequency corresponding to shaft speed multiplied by the number of rotating pumping chambers (typically pistons or gear teeth). This frequency is referred to as the fundamental frequency of the pump or motor. Flow fluctuations due to geometric design are generally close to sinusoidal. Flow fluctuation becomes pressure fluctuation because of the inertia of the fluid volume moving around the system, and it is the resultant pressure fluctuation that generates loads that excite system components and structures. Piston units generally have six or more pistons, and flow fluctuations can be low with careful element design.

Of greater concerns are the pressure fluctuations due to fluid compressibility and valve timing. These occur at multiples of the fundamental frequency (harmonics) because they are due to short duration "spikes" corresponding to opening of individual pumping chambers to the high-pressure passage of the pump or motor. The combination of harmonics gives rise to the characteristic whine noise heard from some hydraulic systems. Pressure spikes may be above or below the prevailing average fluid pressure, and the perturbations propagate sonically through the fluid system. Both are high-frequency noise generators.

Valve opening occurs as the rotating chamber of the axial piston unit becomes exposed to the opening in the port plate. At this time, the cylinder pressure can be above or below the system pressure, and a pressure spike occurs. The pressure in the cylinder is a function of geometry, leakage, rotor speed, and compressibility of the fluid. For a variable-capacity axial piston unit, it is theoretically possible to eliminate pressure spikes by precise geometry for one operating condition, but not for the wide range of speeds, capacities, and pressures encountered during vehicle operation, particularly if the unit operates in both directions, and as a pump and as a motor.

Measures to reduce spikes include provision of small grooves in the port edges to allow some measure of pressure equalization at the cost of some increase in leakage. It is also important to have careful control of component location accuracy.

The variable radial piston machine described in Section 8.2.6.2 has valves on each cylinder with timing controlled electronically. Here it is possible to program the valve opening so that it occurs when pressure is equalized over a wide range of speeds and pressures, and virtually eliminate pressure spikes. Systems using these machines have demonstrated very quiet operation in vehicles.

8.3 Valves

Robust and well-engineered valve components are available from suppliers to the transportation and industrial fluid power market. Valves can be specified for flows, pressures, and speed of response appropriate for vehicle hydraulic hybrid systems. Valve manufacturers have developed components that perform stably and reliably over a wide range of fluid viscosity, flow velocity, and control parameters. Some minor reconfiguration may be appropriate for on-highway use when requirements differ and production volumes are larger.

Valve functions can be combined into manifolds, or valves can be installed into other system components. Valves can have automatic operation (e.g., non-return or pressure relief valves), operate manually or with solenoids, or have servo operation under electronic control.

Valve manifolds are cast or machined assemblies containing multiple valves and the appropriate fluid connection passages. This can considerably simplify the installation and reduce the number of pipes and hoses.

There is some industry standardization. "Cartridge valve" components are designed to fit into standardized cavities machined into manifolds or other assemblies. For automotive use (high production volume) more integrated valve assemblies would be designed.

8.3.1 Automatic Valves

Non-return valves and pressure relief valves are used to ensure correct system operation and basic safety, particularly when faults occur. A schematic of a pressure relief valve is shown in Fig. 8.13. The valve operates completely automatically, for example where flow tries to reverse direction or a pressure is exceeded. Additional lines may be required to return relief flow to the fluid reservoir. It is important that they operate predictably, without oscillation, and a number of design alternatives are available to achieve this in certain installations. Care is required to specify possible transient pressure peaks to ensure that a robust enough valve is selected.

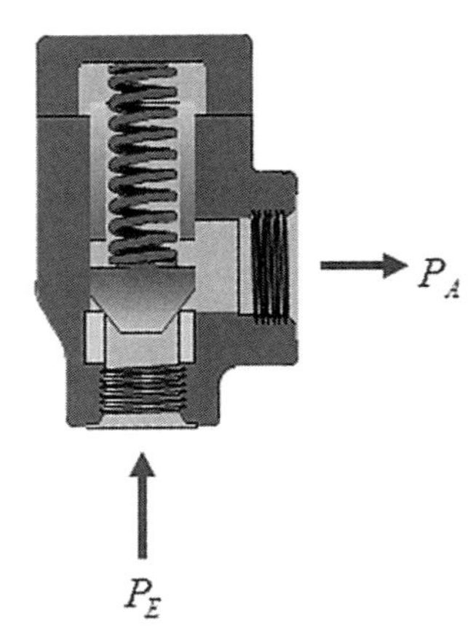

FIG. 8.13 Schematic of pressure relief valve.

A more-complex example of an automatic valve is the so-called "flow fuse," which is designed to isolate the fluid passage when flow exceeds a certain value, for instance if a line ruptures. This operates by activating shut-off when pressure drop ($\Delta p \propto Flow^2$) reaches a predetermined value.

8.3.2 Isolation Valves

These valves are used to open or close fluid passages according to the system requirements. A hand- or wrench-operated valve may be used, for example to isolate part of the system for safety or to minimize fluid loss during maintenance. A solenoid-operated valve may be specified where a routine control function is required, for example to isolate the accumulator. The required time constant and transient overpressure should be specified during component selection. There should be minimal pressure drop in these valves in the open position. These units may also give status feedback to the system control unit, and may be designed to fail closed or open when power is lost.

8.3.3 Control Valves

These valves are used to modify flow or pressure by varying the flow area in a fluid passage. There are many designs with operation to give constant flow, pressure, or power, and examples of these may appear in pump control servomechanisms. However, for precise control of hybrid system components, the proportional control valve with electronic signaling and feedback to achieve target value is used most widely.

The proportional control valve facilitates the achievement of a robust, power-dense hydraulic drive. The valves are not used to throttle the hydrostatic drive system primary fluid flow, as this would give rise to unacceptable energy loss. The primary use is to control pump and motor flow capacity to achieve efficient and smooth power transfer to satisfy driver demand.

These valves are, therefore, built into the swash plate or bent-axis pump/motor mechanisms and are used to feed oil to actuators that adjust the angles to determine machine output. There is usually angular or positional feedback from the pump mechanism to the control unit to improve precision and reduce hysteresis.

The required control, including transient response, for industrial and off-highway pump systems is usually provided by a discrete electronic control unit, but for vehicle hybrid systems the required functionality is built into the hybrid system controller, which operates all of the controllable system components.

8.4 Pipes and Fittings

Mobile and industrial suppliers have well-engineered ranges of hydraulic hose, pipe, and connector products. Fabrication methods for hose/pipe assemblies are well established, and fabrication equipment is readily available for workshop use. Supplier regional distributors and agents also have facilities for fabricating and testing assemblies.

Hoses are made from synthetic elastomers reinforced with polymer or steel fibers. Pipes and fittings are made from steel, which may be stainless or coated for corrosion protection. Pipe ends are normally flared to locate threaded end fittings which incorporate seals. Sometimes welded connectors are specified.

8.4.1 High-pressure Lines

Pipes and hoses suitable for use at maximum pressures of 275 bar, 345 bar, and 414 bar (4000, 5000, and 6000 psi) are available from catalogs. Equipment for higher pressures still is becoming generally available. Hose is specified by internal diameter, and sizes appropriate to vehicle hydraulic hybrid systems are readily available.

Hoses are specified by internal diameter, but pipes are specified by a "nominal" internal diameter. This is correct for the commonly used metal thickness, but because common dies are used in manufacture, the internal diameter will change slightly when other material thickness is specified. Tables are available from pipe manufacturers (as shown in Section 7.3.3).

Assemblies are usually pressure tested to 1.5 times the maximum working pressure, and samples should be pulse tested to demonstrate process consistency. For high-pressure lines in hydraulic hybrid systems, a maximum flow velocity of 6–9 m/s (20–30 ft/s) is recommended to minimize pressure losses.

8.4.2 Low-pressure/Return Lines

Maximum pressures of 7–28 bar (100 to 400 psi) are used, depending on whether a pressurized reservoir is used or if transient pressure peaks are expected.

For low-pressure pipe/hose assemblies, the following maximum fluid velocities are recommended:

Pressurized return line—3 m/s (10 ft/s)

Suction line (non pressurized)—1.2 m/s (4 ft/s)

The lower velocities indicated are to ensure that pressure drops are very low to prevent cavitation in the fluid at entry to the pump or valve components, particularly under cold conditions at which oil viscosity is high. It follows that particular attention should be paid to avoid sharp bends and restrictions in these lines, and low pressure lines should be as short as possible.

It is important to indicate that under some transient and cold ambient conditions, it may be possible to experience pressures below atmospheric in the hose. The hose must be stiff enough to prevent collapse under these conditions.

8.4.3 Hydraulic Fittings

A large range of hydraulic fittings is available for connecting fluid lines to components [8.8]. These are designed to cater for many geometrical arrangements in a range of sizes to suit various flow and pressure requirements. Where required fittings incorporate or mate with seals, use of these (according to specification) will ensure leak-free operation. Standard fittings are recommended for commercial vehicle applications. For high-volume automotive, use of custom fittings may be specified, functionally based on the above standards.

8.5 Hydraulic Oils and Related Systems

8.5.1 Oil Characteristics

Existing hydraulic components are designed for use with premium hydraulic oils. Viscosity affects the fluid's ability to be pumped, transmitted through the system, carry a load, and lubricate the moving surfaces. The system must be able to start effectively without damage under cold conditions and operate efficiently without damage at a practical maximum oil temperature.

Viscosity increases progressively rapidly as temperature drops. It is usual to plot a graph of log log Kinematic Viscosity (cSt) versus temperature for various grades of hydraulic oil (Fig. 8.14). These plots are almost linear, and high viscosity index (VI) fluids have a flatter curve. DIN 51564 describes the classification of fluid viscosity. Premium Hydraulic oils are specially formulated oils with high VI and are the result of extensive and ongoing development by major oil suppliers.

The arbitrary VI scale was set up by SAE International. The original scale only stretched between 0 and 100 and applied to mineral oils, but higher values

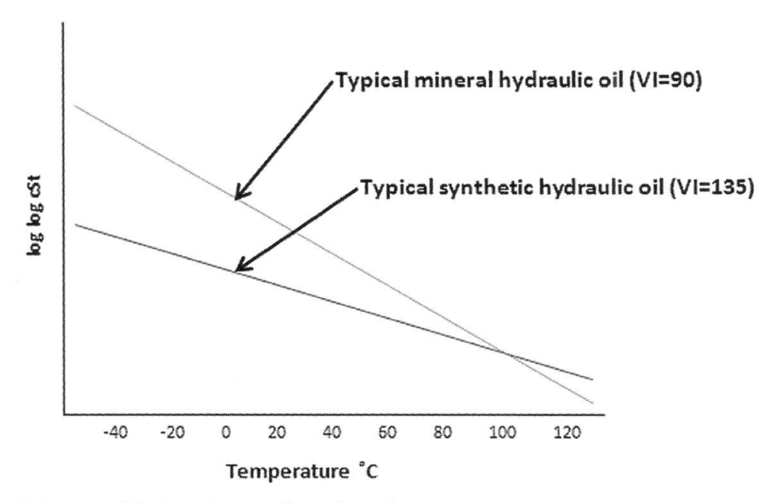

FIG. 8.14 Oil viscosity as a function of temperature.

can be obtained with oil additives. Modern high-VI oils are formulated from synthetic oils or a mix of synthetic and mineral oils with additives and have VIs from 80 to more than 300.

Premium hydraulic oils are used extensively in off-highway machines and in on-road implement systems such as garbage truck body hydraulics. A range of grades of different viscosity are available.

The oils have good anti-wear characteristics and anti-corrosion, anti-foaming, and self-cleaning characteristics. These properties are retained for long lives if contamination is minimized by prevention and filtration. The design target for hydraulic drive systems is "fill for life." Other oils such as transmission fluids or newer oils with even higher VI could be specified after extensive component and system testing. Pump/motor manufacturers validate their products after extensive testing and typically warrant their products to be used with a relatively limited range of hydraulic oils.

The selection of the appropriate grade of oil is often a compromise to optimize system performance. Figure 8.15 shows the effect of oil viscosity on efficiency of hydraulic pumps and motors. Too high, or too low, a viscosity for a given system could present problems of poor durability, low performance, high leakage and energy usage, etc. [8.3].

Oil viscosity should not fall below 8 cSt at maximum operating temperature to prevent component damage and minimize internal fluid leakage. Limited operation should be available for system start between −20 and −40°F (−30 to −40°C) without damage to components. Under these conditions, provided oil can flow, friction is high and heating of the oil is rapid.

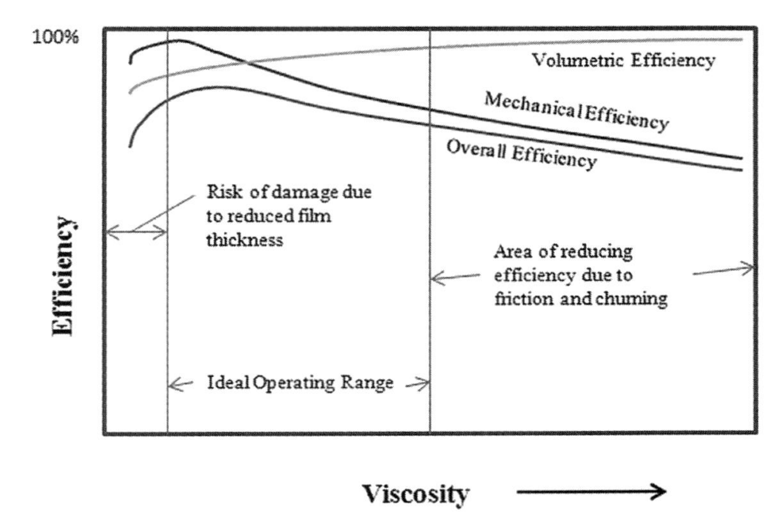

FIG. 8.15 Effect of oil viscosity on efficiency of hydraulic pumps and motors.

There are numerous methods for defining viscosity. Among these are:

- Absolute dynamic viscosity (Centipoise/cP)
- Kinematic viscosity (Centistoke/ cSt)
- Relative viscosity as measured in Saybolt Universal Seconds (SUS or SSU)

Most hydraulic systems operate with oil in the range of 150 to 300 SUS with typical ISO viscosity grade (ISOVG) range of ISOVG-22 to -68 [8.3].

8.5.2 Fluid Conditioning

Fluid conditioning is an important factor in assuring proper function of hydraulic systems in hydraulic hybrid vehicles. Fluid conditioning includes:

- Filtration to ensure adequate cleanliness of the fluid
- Maintenance of an acceptable range of fluid temperature
- Elimination or control of shock in the system
- Proper maintenance practices and procedures

8.5.2.1 Contamination

Contamination of hydraulic fluids is the major contributor to failures of hydraulic components and systems. Upward of 80% of hydraulic system failures may be caused by contamination of the fluid and thus the system [8.3]. Although contamination is often defined as solid particles, serious damage may be caused by oil contamination with air or other gases, fluids such as water, cleaning solvents, or incompatible hydraulic fluids.

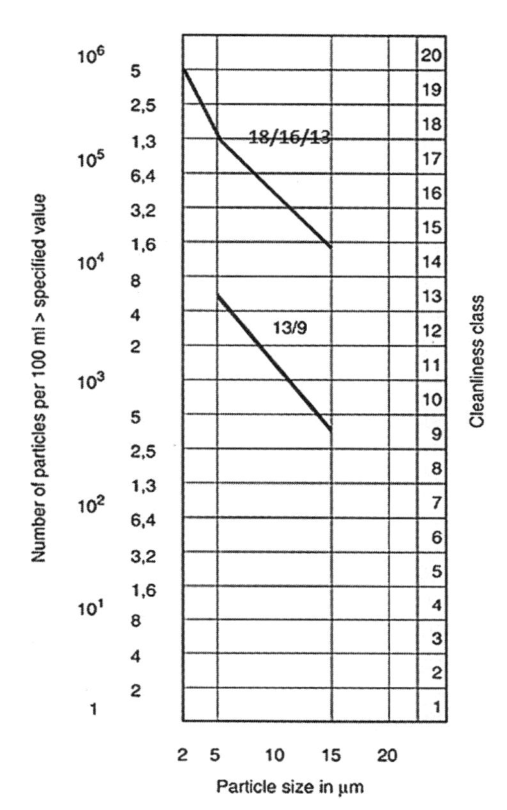

FIG. 8.16 Fluid cleanliness class to ISO DIS 4406.

Solid contaminants are generated by wear within components and unfortunately can be found inside new components and even in the oil supplied. Considerable study has taken place, and standards have been developed. ISO DIS 4406 describes a classification system for particles analyzed from samples of contaminated hydraulic fluid, as shown in Fig. 8.16.

Figure 8.17 illustrates typical clearances in hydraulic components in micrometers and inches. From these dimensions, it is readily observable that very small solid contamination, i.e., between 2μ and 15μ, can cause significant problems in hydraulic applications, depending on the type of contaminant material.

8.5.2.2 Cleanliness Requirements

ISO code 4406 1999 [8.9] established a range code of particle contamination concentration, independent of particle size, and a classification to indicate the concentration of particles larger than 2μ to 15μ [8.9]. The cleanliness level range codes are as $\geq 4\mu_{(c)}$, $\geq 6\mu_{(c)}$, $\geq 14\mu_{(c)}$. The classification class is assigned to the number of particles found in a 1-ml sample. It is usual to quote three numbers resulting from analysis. The first number is the Cleanliness Class for particles

COMPONENT	CLEARANCE LOCATION	MICRONS	INCHES
Gear Pump	Gear to side plate	1/2 - 5	0.00002-0.0002
	Gear tip to case	1/2 - 5	0.00002-0.0002
Vane Pump	Tip of vane	1/2 - 1	0.00002-0.00004
	Sides of vane	5 - 13	0.0002-0.0005
Piston Pump	Piston to bore	5 - 40	0.0002-0.0015
	Valve plate to cylinder	1/2 - 5	0.00002-0.0002
Servo Valve	Orifice	130 - 450	0.005-0.018
	Flapper wall	18 - 63	0.0007-0.0025
	Spool sleeve	1 - 4	0.00005-0.00015
Control Valve	Orifice	130 - 10,000	0.005-0.4
	Spool sleeve	1 - 23	0.00005-0.0009
	Disk type	1/2 - 1	0.00002-0.00004
	Poppet type	13 - 40	0.0005-0.0015
Actuators		50 - 250	0.002-0.01
Hydrostatic Bearings		0 - 25	0.00005-0.001
Antifriction Bearings		1/2	0.00002
Slide Bearings		1/2	0.00002

FIG. 8.17 Typical hydraulic system design clearance.

larger than 2 μm, the second number is the Cleanliness Class for particles larger than 5 μm, and the third number is the Cleanliness class for those larger than 15 μm. The example in Fig. 8.15 shows a contamination level of 18/16/13. This indicates that 1300 to 2500 particles above 2 μm, 320 to 640 particles above 5 μm, and 40 to 80 particles above 15 μm were found in the 1-ml fluid sample. Figure 8.18 provides general component category information for cleanliness recommendations for Vickers components at typical pressures [8.3].

Robust components have been designed to tolerate low levels of contamination, and reliability is excellent when good filtration and cleanliness, and fluid conditioning practices are employed.

8.5.2.3 *Filtration*

Filtration of the working fluid is essential to long life and robust operation of hydraulic system components. A filter system can be specified to achieve a target cleanliness in a hydraulic system.

There are various filter types:

1. Strainers—Mesh devices that prevent ingress of large particles or material (e.g., fibers) to the system or sensitive components such as control valves.

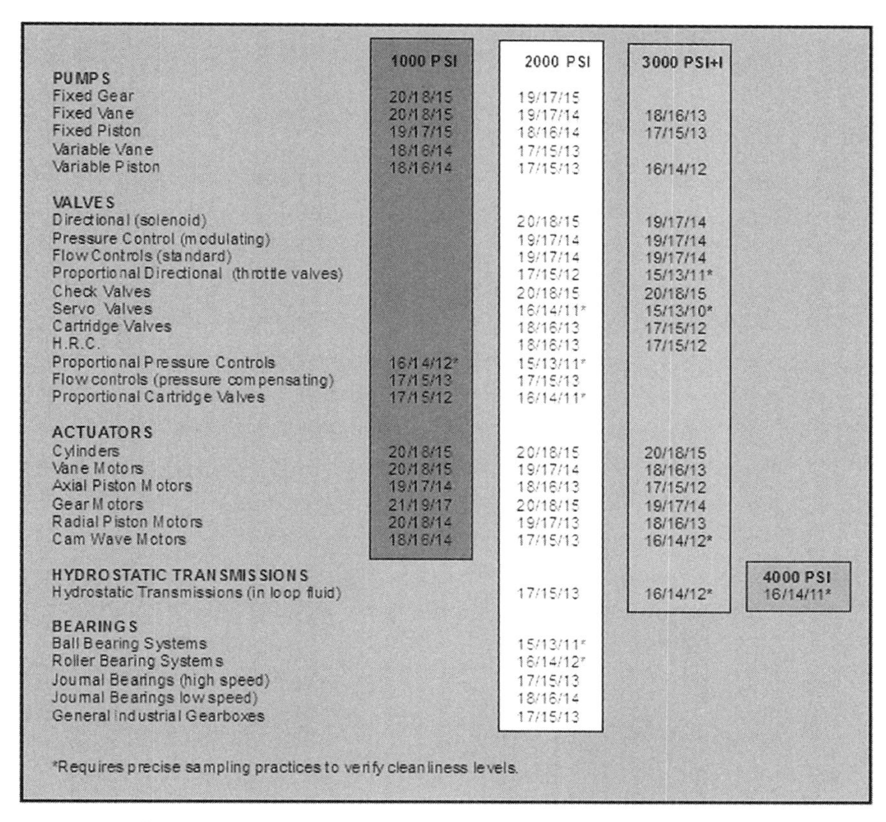

	1000 PSI	2000 PSI	3000 PSI+I	4000 PSI
PUMPS				
Fixed Gear	20/18/15	19/17/15		
Fixed Vane	20/18/15	19/17/14	18/16/13	
Fixed Piston	19/17/15	18/16/14	17/15/13	
Variable Vane	18/16/14	17/15/13		
Variable Piston	18/16/14	17/15/13	16/14/12	
VALVES				
Directional (solenoid)		20/18/15	19/17/14	
Pressure Control (modulating)		19/17/14	19/17/14	
Flow Controls (standard)		19/17/14	19/17/14	
Proportional Directional (throttle valves)		17/15/12	15/13/11*	
Check Valves		20/18/15	20/18/15	
Servo Valves		16/14/11*	15/13/10*	
Cartridge Valves		18/16/13	17/15/12	
H.R.C.		18/16/13	17/15/12	
Proportional Pressure Controls	16/14/12*	15/13/11*		
Flow controls (pressure compensating)	17/15/13	17/15/13		
Proportional Cartridge Valves	17/15/12	16/14/11*		
ACTUATORS				
Cylinders	20/18/15	20/18/15	20/18/15	
Vane Motors	20/18/15	19/17/14	18/16/13	
Axial Piston Motors	19/17/14	18/16/13	17/15/12	
Gear Motors	21/19/17	20/18/15	19/17/14	
Radial Piston Motors	20/18/14	19/17/13	18/16/13	
Cam Wave Motors	18/16/14	17/15/13	16/14/12*	
HYDROSTATIC TRANSMISSIONS				
Hydrostatic Transmissions (in loop fluid)		17/15/13	16/14/12*	16/14/11*
BEARINGS				
Ball Bearing Systems		15/13/11*		
Roller Bearing Systems		16/14/12*		
Journal Bearings (high speed)		17/15/13		
Journal Bearings low speed)		18/16/14		
General industrial Gearboxes		17/15/13		

*Requires precise sampling practices to verify cleanliness levels.

FIG. 8.18 Vickers component cleanliness recommendations.

2. Full Flow Filters—Engineered filters that contain contaminant larger than the rated size. They are usually situated inside a housing in the return line from the hydraulic motor to the reservoir. Ideally, the filter assembly is located within the reservoir assembly if this is not a pressure vessel. The filter element is designed to be replaced at a service interval or as a result of a pressure differential measurement. The filter housing usually contains a bypass valve for low temperatures (high pressure drop).
3. Partial Flow Filters—Engineered filters that contain fine contaminant larger than the rated size. These are designed to filter a small fraction of the oil on a continuous basis. A circulating motor may be used or a suitable system pressure drop to pass the flow fraction through the filter.

Filter elements are usually given a rating in μm based on the pore size in the filter medium, which nominally indicates the maximum size of particle not contained in the element. This is a statistical process, however, as there is variability both in the media and in the shape of the contaminant. Components used in hydraulic hybrid drives usually require filter elements of 10 μm or better, which should give sample measurements as shown in Fig. 8.15.

8.5.3 System Thermal Management

All pumps and motors generate heat because they are less than 100% efficient. This is manifest as a rise in oil temperature between inlet and exhaust (typically 20 to 40°F maximum). A stable system rejects heat to the environment at the same rate that it is generated.

The oil circulation in a hydraulic system gives ample opportunity for heat rejection from the many exposed surfaces. High thermal inertia of the system (fluid and components) allows for significant transient heat generation without overheating. This is a major advantage of hydraulic systems over batteries and electric power management systems. The target is to contain maximum oil temperature to 180–200°F (80–95°C) and, depending on oil used, to achieve minimum oil viscosity of 8 cSt to maintain adequate oil film thickness.

The maximum load condition at high ambient temperature may dictate the use of a system cooler similar to a conventional transmission cooler. This is usually fitted in the return line from the motor to the reservoir. The unit may admit full or partial flow to achieve the required cooling and may have a bypass valve for cold operation or a thermostatic valve.

8.6 Control Units and Transducers

8.6.1 System Control Units

Electronic control units enable effective and safe operation of hydraulic hybrid systems. Figure 8.19 shows a possible control architecture for a hydraulic hybrid vehicle.

The local control units provide inputs to the hydraulic components (variable pumps, valves, etc.) in response to sensor signals and other vehicle control signals.

System controllers provide inputs to the local controllers in response to vehicle demand (required state) and system status. Supervisory control, usually within the system controller, ensures safe operation of the system under all conditions, including component or system failure.

In practice, the various control functions may be combined into a reduced number or even a single control box, specifically designed for vehicle use, during system integration. Such a unit would contain embedded control strategy, input and output devices for transducers and vehicle information (J1939), and power outputs for solenoids and pump/motor control valves [8.10]. For series drive systems, this unit is the primary vehicle drive control system. The engine

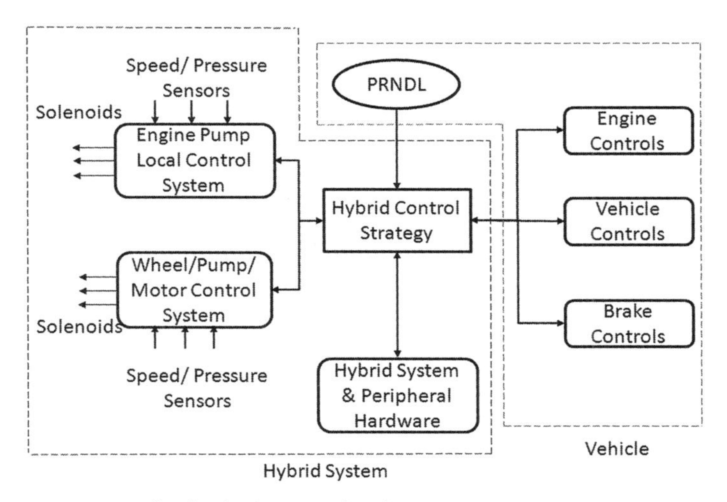

FIG. 8.19 Hybrid vehicle control architecture.

control unit is secondary, taking instructions to operate the engine safely with high efficiency and low emissions.

8.6.2 Transducers

Transducers for vehicle use are available to measure oil pressure and temperature, RPM, and solenoid displacement with required precision and responsiveness. The specifications are dictated by system operational requirements and the system FMEA. Vehicle parameters are available according to J 1939 protocols. Direct torque measurement is possible, but this parameter is usually derived from other data for control purposes because such transducers are expensive and difficult to install with acceptable reliability.

8.7 References

8.1 "A brief history of pumps," http://www.worldpumps.com/view/813/a-brief-history-of-pumps-/, assessed December, 2011.

8.2 Rothbart, H. A., *Mechanical Design Handbook*, pp. 30.1–30.61, McGraw-Hill, 1985.

8.3 "Mobile Hydraulics Manual—Your Comprehensive Guild to Mobile Hydraulics," Eaton Fluid Power Training, Eaton Corporation, 2006.

8.4 Kargul, J. J., " Hydraulic Hybrid Vehicles 101—Delivering Efficiency to Commercial Vehicles," South Coast AQMD Hydraulic Hybrid Forum, 2007.

8.5 Rampen, W., "Hydraulic Transmissions for Hybrid Vehicles," Artemis Intelligent power LTD. Presentation at IFS meeting in Eskilstuna, November 5, 2008.

8.6 Ivantysyn, J., Ivantysynova, M., *Hydrostatic Pumps and Motors*, Tech Books International, 2003 (English Edition).

8.7 Rydberg, K-E, "Energy Efficient Hydraulic Hybrid Drives," The 11th Scandinavian International Conference on Fluid Power, SICFP'09, Linköping, Sweden, June 2–4, 2009.

8.8 SAE J517: "Hydraulic Hose Fittings," SAE International, October 2011.

8.9 ISO 4406: "Hydraulic fluid power—Fluids—Method for coding the level of contamination by solid particles," the International Organization for Standardization, 12, 1999.

8.10 SAE J1939 "Recommended Practice for a Serial Control and Communications Vehicle Network," SAE International, October 2010.

8.11 Bosch Rexroth, "Hydraulic Trainer Volume 1" (R900018614), Website www.boschrexroth-us.com/training, 2007.

Chapter 9

Fuel Cell Hybrid Powertrain Systems

9.1 Fuel Cell Hybrid Powertrain

Fuel cells convert the chemical energy contained in a fuel electrochemically into electrical energy, and were first discovered in the 1800s. There are many types of fuel cells, which are distinguished by the electrolytes and operating temperatures. They include proton exchange membrane (PEMFC), molten carbonate (MCFC), solid oxide (SOFC), phosphoric acid (PAFC), and alkaline (AFC). The major activities in transportation fuel cell developments have focused on the PEMC and SOFC in the past 30 years [9.1]. A comparison of different fuel cell technologies is summarized in Appendix 9.6.

Fuel cells can operate on conventional fuels (gasoline, diesel) as well as renewable and alternative fuels (hydrogen, methanol, ethanol, natural gas, and other hydrocarbons).

A fuel-cell-powered vehicle offers the potential of being a zero-emission vehicle if hydrogen is used as the onboard fuel. Fuel cells also offer unique performance benefits with electrically driven accessories to enhance the overall fuel efficiency. Because fuel cells are electric power plants, it is easier to integrate electronic components such as air conditioning, GPS tracking, laptop access to the Internet, paging/fax devices, traffic information devices, climate-controlled seats, and more, into vehicles.

Since late 1980, the development of fuel-cell-powered vehicles is being sponsored by governments in North America, Europe, and Asia, as well as by major vehicle manufacturers worldwide. In 1993, Ballard Power Systems demonstrated a 10-m (32-ft) light-duty transit bus with a 120-kW fuel cell system and a 200-kW, 12-m (40-ft) heavy-duty transit bus in 1995. These buses operate with onboard compressed hydrogen as the fuel. In 1997, Ballard provided 205-kW (275-hp) units for a small fleet of hydrogen-fueled, full-size transit buses for demonstrations in Chicago, Illinois, and Vancouver, British Columbia. Table 9.1 lists many active programs in the United States, Europe, Japan, and China as of 2011.

TABLE 9.1 Current and Planned Fuel Cell Heavy-duty Vehicle Programs [9.2]

Country	Vehicle Type	Units	General System Description: Fuel	General System Description: Type	Fuel Cell Type
USA	Bus	48	Hydrogen/ Methanol	Hybrid FC	PEM
EU	Bus	44	Hydrogen	FC-battery-ultra cap hybrid system	PEM
Brazil	Bus	6	Hydrogen	Fuel Cell hybrid	PEM
South Korea	Bus	6	Hydrogen	Upgraded hybrid system from Gen 1 bus	PEM
India	Bus	1	Hydrogen	Hybrid FC Bus	PEM
China	Bus	9	Hydrogen	Next generation system from Phase I, hybrid FC bus	PEM
Singapore	Bus	1	Hydrogen	Hybrid FC bus	PEM
Canada	Bus	35	Hydrogen/ CNG blend	Upgraded Cummins-Westport C Gas Plus NG engine	PEM

9.2 Operating Principles of Fuel Cells

The basic physical structure of a fuel cell consists of an electrolyte layer in contact with an anode and a cathode on either side. A schematic representation of a unit cell with the reactant/product gases and the ion conduction flow directions through the cell is shown in Fig. 9.1.

In a typical fuel cell, fuel is fed continuously to the anode (negative electrode), and an oxidant (often oxygen from air) is fed continuously to the cathode (positive electrode). The electrochemical reactions take place at the electrodes to produce an electric current through the electrolyte, while driving a complementary electric current that performs work on the load.

The maximum electrical work (W_{el}) obtainable in a fuel cell operating at constant temperature and pressure is given by the change in Gibbs free energy (ΔG) of the electrochemical reaction:

$$W_{el} = \Delta G = -nFE \tag{9.1}$$

where n is the number of electrons participating in the reaction, F is Faraday's constant (96,487 coulombs/g-mole electron), and E is the ideal potential of the cell.

The Gibbs free energy change is also given by the following state function:

$$\Delta G = \Delta H - T\Delta S \tag{9.2}$$

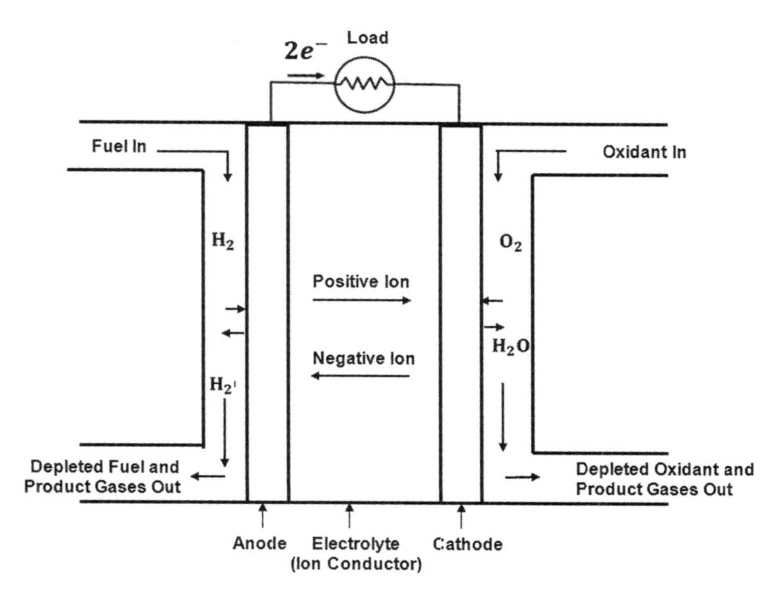

FIG. 9.1 Basic fuel cell structures.

where ΔH is the enthalpy change and ΔS is the entropy change. The total thermal energy available is ΔH. The available free energy is equal to the enthalpy change less the quantity TΔS, which represents the unavailable energy resulting from the entropy change within the system. The amount of heat that is produced by a fuel cell operating reversibly is TΔS. The ideal efficiency of a fuel cell, operating reversibly, is then defined as:

$$\eta_{ideal} = \frac{\Delta G}{\Delta H} = 1 - \frac{\Delta F}{\Delta H} T \tag{9.3}$$

Typical thermodynamics data for different reactions at 77°F (25°C) and 1 atm pressure are listed in Table 9.2.

The most widely used efficiency of a fuel cell is based on the change in the standard free energy for the cell reaction:

$$H_2 + \frac{1}{2} O_2 \rightarrow H_2O(l) \tag{9.4}$$

At standard conditions of 25°C (298 K) and 1 atmosphere, the thermal energy (ΔH) in the hydrogen/oxygen reaction is 285.8 kJ/mole, and the free energy available for useful work is 237.1 kJ/mole. Thus, the thermal efficiency of an ideal fuel cell operating reversibly on pure hydrogen and oxygen at standard conditions is calculated using Eq. 9.5.

TABLE 9.2 Typical Thermodynamics Data for Different Reactions at 25°C and 1 atm Pressure

	ΔH^0_{298} (kJ/Mol)	ΔH^0_{298} (kJ/Mol)	ΔG^0_{298} (kJ/Mol)	E(V)	η_{ideal} (%)
$H_2 + \frac{1}{2}O_2 \rightarrow H_2O(l)$	-285.8	-0.1641	-237.1	1.23	83
$H_2 + \frac{1}{2}O_2 \rightarrow H_2O(g)$	-242	-0.045	-228.7	1.19	94
$C + \frac{1}{2}O_2 \rightarrow CO(g)$	-116.6	0.087	-137.4	0.71	124
$C + O_2 \rightarrow CO(g)$	-393.8	0.003	-394.6	1.02	100
$CO + \frac{1}{2}O_2 \rightarrow CO_2(g)$	-279.2	-0.087	-2.53.3	1.33	100

$$\eta_{ideal} = \frac{237.1}{285.8} = 0.83 \tag{9.5}$$

For some direct electrochemical oxidation of carbon, ΔG is larger than ΔH, and consequently the ideal efficiency is slightly greater than 100% when using this definition of ideal efficiency.

For a general cell reaction,

$$\alpha A + \beta B \rightleftharpoons \chi C + \delta D \tag{9.6}$$

where α, β, χ, and δ have the units of moles. A and B are reactants, and C and B are products. The Gibbs free energy change of reaction can be expressed by the equation:

$$\Delta G = \Delta G^0 + RT\ \ln \frac{f_C^{\chi} f_D^{\delta}}{f_A^{\alpha} f_B^{\beta}} \tag{9.7}$$

where ΔG^0 is the Gibbs free energy change of reaction at the standard state pressure (1 atm) and at temperature T, and f_i is the fugacity of species i. Substituting Eq. (9.1) in Eq. (9.7) gives the relation

$$E = E^0 + \frac{RT}{nF} \ln \frac{\prod \text{[reactant fugacity]}}{\prod \text{[product fugacity]}} \tag{9.8}$$

which is the general form of the Nernst equation. The reversible potential of a fuel cell at temperature T, E^0, is calculated from ΔG^0 for the cell reaction at that temperature.

Fuel cells generally operate at pressures low enough that the fugacity can be approximated by the partial pressure.

The ideal standard potential (E^0) at 298 K for a fuel cell in which H_2 and O_2 react is 1.23 V with liquid water product, or 1.19 V with gaseous water product, as shown in Table 9.2. The potential is the change in Gibbs free energy resulting from the reaction between hydrogen and oxygen. The difference between 1.23 V and 1.19 V represents the Gibbs free energy change of vaporization of water at standard conditions.

The efficiency of an actual fuel cell is often expressed in terms of the ratio of the operating cell voltage to the ideal cell voltage as

$$\eta = \frac{\text{useful energy}}{\Delta H} \tag{9.9}$$

The thermal efficiency of a hydrogen/oxygen fuel cell can then be written in terms of the actual cell voltage:

$$\eta = \frac{0.83 V_{cell}}{E_{ideal}} = 0.675 V_{cell} \tag{9.10}$$

The actual cell potential is decreased from its ideal potential because of several types of irreversible losses, as shown in Fig. 9.2. Multiple phenomena

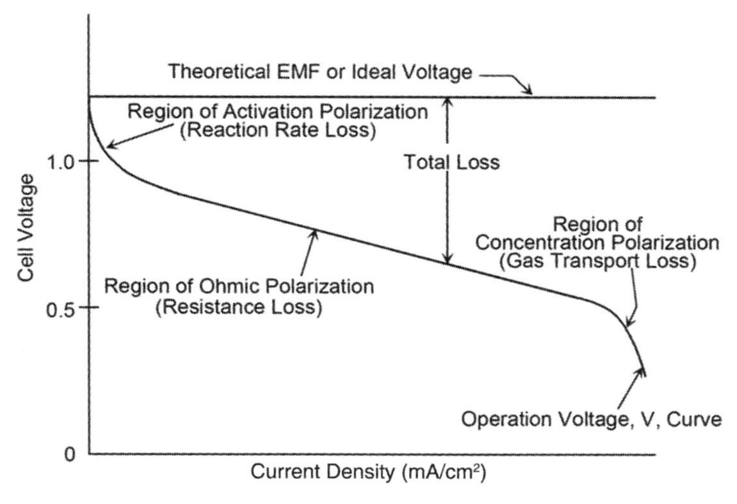

FIG. 9.2 Ideal and actual fuel cell voltage/current characteristic [9.1].

contribute to irreversible losses in an actual fuel cell. These losses are often referred to as polarization, over-potential, or over-voltage, though only the ohmic losses actually behave as resistances [9.1].

9.3 Polymer Electrolyte Membrane (PEM) Fuel Cells for Commercial Vehicles

PEM fuel cells use a solid polymer as an electrolyte and porous carbon electrodes containing a platinum catalyst, and operate at relatively low temperatures of approximately 176°F (80°C) with hydrogen, oxygen from the air, and water. Due to their fast startup time, low sensitivity to orientation, and favorable power-to-weight ratio, the PEM fuel cells are used primarily for transportation applications, such as cars and buses, and some stationary applications.

The PEM fuel cell was first developed by General Electric in the United States in the 1960s for use by NASA for the first manned space vehicles. In the late 1980s and early 1990s, Ballard Power Systems of Canada and Los Alamos National Laboratory of the United States led the technical advancements of PEM fuel cells. The development over the recent years have brought the current densities up to around 1 A cm^{-2} or more, while reducing the use of platinum by a factor of over 100. These improvements have led to huge reductions in cost per kilowatt of power, and power density improvement for use in commercial applications.

9.3.1 How the PEM Fuel Cell Works

Typical PEM fuel cells consist of a membrane electrode assembly (MEA), which is placed between two flow-field plates as shown in Fig. 9. 3. The MEA consists of two electrodes, the anode and the cathode, which are each coated on one side with a thin catalyst layer and separated by a proton exchange membrane. The most used membranes are perfluorosulfonic acid membranes manufactured by DuPont and sold under the trademark of Nafion®. The flow-field plates direct hydrogen to the anode and oxygen (from air) to the cathode.

Hydrogen flows through channels in flow-field plates to the anode, where the platinum catalyst promotes its separation into protons and electrons. Hydrogen can be supplied to a fuel cell directly or may be obtained from natural gas, methanol, or petroleum using a fuel processor, which converts the hydrocarbons into hydrogen and carbon dioxide through a catalytic chemical reaction. When hydrogen reaches the catalyst layer, it separates into protons (hydrogen ions) and electrons. The free electrons, produced at the anode, are conducted in the form of a usable electric current through the external circuit. At the cathode, oxygen from the air, electrons from the external circuit, and protons combine to form water and heat.

(a)

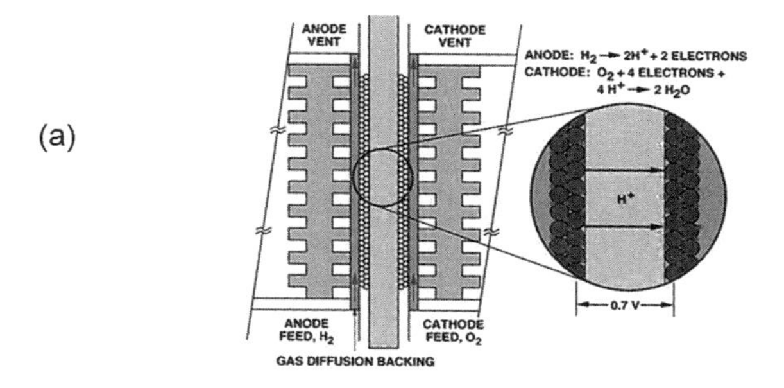

(b)

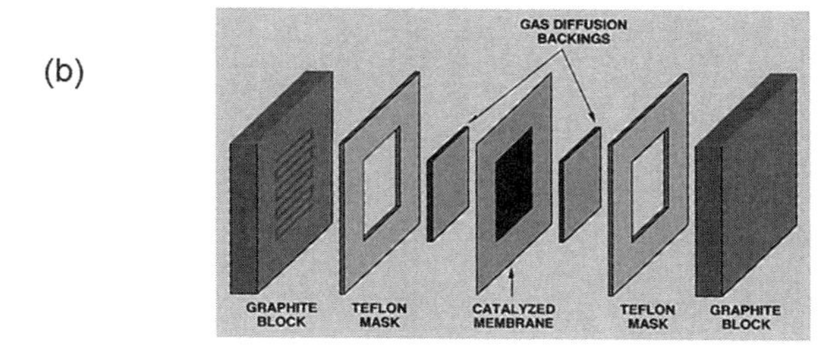

FIG. 9.3 (a) Schematic of PEM fuel cell; (b) Single cell structure of PEM fuel cell [9.1, 9.4].

To obtain the desired amount of electrical power, individual fuel cells are combined to form a fuel cell stack. Increasing the number of cells in a stack increases the voltage, while increasing the surface area of the cells increases the current.

There are must be sufficient water content in the polymer electrolyte to assure high proton conductivity. Reliable forms of water management have been developed based on continuous flow-field design and appropriate operating adjustments. If more water is exhausted than produced, then humidification of the incoming anode gas becomes important. If there is too much humidification, however, the electrode floods, which causes problems with gas diffusion to the electrode. A temperature rise between the inlet and outlet of the flow field increases evaporation to maintain water content in the cell.

Fig. 9.4 shows a schematic of a fuel cell engine with the water and thermal management systems used in a city bus at the 2010 Shanghai World Expo [9.4]. Hydrogen is supplied from a hydrogen tank with working pressure between 1.76 and 3.2 bar into the cell stack inlet, and the hydrogen gas is humidified with a hydrogen circulation pump. Air into the cell stack inlet is humidified with a spray-type humidifier; water content in the tail gas from the cell stack outlet is condensed and recovered to ensure the balance of the humidifier water system.

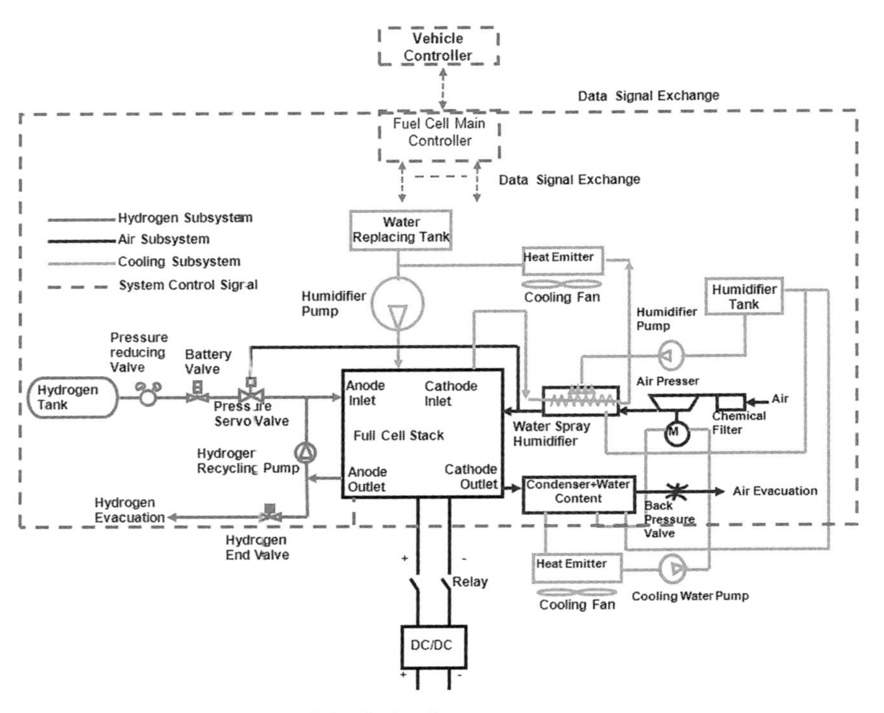

FIG. 9.4 Schematic diagram of the fuel cell engine system.

9.3.2 PEM FC System for Commercial Vehicles

The fuel cell as an alternative propulsion system was first used in transit bus applications because the transit buses have limited daily operation ranges and are usually refueled at the bus service stations. Figure. 9.5 shows a schematic of a PEM fuel cell powertrain for the transit bus application. It consists of two fuel cell systems. Each system has one fuel cell stack, a hydrogen delivery system, and an air supply system.

Hydrogen is supplied to the fuel cell stack from compressed storage tanks through a pressure regulator. Surplus hydrogen from the stack is recirculated through a hydrogen pump to improve the fuel cell performance.

Air is compressed and pumped into the stack through a humidifier. The humidifier takes water vapor from stack exhausted air and uses it to humidify the incoming compressed air. A regulator between the stack and the humidifier is electrically controlled to keep air pressure at a certain value.

The fuel cell powertrain system consists of fuel cell stacks, DC/DC converters; energy storage secondary batteries, inverters, and traction motors. The fuel cell stack and the traction motor/inverter are connected directly to avoid loss of energy through a converter. Each secondary battery is connected through a DC/DC converter, respectively.

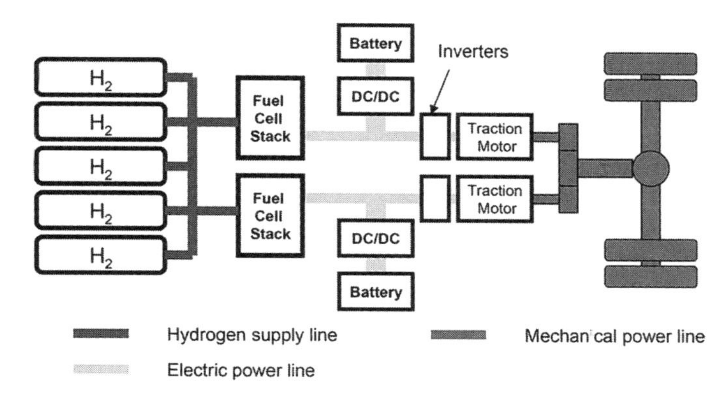

FIG. 9.5 Schematics of PEM fuel cell propulsion system for city buses.

The DC/DC converters manage not only the battery input/output power, but also fuel cell output power by means of controlling its output voltage at the desired operational point. Additionally, DC/DC converters keep two energy storage secondary batteries' state of charge at almost the same level. Each fuel cell stack is electrically separated from each other to control the stacks independently. Two traction motors are mechanically connected and jointly drive the rear axle of the vehicle.

Comparative tests show that the fuel cell bus is about a 66% improvement in fuel economy than that of the conventional diesel IC engine bus. In addition to the high efficiency of the fuel cell system itself, this high vehicle efficiency is realized by the power regeneration during deceleration and the proposed power distribution algorithm. The power regeneration and the power distribution algorithm improved the vehicle efficiency by 17.6% and 5.0%, respectively [9.5].

Figure 9.6 shows a PEM fuel-cell-powered Class 8 heavy-duty truck that was displayed at the 2011 HTUF conference in Baltimore, Maryland. The

FIG. 9.6 PEM fuel-cell-powered heavy-duty commercial vehicle.

vehicle is an electric truck with a battery recharged by a 65-kW fuel cell, which generates electricity from a reaction of hydrogen and oxygen. The vehicle is driven by a 400-kW and 4500 N·m electric motor. The fuel-cell-powered truck operates under typical short-haul conditions with a maximum speed of 65 mph. There is no combustion and no air pollution. Pure H_20 is the only byproduct.

9.3.3 PEM Fuel Cell Vehicle Performance and Development

In 2010, NREL, a national laboratory of the U.S. Department of Energy, conducted extensive evaluations of hydrogen-powered fuel cell vehicles. These demonstrations focus on identifying improvements to optimize reliability and durability. As of August 2010, 15 fuel cell buses were in service at seven locations in the United States [9.6, 9.7].

The fuel economy in diesel energy equivalent gallons (DGE) for the fuel cell and baseline buses was evaluated during the tests. The fuel cell buses at three test locations showed fuel economy improvement ranging from 53 to 141% when compared to diesel and CNG baseline buses. During the evaluation, the fuel cell buses reached more than twice the fuel economy of the diesel buses.

The sponsored tests by the U.S. Department of Energy (DoE) also show that future developments are required for commercialization of fuel-cell-powered transit buses. Availability of the fuel cell bus is one of the major challenges. Availability is the percentage of days that buses are planned for operation compared to the percentage of days the buses are actually available. For this evaluation period, the fuel cell buses were available from 52 to 69% compared with over 95% for conventional powered transit buses. Figure 9.7 categorizes the reasons that the buses were not available for operations. About

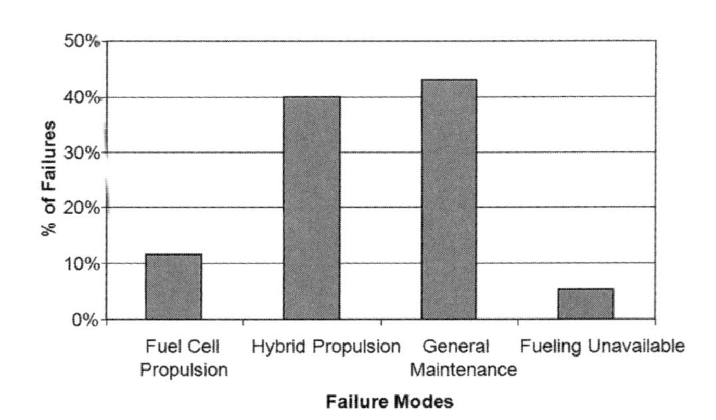

FIG. 9.7 Reasons for unavailability of the fuel cell buses [9.7].

12% of the unavailability was caused by the fuel cell propulsion system, while over 80% was caused by the hybrid propulsion system and general maintenance of the fuel-cell-powered vehicles.

Another major challenge is the cost of fuel-cell-powered city buses. Fuel cell buses are following the typical trend of all prototype technology: capital costs are high in the early stages and begin to fall with increased production and further product development. However, purchase price has little relevance if the buses cannot meet performance standards. After fuel cell bus designs have proven performance and durability, the industry can investigate ways to reduce the cost of the buses and replacement components.

The operating costs are also higher than those of conventional technology. Operating costs can be lower than expected in the first year while the buses are under warranty and maintenance is handled by the manufacturer's on-site technicians. Then costs rise as the transit agency staff takes over more maintenance and undergoes a steep learning curve. Once the staff becomes more familiar with maintenance, these costs are expected to drop.

9.4 Solid Oxide Fuel Cells Auxiliary Power Unit (APU) and Range Extender

Solid oxide fuel cells (SOFCs) are a class of fuel cell characterized by the use of a solid oxide material as the electrolyte. In contrast to PEM fuel cells, which conduct positive hydrogen ions (protons) through a polymer electrolyte from the anode to the cathode, the SOFCs use a solid oxide electrolyte to conduct negative oxygen ions from the cathode to the anode. The electrochemical oxidation of the oxygen ions with hydrogen or carbon monoxide thus occurs on the anode side.

SOFCs operate typically between 600 and 1000°C. At these temperatures, SOFCs do not require expensive platinum catalyst material, as is currently necessary for lower-temperature fuel cells such as PEM fuel cells, and they are not vulnerable to carbon monoxide catalyst poisoning. SOFCs are also the most sulfur-resistant fuel cell type; they can tolerate several orders of magnitude more of sulfur than other types of fuel cells. In addition, they are not poisoned by carbon monoxide (CO), which can even be used as fuel. Many attributes of SOFCs are listed in Table 9.3.

High-temperature operation has disadvantages. It results in a slow startup, which may not be acceptable for transportation applications. The high operating temperatures also place stringent durability requirements on materials. The development of low-cost materials with high durability at cell operating temperatures is the key technical challenge facing this technology.

TABLE 9.3 Solid Oxide Fuel Cells—Attributes [9.13]

High electric conversion efficiency	Demonstrated—47% Achievable—55% Hybrid—65% CHP—80%
Superior environmental performance	No NO_x Lower CO_2 emissions Sequestration capable Quiet; no vibrations
Co-generation—combined heat and power	High quality exhaust heat for heating, cooling, hybrid power generation, and industrial use Co-production of hydrogen with electricity Compatible with steam turbine, gas turbine, renewable technologies, and other heat engines for increased efficiency
Fuel flexibility	Low or high purity H_2 Liquefied natural gas Pipeline natural gas Diesel Coal synthesis gas Fuel oil Gasoline Biogases
Size and siting flexibility	Modularity permits wide range of system sizes Rapid siting for distributed power
Transportation and stationary applications	Watts to megawatts

9.4.1 How Solid Oxide Fuel Cells Work

The cell of an SOFC is constructed with two porous electrodes that sandwich an electrolyte. Air flows along the cathode. When an oxygen molecule contacts the cathode/electrolyte interface, it catalytically acquires four electrons from the cathode and splits into two oxygen ions. The oxygen ions diffuse into the electrolyte material, encounter the fuel at the anode/electrolyte interface, and react catalytically to generate electrons as well as water, carbon dioxide, and heat. The electrons transport through the anode to the external circuit and back to the cathode, providing a source of useful electrical energy in an external circuit.

Figure 9.8(a) shows a planar SOFC cell, which consists of two electrodes sandwiched around a hard ceramic electrolyte. The components are assembled in flat stacks, as shown in Fig. 9.8(b), with air and fuel flowing through

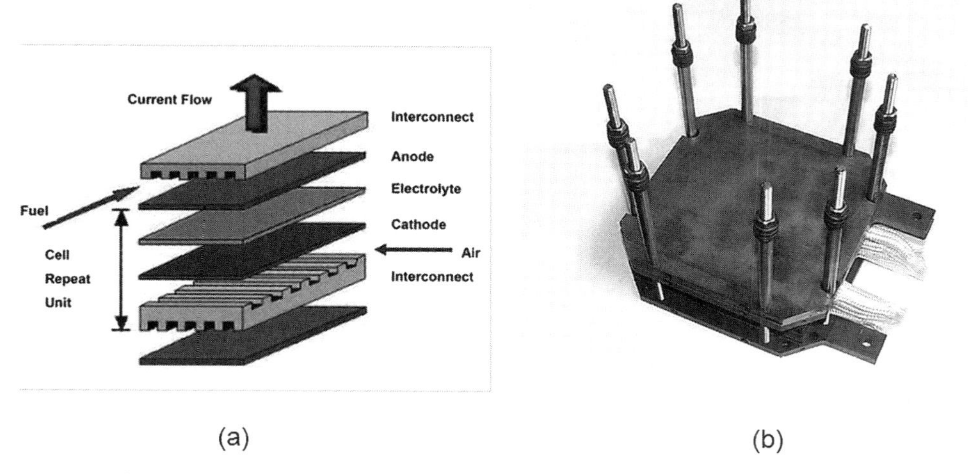

FIG. 9.8 Schematics of (a) SOFC planar stack [9.1]; and (b) 25-cell 400-W SOFC stack.

channels built into the cathode and anode. Hydrogen fuel is fed into the anode of the fuel cell, and oxygen, from the air, enters the cell through the cathode.

The SOFC reactions in the anode side include:

$$H_2 + O^{2-} \leftrightarrow H_2O + 2e^- \tag{9.11}$$

and on the cathode:

$$O_2 + 4e^- \rightarrow 2O^{2-} \tag{9.12}$$

The overall reactions are

$$2H_2 + O_2 \rightarrow 2H_2O \tag{9.13}$$

The corresponding Nernst equation for the reaction is

$$E = E^0 + \frac{RT}{2F} ln \frac{P_{H_2} P_{O_2}^{1/2}}{P_{H_2O}} \tag{9.14}$$

In addition to hydrogen, carbon monoxide (CO) and other hydrocarbons such as methane (CH_4) can be used as fuels by either electrochemical oxidation or reforming within the fuel cell (internal reforming). The direct electrochemical oxidation reaction of CH_4 at the SOFC anode can be described as follows:

$$CH_4 + 4O^{2-} \rightarrow CO_2 + 2H_2O + 8e^- \tag{9.15}$$

Internal reforming involves conversion of hydrocarbons to hydrogen (and carbon monoxide), followed by electrochemical reactions:

$$CH_4 + H_2O \rightarrow 3H_2 + CO \tag{9.16}$$

$$CH_4 + 2H_2O \rightarrow 4H_2 + CO_2 \tag{9.17}$$

followed by

$$H_2 + O^{2-} \rightarrow H_2O + 2e^- \tag{9.18}$$

$$CO + O^{2-} \rightarrow CO_2 + 2e^- \tag{9.19}$$

An example of single-cell performance at reduced temperature (Fig. 9.9) shows that high electrochemical performance can be achieved for reduced-temperature SOFCs.

In addition to the fuel cell stack, an SOFC system includes the fuel processor and the thermal management system, and several balance-of-plant (BOP) components, such as pumps, controls, insulations, packaging, etc. The air for reformer operation and cathode requirements is compressed and then split between the unit operations. Unreacted anode tail gas is recuperated in a tail gas burner. Additional energy is available in an SOFC system from enthalpy recovery from tail gas effluent streams that are typically 400–600°C.

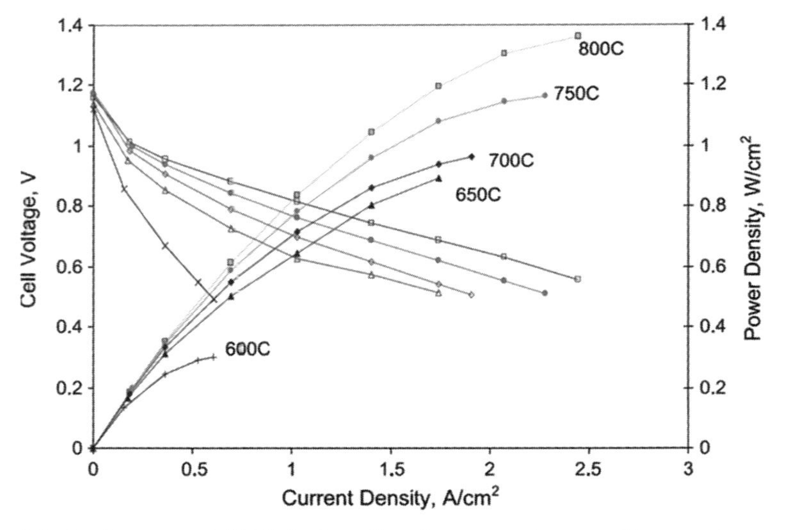

FIG. 9.9 Example of SOFC single-cell performance at reduced temperatures (fuel: hydrogen, oxidant: air) [9.1].

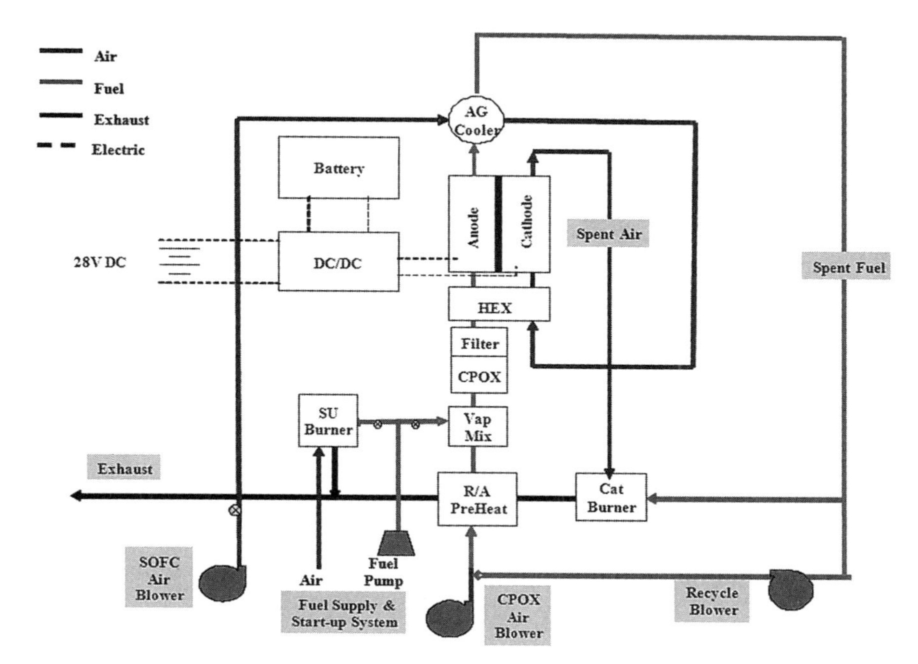

FIG. 9.10 Layout of SOFC systems.

The detailed descriptions of the fuel cell subsystem, as illustrated in Fig. 9.10, are as follows:

Anode Gas (AG) Cooler: Airflow through the anode gas heat exchanger cools the stack spent fuel to a temperature acceptable for the anode gas recycle blower. Higher operating temperature (up to 600°C) of the recycle blower can reduce the size requirement of the anode gas heat exchanger.

Heat Exchanger (HEX): The HEX is located immediately upstream of the stack and is used to minimize the potential for thermal shock at the stack inlet by bringing the air and reformate flows to a uniform temperature prior to entering the stack.

Startup Burner (SU Burner): The SU burner is used to 1) provide a source of heat to raise the fuel cell system and stack temperatures to those at normal operating conditions needed for power production, and 2) provide a water vapor source for reformer startup and oxygen-free blanket gas for the stack anode. Air is supplied to the startup burner from the SOFC air blower. The startup time will be less than 10 minutes (stack temperature rise from ambient to 400°C).

Fuel Reformer: The feed to the fuel reformer is comprised of a) the hydrocarbon stream, b) heated air from the Air HEX, and c) heated anode exhaust gas from the Recycle HEX. The highly efficient and compact catalytic partial oxidation (CPOX) unit produces reformate that can be fed directly to the SOFC Stack subsystem, greatly simplifying the balance-of-plant and thermal integration requirements. The use of stack anode exhaust recycle to supply process steam eliminates the need for onboard water storage and lowers the peak operating temperature in the reformer, thereby improving catalyst durability and inhibiting carbon formation in the reformer and downstream components. CPOX reformers are also an order of magnitude smaller than comparable steam reformers. While steam reformers are more efficient, their design, operation, and cost are negatively impacted by the high heat transfer and process steam requirements.

Filter: Reformate from the CPOX reformer passes through the filter to protect the stack from any particulate materials entrained in the gas stream.

SOFC Air Blower: The SOFC air blower supplies air to the fuel cell stack cathode, anode gas cooler (AG Cooler), startup burner, and emergency air flow at the Catalytic Burner exit to limit the burner exhaust gas temperature into the recycle heat exchanger (Recycle HEX).

Inlet Air HEX: The inlet air heat exchanger provides the heat exchange between the hot exhaust gas and the incoming airflow to the fuel processor.

Catalytic Burner: Cathode air and anode gas mix and combust in the catalytic burner. The heat associated with this exhaust stream is transferred to cooler inlet gas streams that supply the fuel processor and fuel cell stacks.

High-Temperature AG Recycle Blower: The AG blower recycles part of the stack anode gas into the CPOX reformer to achieve the required water vapor needs of the CPOX reformer, with the balance of the anode gas being combusted in the Catalytic Burner.

Recycle Heat Exchanger (Recycle HEX): The Recycle Heat Exchanger heats the spent fuel recycle mixture with water vapor content needed for the reformer.

DC/DC Converter: A DC/ DC converter will perform the main power conversion function as required by the load.

9.4.2 SOFC System for Vehicle Applications

SOFC can provide a variety of future integration strategies for vehicle applications. Figure 9.10 shows three examples of integration of an SOFC system into a vehicle powertrain: (a) and (b) show an SOFC auxiliary power unit (APU), (c) is an SOFC/ICE hybrid powertrain, and (d) is a plug-in SOFC range extender EV.

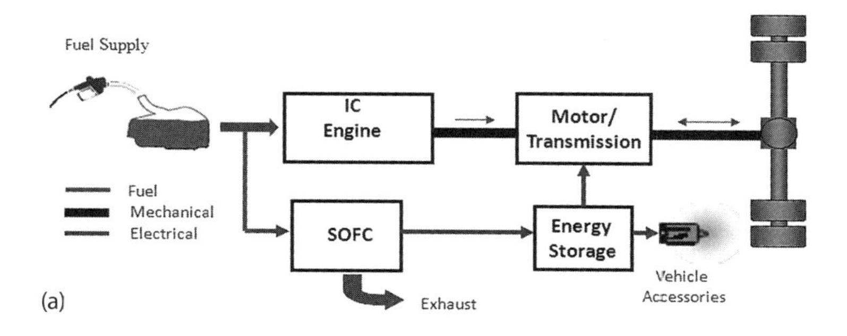

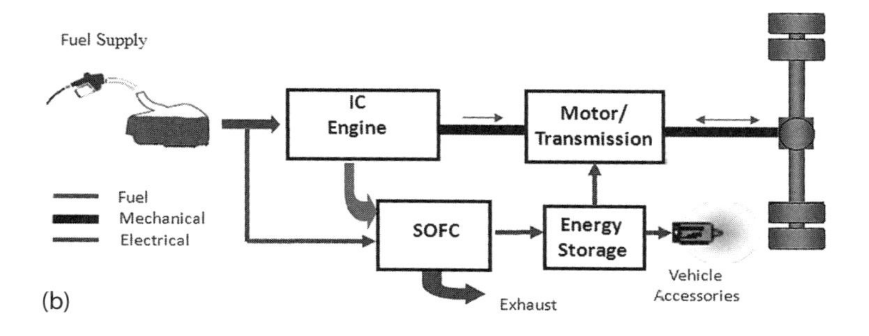

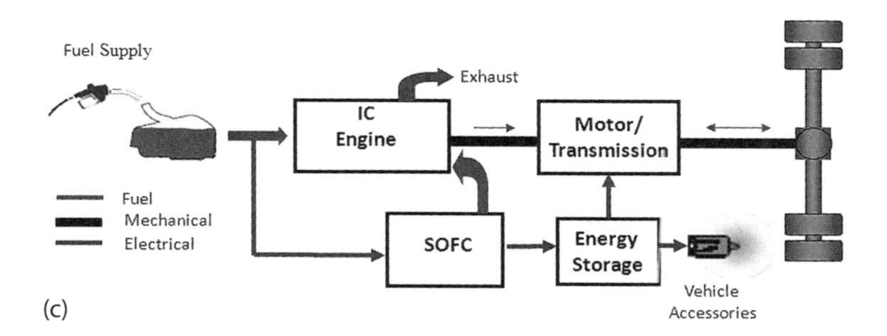

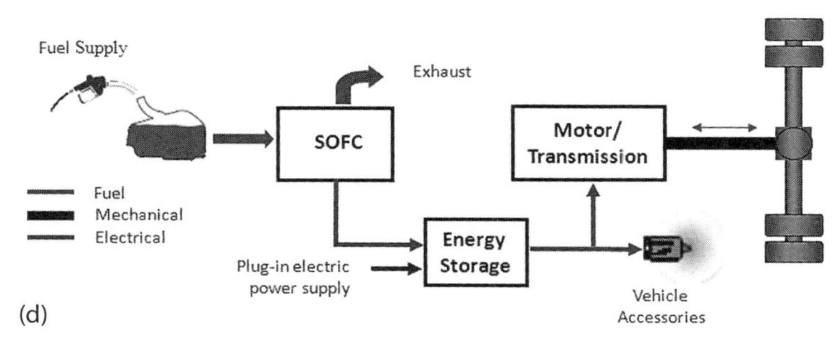

FIG. 9.11 (a) and (b) SOFC auxiliary power unit (APwU); (c) Hybrid powertrain of IC engine and SOFC; (d) Plug-in electric powertrain with SOFC as range extender.

9.4.2.1 SOFC APU

The first application of the solid oxide fuel cell is as the auxiliary power units (APU) for heavy-duty commercial vehicles (Figs. 9.11(a) and (b)). Figure 9.11(a) is a standard SOFC APU system. Figure 9.11(b) is an integrated system with the engine. The exhaust of the SOFC is connected with the intake system of the IC engine. The exhaust of the SOFC will be a part of the EGR system of the IC engine for emission reduction and thermal energy recovery purposes. Integration of an SOFC APU system allows notably higher efficiency for electrical generation compared to traditional onboard systems [9.17, 9.18, 9.19].

An SOFC APU consists of fuel cell stack, a fuel reformer, a start-up burner, a cathode heat exchanger, an anode blower and an air compressor, DC/DC inverter, and other subsystems. In addition to the electric power, the SOFC APU can also provide fuel reformate and high-grade waste heat.

Low-sulfur diesel fuel is injected into circulated hot anode exhaust gas. The vaporization heat of diesel is used to cool down this gas stream to tolerable values for the anode blower (around 250°C). After the diesel injection, the required amount of air is added to reach the targeted reforming temperature. This gas mixture, mainly consisting of diesel fuel, CO_2, N_2, and H_2O enters the reforming catalyst, where all hydrocarbons are split up to CO and H_2. The molar fractions of the anode inlet gas are: 21.5% H_2, 16.9% CO, 8.3% CO_2, 6.7% H_2O, and balance N_2 [9.16, 9.17]. After the electrochemical reaction in the anode, around 30% of the anode off gas is recycled and the rest is directed to the vehicle's after-treatment system to reduce stored NO_x or to be oxidized in the oxidation catalyst.

The cathode air is delivered by a blower. After entering the fuel cell system, this air stream is heated up in a heat exchanger to about 600°C. After leaving the cathode, the temperature has reached the normal operating temperature range of 700 to 800°C, and some oxygen is consumed by the electrochemical reaction. After heat exchange with the cold incoming air, this gas stream leaves the fuel cell system and is fed into the oxidation catalyst of the vehicle.

The start-up of the system is performed by injection and combustion of diesel fuel in the start-up burner. By thermal coupling of the burner with the fuel reformer, the reforming process can be started immediately, which ensures an oxygen-free environment for the anode during the heat up and also transfer of additional heat to the stack.

The SOFC APU will allow full electrification of accessories such as fans, pumps, blowers, and compressors. This permits HVAC, lighting, entertainment, communication, and battery charging functions to be independent of the internal combustion engines. The basic functions of the SOFC APU for heavy-duty commercial vehicles include:

- Idling electrification
- Auxiliary heating and engine preheating
- Electrification of auxiliaries (cooling pump, A/C compressor)
- Substitution of the alternator
- Support of the exhaust gas after-treatment of the main diesel engine

The SOFC APU can be engineered to use many types of fuels that are currently available, including natural gas, diesel, biodiesel, propane, gasoline, coal-derived fuel, and military logistics fuel. The SOFC APU can operate at a higher efficiency than traditional internal combustion engines, while eliminating nearly all pollutants and noise. SOFC APUs have been developed in the precommercialization stage by several companies [9.16, 9.20]. In January 2011, The U.S. EPA approved the SOFC APU for use as an emerging technology under the Energy Policy Act of 2005 [9.10]. The benefits of a 5-kW SOFC APU developed by Delphi include [9.17]:

- 5-kW electrical output power
- High-quality, reliable power: 110 VAC and/or 12 VDC
- High fuel efficiency: 40 to 50% higher than current non-fuel cell APUs
- Low noise: <60 dBa at 3 m
- Ultra-clean, near zero emissions:
  - Meets Tier 4 Emissions standards for non-road diesel engines
  - For CO, less than 8 g/kW·h
  - For NMHC and NO_x, less than 0.2 g/kW·h
- Onboard reforming capability

9.4.2.2 Hybrid ICE/SOFC Powertrain

The hybrid ICE/SOFC powertrain system consists of a solid oxide fuel cell and a specially designed IC engine (Fig. 9.11(c)). The IC engine operates in a fuel-rich, low-temperature combustion mode and functions as a power-generating fuel reformer [9.21, 9.22]. Unlike a conventional diesel engine, the specially designed IC engine operates in low-temperature combustion (LTC) mode to generate higher concentrations of hydrocarbons (HC) and carbon monoxide (CO) in the exhaust stream, as well as delivers mechanical power to the vehicle drivetrain. Additional fuel can also be injected into the SOFC system by a fuel injection system to control electric power output. The electrical power can be used to directly power the vehicle, stored in a battery, or used as an AC/DC auxiliary power source.

Simulated exhaust energy of the IC engine operating in LTC mode is shown in Fig. 9.12. The exhaust chemical energy in the form of CO, H_2, and hydrocarbons will feed to the solid oxide fuel cell to generate electricity. In

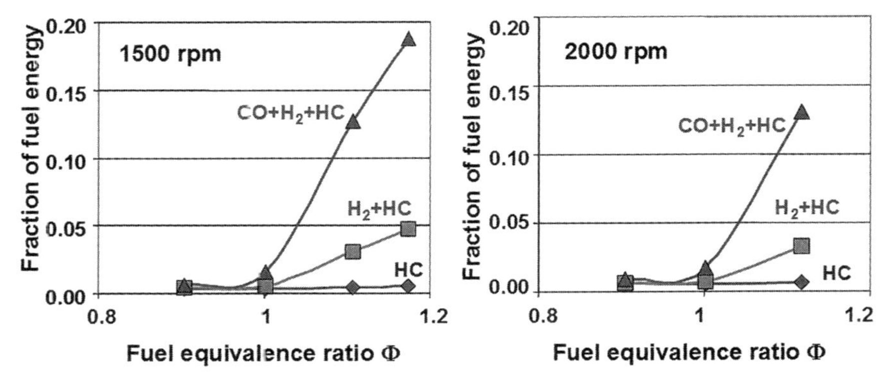

FIG. 9.12 Simulated results of exhaust chemical energy of 2.3-L IC engine [9.11].

other words, the SOFC functions as an energy recuperating device to recover the energy from the engine exhaust, which otherwise will be wasted. Fig. 9.13 shows a comparison of powertrain efficiency. Two factors contributed to the improved fuel conversion efficiency of the hybrid ICE/SOFC powertrain: a) SOFC has high fuel-to-power conversion efficiency compared with a conventional IC engine, and b) SOFC recuperates the energy in the IC engine exhaust gas species such as CO and unburned hydrocarbons. At low load, the hybrid ICE/SOFC powertrain can improve fuel-power efficiency from 40 to 80%.

In addition to the high energy conversion efficiency, the hybrid ICE/SOFC powertrain can also overcome the long warm-up time of the SOFC-only powertrain for transportation applications. It will take up to 30 minutes to bring an SOFC from ambient temperature to the operating temperature of 500°C. With the hybrid ICE/SOFC powertrain, the ICE and the onboard battery can provide power for the initial operation of the vehicle instead of the SOFC system.

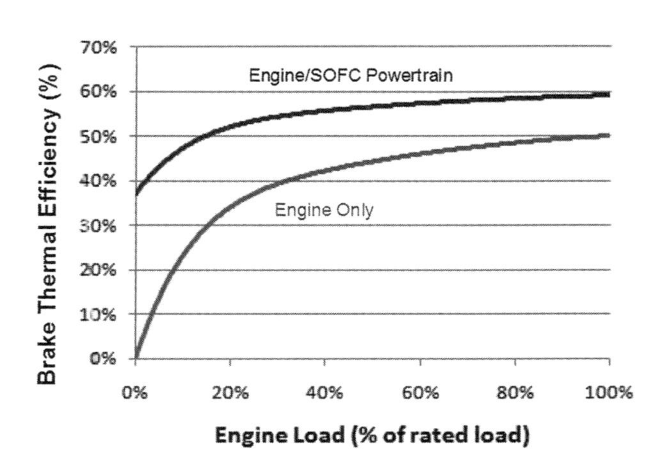

FIG. 9.13 Efficiency comparison of the hybrid ICE/SOFC powertrain and engine-only powertrain.

9.4.2.3 SOFC Range Extender EV

An SOFC range extender EV is an electric vehicle with a solid oxide fuel cell power unit to recharge the battery pack and to power the vehicle accessories during driving. As shown in Fig. 9.11(d), the SOFC range extender EV consists of the SOFC system, battery pack and charger, transmission, electric motor, and controls.

The attraction of this configuration is the use of a relatively smaller battery pack while increasing vehicle range and functionality. Another attraction is to allow the deployment of EV transit buses, which can be charged with grid connecting charging stations instead of depending on the construction of high-cost battery swap stations. Compared with a conventional EV, an SOFC range extender EV should display advantages in the following attributes [9.8]:

Weight and Range:
For commercial vehicles, weight is directly related to the productivity of the vehicles. For the same driving range, adding an SOFC power unit and fuel adds much less weight to an electric vehicle than doubling or tripling the capacity of a battery pack. In another words, the chemical energy of fuel provides additional range with minimal weight and volume compared to the battery pack.

When operating in the range extender mode, the SOFC power unit provides power for the average propulsion and accessory load, and the battery handles peak acceleration and regenerative braking functions.

HVAC Functions:
Accessories such as heating, ventilation, and air conditioning are about 30% of electric power consumption in conventional EVs. Waste heat from an SOFC system is readily available and allows operation of the vehicle in cold climates without compromising range and efficiency. The increased chemical energy (fuel) on board allows electric air conditioning to be used without an unacceptable loss of vehicle range.

Fast Fill/Slow Charge:
The ability to refuel with available fuels (gasoline or diesel in the short term) in a few minutes solves the fast charge requirement for EVs. For a city transit bus, even the 30 min fast charging will be too long. Therefore, spare buses are required for daily operations to allow the charging time when the battery is depleted. The ability to refuel at the end of trip (with SOFC Power Unit run on) or at specified charging stations avoids the constant fear of being stuck with a depleted battery.

TABLE 9.4 EV Transit Bus Operating Conditions

Vehicle Length	M	11.8
Battery Capacity	kW·h	105
Charging C rate	C	0.3
Fully Charged	h	3.0
Round trip Distance	Km	33.6
Total Miles Traveled per Day	kM	3528.0
Average Speed (China city bus drive cycle w/o Idle)	Km/h	23.5
Required Round Trip per Day		8.5

Table 9.4 shows an example of an 11.8-m city transit EV bus in Southern China. The energy capacity of the EV battery pack will be in the range of 105 kW·h with the charging C rate of 0.3. The transit EV bus is required to travel 33.6 km on a round trip with average speed of 23.5 km/h. On average, the buses are required to travel a total 3528 km, and about 30% of the battery energy will be used for air conditioning and other accessories.

A study of an SOFC power unit with output of 10 kW, 20 kW, and 30 kW is shown in Table 9.5. A 10-kW SOFC power unit can extend the city transit bus travel range for 19%, a 20-kW for 58%, and a 30-kW unit for more than double the travel distance to 117%.

Additional benefits of the SOFC power unit include the capability of allowing the vehicle to operate as a back-up generator for buildings and facilities (Vehicle to Grid), the capability of using renewable fuels such as biofuels, and enabling flexible charging intervals for commercial vehicles.

TABLE 9.5 EV Transit Bus with SOFC as a Range Extender

			With Air Conditioning (A/C)			
SOFC (kW)		W/O A/C	0	10	20	30
Energy Consumed per Km	kW·h/Km	1.2	0.85	0.85	0.85	0.85
Depth of Discharge	%	70%	70%	70%	70%	70%
Usable Energy	kW·h	73.5	73.5	87.8	116.4	159.3
Distance Traveled with Usable Energy	Km	61	86	103	137	187
Round trip Distance	Km	33.6	33.6	33.6	33.6	33.6
Total round trip with Usable Energy		1.8	2.6	3.1	4.1	5.6
Percent of Range Increase				19%	58%	117%

9.5 References

9.1 "Fuel Cell Handbook (Seventh Edition)," EG&G Technical Services, Inc., Under Contract No. DE-AM26-99FT40575, 2004.

9.2 Larminie, J. and Dicks, A. *Fuel Cell Systems Explained*, 2nd ed., Chichester; Hoboken, NJ: Wiley, 2003. pp. 45–66.

9.3 "Fuel Cell Vehicles," http://www.afdc.energy.gov/afdc/vehicles/fuel_cell.html, assessed, Sept. 2011.

9.4 Zhang, T., "The Development of Fuel Cell Vehicles in China," www.iphe.net/docs/...9.../1-2_FCV in China-2020100921.pdf, 2010.

9.5 Yokoyama, T. et al., "Development of Fuel cell Hybrid Bus," SAE Paper No. 2003-01-0417, SAE International, Warrendale, PA, 2003.

9.6 "Fuel Cell Transit Bus Evaluations," Technical Report, NREL/TP-5600-49342-1, November 2010.

9.7 Eudy, L. et al., "Fuel Cell Buses in U.S. Transit Fleets: Current Status 2010," NREL/TP-5600-49379. Golden, CO: National Renewable Energy Laboratory, 2009.

9.8 Velasco, E., "FCV's for a More Sustainable Mobility in 2050," SAE Paper No. 2010-01-0850, SAE International, Warrendale, PA, 2010.

9.9 Spitzley, D. et al., "Assessing Fuel Cell Power Sustainability," SAE Paper No. 2000-01-1490, SAE International, Warrendale, PA, 2000.

9.10 US EPA Letter regarding Delphi SOFC APU, www.epa.gov/cleandiesel/documents/et-letter-delphi-sofc-apu.pdf, assessed, Sept. 2011.

9.11 Hahn, T., "Study of Low Temperature Combustion Diesel Engine for Intermediate Temperature Solid Oxide Fuel Vehicle," Master Thesis. MIT, 2006.

9.12 Surampudi, B., Redfield, J., Ray, G., Montemayor, A. et al., "Electrification and Integration of Accessories on a Class-8 Tractor," SAE Paper No. 2005-01-0016, SAE International, Warrendale, PA, 2005.

9.13 Williams, M. C. et al., "Solid oxide fuel cell technology development in the U.S.," Solid State Ionics 177, pp. 2039–2044, 2006.

9.14 Rechberger, J. and Prenninger, P., "The Role of Fuel Cells in Commercial Vehicles," SAE Paper No. 2007-01-4273, SAE International, Warrendale, PA, 2007.

9.15 Chaudhari, A., "Development of Model Predictive Controller for SOFC-IC Engine Hybrid System," SAE Paper No. 2009-01-0146, SAE International, Warrendale, PA, 2009.

9.16 Norrick, D., "Diesel-Fueled SOFC System for Class 7/Class 8 On-Highway Truck Auxiliary Power," FY 2009 Annual Progress Report, US DOE Hydrogen Program, 2009.

9.17 Shaffer, S., "Solid Oxide Fuel Cell Development for Auxiliary Power in Heavy Duty Vehicle Applications," FY 2009 Annual Progress Report, US DOE Hydrogen Program, 2009.

9.18 Botti, J., Grieve, J., and MacBain, J., "Electric Vehicle Range Extension using an SOFC APU," SAE Paper No. 2005-01-1172, SAE International, Warrendale, PA, 2005.

9.19 Botti, J. and Speck, C. E., "Solid Oxide Fuel Cell Power Unit for Hybrid Vehicle and Electric Vehicles," SAE Paper No. 2005-24-094, SAE International, Warrendale, PA, 2005.
9.20 Hu, H. et al., "Mechanism and method of combined fuel reformer and dosing system for exhaust aftertreatment and anti-idle SOFC APU," US Patent No. 7,213,397, May 8, 2007.
9.21 Hu, H. et al., "Clean Power System," US Patent No. 7,818,959, October 26, 2010.
9.22 Hu, H. et al., "Clean Power System," US Patent No. 7,648,785, January 19, 2010.
9.23 Ebner, A. et al., "Study of Possible Range Extender Concepts with Respect to Future Emission Limits," SAE Paper No. 2010-32-0129, SAE International, Warrendale, PA, 2010.

9.6 Appendix: Comparison of Fuel Cell Technologies

Fuel Cell Type	Common Electrolyte	Operating Temperature	Typical Stack Size	Efficiency	Applications	Advantages	Disadvantages
Polymer Electrolyte Membrane (PEM)	Perfluoro sulfonic acid	50–100°C 122–212°F typically 80°C	< 1kW–100kW	60% transportation 35% stationary	Backup power Portable power Distributed generation Transportation Specialty vehicles	Solid electrolyte reduces corrosion & electrolyte management problems Low temperature Quick start-up	Expensive catalysts Sensitive to fuel impurities Low temperature waste heat
Alkaline (AFC)	Aqueous solution of potassium hydroxide soaked in a matrix	90–100°C 194–212°F	10–100 kW	60%	Military Space	Cathode reaction faster in alkaline electrolyte, leads to high performance Low cost components	Sensitive to CO_2 in fuel and air Electrolyte management
Phosphoric Acid (PAFC)	Phosphoric acid soaked in a matrix	150–200°C 302–392°F	400 kW 100 kW module	40%	Distributed generation	Higher temperature enables CHP Increased tolerance to fuel impurities	Pt catalyst Long start up time Low current and power
Molten Carbonate (MCFC)	Solution of lithium, sodium, and/ or potassium carbonates, soaked in a matrix	600–700°C 1112–1292°F	300 kW–3 MW 300 kW module	45–50%	Electric utility Distributed generation	High efficiency Fuel flexibility Can use a variety of catalysts Suitable for CHP	High temperature corrosion and breakdown of cell components Long start up time Low power density
Solid Oxide (SOFC)	Yttria stabilized zirconia	700–1000°C 1202–1832°F	1 kW–2 MW	60%	Auxiliary power Electric utility Distributed generation	High efficiency Fuel flexibility Can use a variety of catalysts Solid electrolyte Suitable for CHP & CHHP Hybrid/GT cycle	High temperature corrosion and breakdown of cell components High temperature operation requires long start up time and limits

Reference: http://www1.eere.energy.gov/hydrogenandfuelcells/fuelcells/pdfs/fc_comparison_chart.pdf. More information on the Fuel Cell Technologies Program is available at: http://www.hydrogenandfuelcells.energy.gov.

Chapter 10

Commercial Vehicle Electrification

10.1 Introduction

Hybrid drive technology offers significant fuel efficiency benefits for commercial vehicles. To date, those benefits have been largely limited to energy recovery from vehicle braking events subsequently used to aid vehicle acceleration. Replacing mechanically driven systems on vehicles with their electric equivalents can lead to additional efficiency improvements and fuel savings. For non-hybrid vehicles, especially light-duty passenger cars, the electrification trend started more than a decade ago with the electrification of brake, steering, and other systems. Efficiency improvements and fuel savings come from the change from mechanically driven always-on systems to electrically actuated on-demand operation. With the advent of hybrid vehicles, electrification is becoming more prevalent. This is not only driven by the need for improved efficiency, but also because one of the primary constraints for electrification is removed, namely the limited power that can be generated by even advanced 12-V alternators. For 12-V electric needs on the vehicle, hybrid systems can provide power via a DC to DC converter from the high-voltage hybrid battery system. High-power-demand systems on board a hybrid vehicle, such as A/C compressors, are designed to run directly off the high-voltage DC power bus. Any hybrid vehicle designed to drive any distance in pure electric mode, as well as range extended and pure electric vehicles, will of course have all-electric systems.

For commercial vehicles, there is another dimension to vehicle electrification. Many commercial vehicles have vocation-specific tools or systems on the vehicle. A good example is the PTO-driven hydraulic boom on utility trucks. These trucks are generally used to repair or service overhead electrical lines, transformers, or telecommunications equipment. On a normal utility boom truck, the engine is always on when the vehicle is on the job site and the boom is operated. The electric PTO feature found on utility trucks employing Eaton Corporation's hybrid-electric drive system are a good example of equipment electrification with significant resulting efficiency improvement [10.1]. With this system, the transmissions PTO output is connected to a hydraulic pump. While the engine is off and the clutch disengaged, the electric traction motor can drive the transmission PTO output and, via the hydraulic pump, power the hydraulic boom and tool chain. Occasionally, the engine will start automatically to recharge the battery. This system has been demonstrated to reduce engine idle fuel consumption by greater than 85% [10.1]. On an annual basis, the fuel savings from job site operations with engine-off are approximately double that of fuel savings from hybrid driving alone [10.2]. This chapter will provide an overview of a range of technologies and electrified components that are either more efficient than their mechanical counterparts or enable more-efficient vehicle operation. The most obvious example of the latter is the avoidance of engine idling while still providing heat or cooling and maintaining a charge in the 12- or 24-volt battery system.

10.2 Idle Reduction Technologies

Commercial trucks of all sizes idle for extended periods (half hour or more) during their workdays for a variety of reasons, such as while drivers wait to pick up or drop off a load, to operate vocation-specific equipment on the truck (booms, digger derricks, compactors, mixers, etc.), or to provide shelter from hot or cold ambient temperatures. To date, attention has been focused primarily on overnight idling by long-haul trucks because these provide an obvious and visible target for state and local regulatory agencies. At up to 2000 gal per year, the amount of diesel fuel used by long-haul truck idling is substantial [10.3]. However, as Table 10.1 shows [10.4], the sheer quantity of vocational trucks results in idle fuel consumption from long-haul trucks accounting for only approximately 20% of combined overnight and job site commercial vehicle idle fuel consumption.

Over the last decade, many States and localities have adopted a variety of no-idling rules and regulations [10.5]. The drivers for this patchwork of rules and regulations vary by region and regulatory body, but they generally encompass one or more of the following: criteria pollutant emissions [10.6], greenhouse gas emissions, noise, and fuel consumption reductions. No-idle

TABLE 10.1 Fuel Used by Commercial Trucks while Idling [10.4]

	Class 8 Sleeper Trucks	All Commercial Body Types—job site idling
Number of Trucks	659,606	18,147,256
Total Fuel Used (gal/yr)	9,709,129,733	36,982,024.82
Fuel Used to Idle (gal/yr)	657,185,760	2,491,143,761

regulations set forth by the State of California [10.7] have had a significant impact on idle reduction technology development for long-haul trucks. The foremost reason, besides size and population, is that the State of California is home to several major ports where imported goods are transferred to road transport for distribution across the nation, and therefore a significant number of long-haul trucks pass through California. Another reason is that many other states and localities have adopted all or portions of California's no-idle regulations. With few exceptions, idling for more than 5 min for any diesel-powered vehicle with a gross vehicle weight rating of 10,000 lb is prohibited within the borders of the State of California. In response to these regulations, a variety of technologies have been developed, and some [10.8]are now readily available. Many of these technologies are focused specifically on long-haul sleeper trucks for reasons mentioned earlier: up to 2000 gal consumed per year from idling on average more than 8 h daily. With diesel fuel prices exceeding $3 per gallon, many idle-reduction technologies have the potential to be cost effective based on economic metrics alone, as shown [10.9] in Table 10.2.

10.2.1 Diesel Auxiliary Power Unit (APU)

Diesel Auxiliary Power Units (APUs) are typically a fraction of the size of the truck primary engine and sized to efficiently handle all the auxiliary loads on the vehicle when it is not moving, consuming as little as one gallon of fuel during an 8-h period. APUs based on diesel internal combustion engines have been commercially available since the 1980s. An early pioneer in this field was the Pony Pack™, but many models are available today [10.8]. These systems typically have a power output range of 3–6 kW and are used in thousands of heavy-duty diesel trucks in the United States. The most popular models provide cabin cooling and heating, can power auxiliary equipment, and have the capability to be plugged into electrified truck stops. A 2002 EPA study [10.6] measured fuel consumption during extended idle periods for several long-haul trucks with either the stock engine idling or with an APU replacing the functions of the main engine. The EPA reports in this study that using an APU, rather than the truck's stock engine, can save up to 73% of the fuel used to idle the main engine and also lead to significant criteria pollutant

TABLE 10.2 Average Capital and Maintenance Costs of On-Board Idle Reduction Technologies

	Capital Cost	Annual Maintenance Cost
Direct-fired heaters	$888	$110
Battery powered A/C	$4,300	$200
APU/Gen Set	$7,750	N/A

reductions. California regulations require that for diesel trucks with engines manufactured in 2007 or later and use a diesel-powered APU, the APU must meet additional equipment requirements [10.7]. With additional emissions and noise requirements, and a general regulatory policy direction toward diesel alternatives, diesel ICE-based APUs will be made less attractive to fleets and truck operators in spite of their cost effectiveness.

10.2.2 Fuel Cell APU

Fuel cells are electrochemical conversion devices that produce a DC voltage. Chapter 9 of this book has discussed the design and operations of the fuel cells. In the late 1990s and early 2000s, fuel cell technology received a great deal of attention [10.10, 10.11, 10.12, 10.13, 10.14] for their potential application in auxiliary power units and eventual replacement of internal combustion engines altogether. Due to economic challenges of the technology, however, fuel cells in transportation applications remain more promise than reality. Depending on the type of fuel cell technology used, the input fuel can be hydrogen, methanol, or a range of select hydrocarbons. An additional challenge in APU applications for long-haul trucks is that diesel fuel already carried on board the truck should be used as the input fuel for the APU. This will require fuel reforming technology that can convert diesel fuel into an input fuel compatible with the selected fuel cell technology [10.13, 10.15].

10.2.2.1 Solid Oxide Fuel Cell (SOFC)

SOFCs typically operate at high temperatures from 700°C to 1000°C. This high-temperature operation presents a number of challenges, namely poor thermal shock properties, limited material selection, and long startup times to operating temperature. SOFCs use a solid oxide material as the electrolyte, hence the name. A positive characteristic of SOFCs compared to Proton Exchange Membrane (PEM) fuel cells is that expensive precious metals are not required. Another benefit is that a variety of fuels—hydrogen, carbon monoxide, and light hydrocarbons—can be used directly in the SOFC. With the help of a fuel reformer, even liquid fuels such as diesel fuel can be used successfully in an SOFC by reforming it into a mixture of hydrogen, carbon monoxide, and gaseous hydrocarbons.

Figure 10.1 shows a schematic overview of an SOFC. Assisted by the high operating temperature, light hydrocarbons are reformed directly on the SOFC anode. In an SOFC, negative oxygen ions travel from the cathode to the anode.

The trucking industry has historically been very resistant to a secondary fuel, thus limiting the potential for hydrogen-fueled SOFC and PEM fuel cells. Because of the SOFC's greater fuel flexibility, it is generally considered a more viable fuel cell technology for APU applications in the transportation

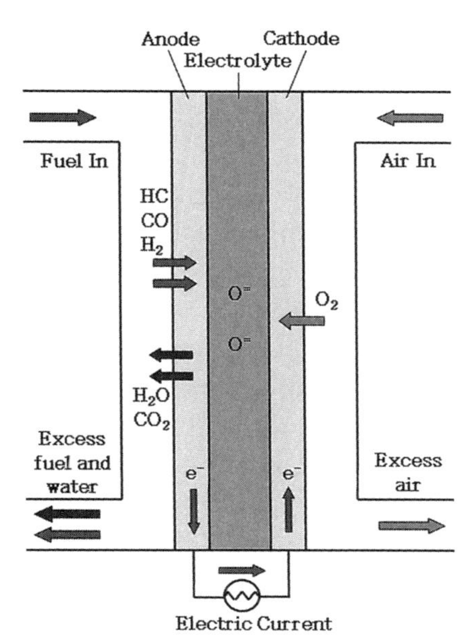

FIG. 10.1 Solid oxide fuel cell schematic.

industry. A number of studies [10.16, 10.17, 10.18, 10.19] suggest that the appropriate sizing of a fuel cell for long-haul truck APU applications is 3–10 kW. While the efficiency of an SOFC can easily exceed 50%, the efficiency of a complete SOFC-based APU is significantly lower. This is mostly due to the losses associated with reforming the onboard diesel fuel into a mixture acceptable as an input fuel to the SOFC. Furthermore, overall SOFC APU efficiency is highly dependent on overall plant design and engineering effort expended to minimize parasitic losses. Both Dobbs et al. [10.17] and Filipi et al. [10.18] describe that careful system integration is essential to acceptable overall SOFC APU efficiency. Even then, overall efficiency is not expected to exceed 30%, which can quite easily be obtained by a conventional diesel ICE-powered APU. Primary drivers for the adoption of fuel cell APUs in class 8 long-haul trucks can therefore not be found in any efficiency advantages over more conventional approaches. Combined with the continuing relatively high cost of fuel cell technology, this explains why broad industry adoption continues to elude this technology.

10.2.2.2 PEM Fuel Cell

A Proton Exchange Membrane (PEM) fuel cell works with a polymer electrolyte in the form of a thin permeable sheet or membrane. Some of the key characteristics of PEM fuel cells are that they operate at relatively low temperatures, around 80°C, and require pure hydrogen for a fuel. Platinum-based catalysts are used to ionize the hydrogen atoms at the anode. The positively charged protons pass through the membrane to the cathode, while the electrons must pass through an external circuit, providing electrical power. Figure 10.2 shows a schematic of a PEM fuel cell.

The primary drawback for using a PEM fuel cell in an APU for long-haul truck applications is the requirement for pure hydrogen as an input fuel. Given the industry requirement of using fuel already onboard the vehicle (i.e., diesel fuel), a fuel treatment system is required to convert diesel fuel to pure hydrogen. Filipi et al. [10.18] describe such a fuel processing system. There are essentially five steps in the fuel reforming process: 1. Fuel vaporization and sulfur removal; 2. Steam reforming, typically above 850°C; 3. High-temperature water gas shift reaction; 4. Low-temperature water gas shift reaction; and 5. Preferential oxidation to remove any remaining carbon monoxide in the end gas. Complexity, cost, and reliability concerns for PEM fuel cells make them a relatively unattractive solution as an APU in commercial vehicles.

10.2.3 Microturbine APU

First introduced in the distributed power generation market [10.20, 10.21] in the mid-nineties, thousands of microturbines are now in service. Microturbines are small combustion turbines with outputs of 25 to 500 kW. They evolved from automotive and truck turbochargers, auxiliary power units (APUs) for airplanes, and small jet engines [10.22]. Most microturbines are composed of a

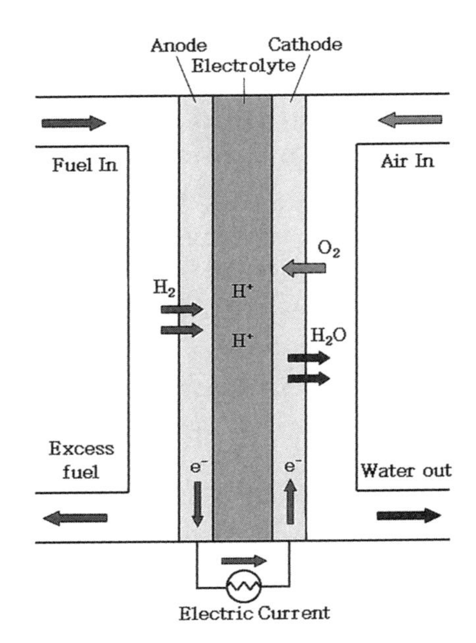

FIG. 10.2 Proton exchange membrane fuel cell schematic.

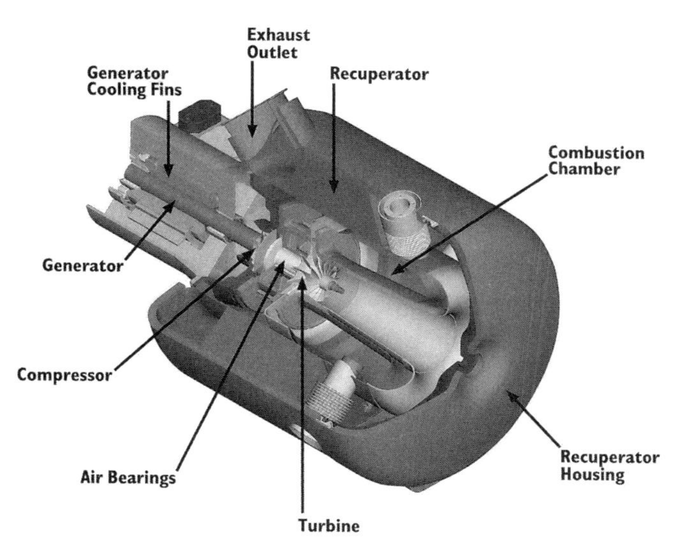

FIG. 10.3 Recuperated 30-kW microturbine. (Courtesy of Capstone Turbine Corporation)

compressor, combustor, turbine, alternator, recuperator (a device that captures waste heat to improve the efficiency of the compressor stage), and generator, as illustrated in Fig. 10.3.

In an unrecuperated or simple cycle turbine, compressed air is mixed with fuel and burned under constant pressure conditions. The resulting hot gas expands through the turbine to generate work. Simple cycle microturbines typically have efficiencies around 15 to 18%. However, these units are lower in capital cost and offer higher reliability. Recuperated microturbines use an air to air heat exchanger to transfer heat from the exhaust air to the incoming air stream. As a result, recuperated microturbine efficiency can be as high as 30%. Advantages of microturbines are their small size, small number of moving parts, good efficiency in cogeneration, low maintenance cost, and fuel flexibility. Disadvantages are a low fuel-to-electricity efficiency and reduced power output with higher ambient temperatures and altitude. More recently, work is progressing toward the development of a 7.5-kW microturbine [10.23] that could be used in commercial vehicle APU applications. Challenges remain to ensure robust starting and operation under the harsh conditions typical for commercial vehicles. Given that the expected cost in high-volume manufacturing for these very small microturbines is estimated to be around $1000 per kW, it is unlikely that microturbines will find a significant market in commercial vehicle APU applications.

10.2.4 Truck Stop Electrification

In addition to mobile idle-reduction technologies discussed in the previous sections, truck idle emissions can also be avoided through Truck Stop Electrification (TSE). TSE can be installed at truck stops, service plazas, or rest areas to provide electric power to a truck parking area. TSE systems can be classified as onboard or off-board, depending on the type of services provided. With onboard TSE, truckers park and connect their trucks to an electric power source. Without idling, they can use the supplied electricity to recharge their onboard batteries and operate electric systems on the truck, such as sleeper cab heating and cooling, microwave ovens, refrigerators, televisions (TVs), telephones, personal computers, and other small appliances. TSE service providers charge truck owners between $1.25 and $1.75 per hour [10.24] for basic cooling and heating services depending on local utility rates, while additional services such as internet or satellite TV are extra. TSE off-board technologies provide heating and air conditioning infrastructure at truck stops. Cooling or heating is provided to the truck cab from an overhead unit to the truck cab via a hose connection. In addition to basic heating and cooling services, TSE users can purchase other services such as internet access, movies, and satellite programming. An example of such a system, manufactured by IdleAire Technologies Corporation, is shown in Fig. 10.4.

A number of studies [10.25, 10.26] have shown that truck stop electrification is an economically viable and effective approach to reducing idle emissions from long-haul trucks. Early demonstration projects were implemented

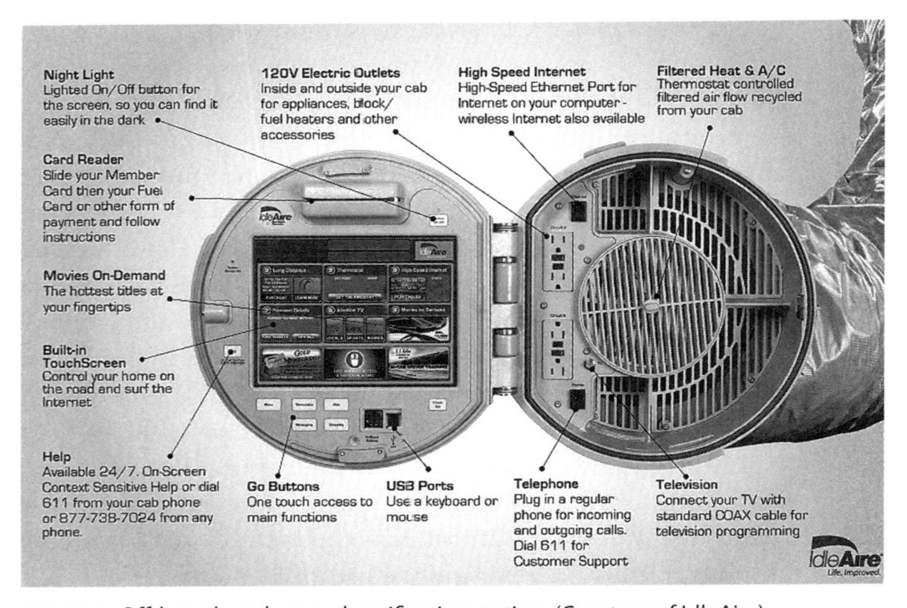

FIG. 10.4 Off-board truck stop electrification station. (Courtesy of IdleAire)

in the Northeast, predominantly New York State, as well as California and Texas. A more recent study [10.27] details an evaluation of major national truck corridors in an effort to identify appropriate distribution of electrified truck stops along these national corridors based on prevailing truck traffic patterns. The Texas Transportation Institute (TTI) maintains a website [10.28] with an interactive map of the United States with major truck routes and locations of electrified truck stops.

The Department of Energy Vehicle Technologies Program reports that a total of 134 electrified truck stops were in operation by the end of 2009 [10.29]. With an average of more than 75 units per truck stop [10.28], that means that more than 10,000 units are available throughout the nation.

10.3 Electrically Driven Boost System and Other Components

Diesel engines incorporate a number of components that are typically driven mechanically, such as water pumps and cooling fans. Replacing such components by electrically driven versions can deliver benefits in several ways. First, removing the always present parasitic losses associated with mechanical drives and only using the electrically driven components when needed can provide energy efficiency improvements. Second, by no longer being tied to engine speed, components can be sized more appropriately to meet the particular engine's design specifications.

Similarly for turbochargers, electrically driven or electrically assisted versions have been under development by a variety of companies over the last decade. Benefits for electrically driven or assisted engine boosting typically relate to improved performance due to the uncoupling of the compressor shaft speed from engine speed. This can lead to improved low-engine-speed torque output as well as better dynamic response.

10.3.1 Electrically Driven Turbocharging

A turbocharger is a radial fan pump driven by the energy of the exhaust gases of the diesel engine, and it shares a common shaft with a compressor. The turbine, when driven by the exhaust gases, spins the compressor. The compressor draws in ambient air and pumps it into the intake manifold at increased pressure, resulting in a greater mass of air entering the cylinders on each intake stroke.

Filling of the intake and exhaust manifolds, as well as consequent increase of the pressure and acceleration of the rotating components of the turbocharger, require a certain period of time. The time it takes from a step change in engine torque demand to delivery of that torque after the turbo spools up is called turbo lag. There are many options to reduce turbo lag: for example, through the use of variable-geometry turbines or two-stage sequential turbocharging, in which a smaller turbine will spool up at lower engine speeds. Another option is through the addition of an electric drive to the turbocharger main shaft. As a result, the dynamic performance of the turbocharger can be significantly improved. With electric assist, the turbocharger can be spooled up much faster, and subsequently turbo lag can be reduced from a typical 2–3 seconds to significantly less than 1 second.

In addition to providing increased engine responsiveness, electrically assisted turbocharging can provide significantly increased torque at low engine speeds. A turbocharger typically does not provide pressurized air to the intake manifold until higher engine speeds and loads are reached. Therefore, the engine obtains little benefit from turbocharging at low engine speeds and loads when the engine's exhaust does not contain sufficient energy. However, when the turbocharger is electrically assisted, it can be spooled up to deliver compressed air to the intake manifold, even with low exhaust energy content. The availability of higher torque at lower engine speeds allows the engine to be operated at lower speeds, while maintaining vehicle performance, resulting in reduced vehicle fuel consumption.

10.3.2 Electrically Driven Accessories

The primary reason to replace mechanically driven accessories with electrically driven ones is the elimination of unnecessary parasitic losses. Intelligent management of electrically driven accessories can provide fuel savings across a range of commercial vehicles. A program sponsored by the Department of Energy called the "More Electric Truck" [10.29] introduced a range of electrified accessories on a class 8 truck to evaluate the fuel saving benefits of electrified accessories. The project realized significant energy savings in evaluating a class 8 truck during a cross country trip with a range of electrified accessories powered by a fuel cell APU. It was found that with the equivalent of 7 L of diesel fuel, all electrical systems could be powered that would normally have taken 422 L of diesel fuel if the accessories had been powered by the OEM engine. As a note, specific efficiency gains of any similarly equipped class 8 tractors will be a function of vehicle load and driving cycle. Other estimates suggest annual fuel savings for a typical class 8 truck can be approximately 1200 gal of fuel [10.30].

Where mechanically driven accessories are linked to engine speed at all times of their operation, electrically operated accessories can be operated independently of engine speed. The benefits in terms of fuel savings then can be derived from avoiding the parasitic load altogether if the particular accessory component is not required to operate or to operate in the most efficient manner, regardless of engine speed. Another benefit is that the electrified component can be sized for its optimum operating condition without the constraint

TABLE 10.3 Conventional Component Power Requirements [10.31]

	Local Haul Power Required (kW)		Line Haul Power Required (kW)	
Component Name	Maximum Power	Average Power	Maximum Power	Average Power
A/C Compressor	4.5	2.2	4.5	2.2
Power Steering	4–11	2.4–6.6	4–11	0.4–1.1
Air Brake Compressor	Pumping 6.0 no Load 2.4	3.5	Pumping 6.0 no Load 2.4	2.3
Engine Fan	15–30	1.5–3.0	15–30	0.8–1.5
Alternator	Variable (1 kW max. / 0.7 kW avg.)			
Oil Pump	4.5	NA	4.5	NA
Coolant Pump	2	NA	2	NA

of engine speed derived requirements. For instance, a pump required to deliver a high flow rate at low engine speeds would be more than likely over-sized at higher engine speeds.

A 2002 study by NREL [10.3] attempted to quantify the potential energy savings of component electrification. The study focused on class 7 and 8 long-haul tractors, and more specifically on local-haul and line-haul segments for this class of vehicles. These two duty cycles represent use patterns for accessory devices and differ in the time that a component is on, idling, or disconnected. For example, in local-haul driving conditions, the truck is assumed to be in an urban environment. In this type of driving environment, it is generally true that power steering demands are higher than they are in a line-haul duty cycle. Therefore, the power steering pump will demand more power in the local-haul cycle compared to the line-haul duty cycle. Table 10.3 shows the typical power requirements for the various belt-driven auxiliary devices for this class of trucks.

Table 10.4 shows the typical duty cycles for the accessory components listed in Table 10.3.

Using the data from Table 10.3 and Table 10.4, simulations performed by NREL researchers shows that these mechanical accessory components can use as much as 15% of total fuel consumed over the City Suburban Heavy Vehicle Route (CSHVR) duty cycle.

10.3.2.1 Coolant Pump and Engine Cooling System

A variable-speed electric pump allows matching between cooling fluid flow and actual engine heat load, rather than engine speed. For example, the engine coolant pump is not required to operate until the engine is at operating

TABLE 10.4 Component Duty Cycles [10.31]

<table>
<tr><th>Component Name</th><th>Line-Haul Duty Cycle (% of time on)</th><th>Local-Haul Duty Cycle (% of time on)</th></tr>
<tr><td>A/C Compressor</td><td>50%</td><td>50%</td></tr>
<tr><td>Power Steering</td><td>10%</td><td>60%</td></tr>
<tr><td>Air Brake Compressor</td><td>5%</td><td>30%</td></tr>
<tr><td>Engine Fan</td><td>5%</td><td>10%</td></tr>
<tr><td>Alternator</td><td colspan="2">100%</td></tr>
<tr><td>Oil Pump</td><td>100%</td><td>100%</td></tr>
<tr><td>Coolant Pump</td><td>100%</td><td>100%</td></tr>
</table>

temperature. An electrically driven coolant pump can be used on demand, whereas a mechanically driven pump would be operational as soon as the engine starts. Decoupling coolant pump speed from engine speed gives the additional benefit of allowing the pump to be sized for coolant flow and pressure specifications independently of engine speed, typically leading to a smaller pump, requiring less power to operate. Similarly, the radiator cooling fans are controlled to operate as little as possible, while still providing the required cooling performance. The radiator fans are typically controlled using radiator output temperature, increasing speed as necessary to maintain a maximum radiator output temperature.

10.3.2.2 *Heating, Ventilation, and Air Conditioning (HVAC)*

As with the electrification of other mechanically driven components, replacing a mechanically driven A/C compressor with an electrical one results in multiple benefits. The direct load on the engine is reduced, as is the parasitic loss from A/C compressor clutch drag. The efficiency penalty of transforming mechanical energy to electrical is more than compensated for by the efficiency gains of decoupling the A/C compressor and engine duty cycles. The A/C compressor can now be controlled to run at its most efficient operating point, regardless of engine speed. The operation of the air conditioning system will thus be just as efficient whether the engine is idling or at different power levels. Electric HVAC systems are typically installed together with, but separate from, a diesel APU system and powered by a generator driven by the APU engine. The diesel APU-driven generator typically provides 3 to 6 kW of electrical power depending on the electric A/C compressor power rating as well as the power requirement of any additional electric accessories. The variable speed control of electric A/C compressors allows for operation at lower speeds once the cabin temperature is at a comfortable level, saving fuel and lowering power demands on the APU. Also, the variable speed feature is very powerful for implementing automatic temperature control, providing additional comfort for the driver.

10.3.2.3 Compressed Air Module

The air module supplies compressed air for braking and ride control. With an electric air compressor, the system uses the same basic components used in a standard truck air compressor system, standard dryer and controls, and dual air tanks. Compressor operation is functionally identical to the conventional belt-driven unit, but in this case, the compressor is driven by an electric motor. When increased air pressure is required, the air module controller instructs the motor to run until the pressure is restored. By decoupling the air compressor from the engine, it is possible to implement more efficient power management strategies for charging the system, saving fuel and lowering power demands on the APU.

Figure 10.5 shows the power requirement of the compressor module for fully loaded and unloaded conditions as a function of engine speed. Power consumed by auxiliary devices varies with engine speed, not load requirements. For this reason, mechanically driven accessories are typically over designed. As Fig. 10.5 shows, even under zero load conditions, the compressor module draws a relatively high load that can be entirely avoided when switching to an electric compressor module.

10.3.2.4 Power Steering

Similarly to the compressor module, a mechanical power steering pump suffers from relatively high parasitic losses under no-load conditions. As Table

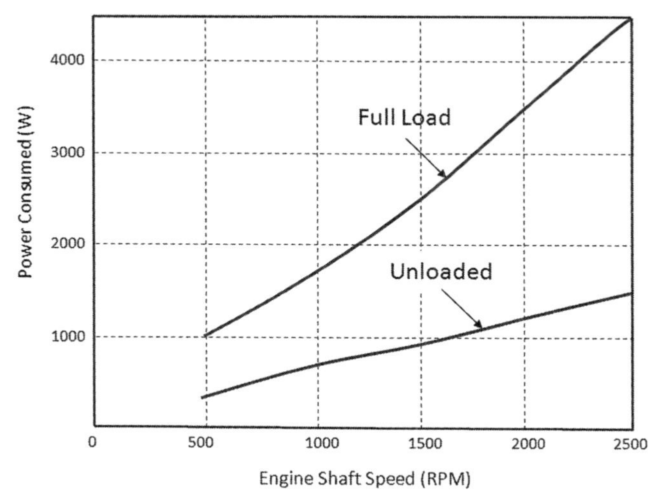

FIG. 10.5 Compressor Module Loading vs. Engine Speed [10.31]

10.3 shows, while the peak power requirements for the power steering pump are similar for local haul and line haul duty cycles, the average power requirements are drastically different. This is another indication that by switching to electrically driven alternatives, significant fuel savings could be accomplished.

10.4 Electrification for Plug-in Hybrid Commercial Vehicles

The previous section discussed accessory component electrification primarily as a means to improve energy efficiency for commercial vehicles. Electrification of previously mechanically driven components is typically achieved using low-voltage power, either 12 V or 42 V. With the advent of hybrid-electric vehicles, a much higher voltage energy source is available to power electric accessories, namely the high-voltage hybrid battery. This battery typically operates anywhere from 250 V DC to 420 V DC. The primary benefit of moving to this much higher voltage is that electric current will be significantly reduced, allowing for thinner wiring and smaller electric components. Accessory component electrification is not required for hybrid-electric vehicles. However, for Plug-in Hybrid Electric Vehicles (PHEV) with a required all-electric range, many mechanically driven components must be replaced with electrically driven alternatives. Those components that purely support engine operation, such as the oil pump, do not need to be electrified, because they are only required to operate when the engine itself is running.

The single greatest challenge for high-voltage electrification of accessory components is the fact that commercial vehicles are designed for a wide variety of applications. Powertrains and accessory components for commercial vehicles span a wide range of vehicle weight and power requirements as well as unique working conditions and duty cycles. Because of this wide range of requirements, it is economically not viable to design custom solutions for each and every application. The expense to engineer a custom solution is disproportionally large compared to the annual sales for a given application. The challenge is twofold:

1. While overall sales of commercial vehicles are quite high, single model/application sales can be quite low due to the high number of unique vehicle applications. For hybrid-electric vehicles in general in this market, the challenge is to build scale on the high-cost components, e.g., batteries and traction drive systems. That objective can lead to compromises. For instance, the electric motor that is right for one application may not have sufficient power or torque for another. Similarly, batteries may not have the right amount of energy storage capacity.
2. For hybrid-electric vehicles with extended all-electric-drive requirements, the challenge extends beyond the basic hybrid drive system to the

high- and low-voltage electric accessories. The design, power, and communication requirements for accessories with the same function (e.g., an electric power steering solution, or electric HVAC) can be vastly different, let alone the wide variety and mix of electrified components that could be integrated in any given application.

To address these challenges, a flexible approach is required that can allow for simple integration of various electrified components onto a wide range of commercial vehicles. For example, Fig. 10.6 shows a diagram of a Hybrid System Flexible Architecture proposed by Eaton Corporation. The Eaton base hybrid drive system has been in production since 2007 with more than 4000 units sold through 2010, into a wide variety of applications. Eaton is adding flexibility to the base hybrid drive system through the development of additional traction drive motors and batteries. This allows for significant commonality of these high-cost components across many different vehicle applications.

Through the introduction of a High Voltage Energy Management System or "junction box," a flexible architecture extension is created. Standardization of the interface between the High Voltage Energy Management System and the electric accessories essentially creates a "plug and play" high-voltage power bus.

10.4.1 Flexible Traction Drive System

On many commercial vehicles, a range of engine and powertrain options are available. However, choices for hybrid versions of those vehicles are very

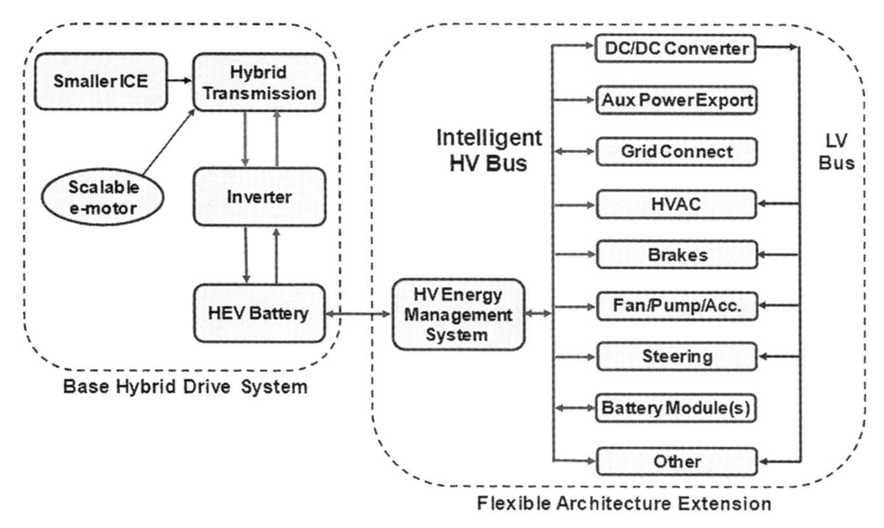

FIG. 10.6 Eaton Hybrid System Flexible System Architecture. (Courtesy of Eaton Corporation)

narrow. Typically only a single hybrid powertrain—internal combustion engine plus hybrid drive system—is available, and then only for select vehicle applications. In part this is due to the very slow growth of hybrid system sales in the commercial vehicle market, which in turn is partially due to the high acquisition cost of hybrid technology developed for commercial vehicles. Another reason is the aforementioned non-recurring engineering cost that is disproportionately high compared to vehicles sales of a given model.

One solution under development by industry is to take a more modular approach for the hybrid transmission or hybrid drive system. Eaton Corporation was one of the early developers of a parallel hybrid drive system that is particularly modular in design. As shown in Fig. 10.7, the Eaton Hybrid Drive Unit consists of an electrically actuated clutch, a standard six-speed automated manual transmission, and between the two a permanent magnet electric motor. Where the original hybrid transmission incorporated a 44-kW and 420-N·m electric motor, recent work has led to an optional higher power and torque motor, 65-kW and 620-N·m, respectively. The ability to match hybrid system performance to the engine and vehicle it is matched to is critical in improving overall powertrain and vehicle performance and fuel economy. Many other commercial vehicle hybrid system manufacturers are taking a similar approach.

10.4.2 Flexible High-voltage System Architecture

High-voltage systems on passenger cars are typically highly engineered and customized to the vehicle. The auto manufacturer can afford to spend more in the development stages because this cost is spread over many tens of thousands of vehicle sales annually. Unfortunately, not many of these high-voltage systems can be readily applied to commercial vehicles because the requirements and environmental conditions for the latter are often quite a bit more severe. As indicated in Section 10.3, a "plug and play" approach is under development to provide the necessary flexibility required to integrate electrified accessories across a wide range of commercial vehicles and platforms. In its simplest form, the High Voltage DC (HVDC) junction box merely provides a means to connect a range of electric accessories to a battery with typically a single high-voltage outlet. "Plug and play" connectivity for the electric accessories is provided through a standardized interface. The interface standard describes both the physical interface (connector and type), as well as electrical (voltage range, current capacity, etc.), and communication. No industry standard has been set yet as of the printing of this book, but efforts are underway in several SAE International standards committees. More highly developed versions of the HVDC junction box could provide additional functionality such as:

FIG. 10.7 Eaton Hybrid Drive Unit. (Courtesy of Eaton Corporation)

- Individual fusing
- Individual contactors and pre-charge circuitry
- Precharge current limiting
- Support for main (battery) contactor control
- Power and energy management across the entire high-voltage power bus
- Diagnostics and prognostics
- Insulation fault detection
- Support for a mixture of high- and low-current devices
- Support to plug in additional battery modules to increase system energy capacity

This approach is applicable in all hybrid vehicles that have a variety of electric accessories, including hybrid-electric vehicles, plug-in hybrid-electric vehicles, and electric vehicles.

10.5 External Charging and Plug-in Hybrid Commercial Vehicles

There are different levels of charging based on the power available at the charging location. SAE International has developed specifications and standards for the different charging levels in the United States. One specification, entitled J1772, defines Level I and Level II charging as well as the interface between the vehicle and the Electric Vehicle Service Equipment, or EVSE. The EVSE is a wall- or pedestal-mounted device that the vehicle connects to while charging. Even though it is commonly referred to as the "charger," the EVSE is not, in fact, a charger. The EVSE is simply an interface device between the vehicle and the electric grid. The vehicle itself will carry a Level I and/or II charger on board. Level I charging is specified for the NEMA outlet that most people are familiar with—the traditional home plug. This charging is relatively slow, with a maximum of 120 V and 12 A. At 1.8 kW, a 25-kW·h battery in a pure EV could take 12.5 h to charge, depending on its initial state of charge. Smaller PHEV batteries would, of course, take less time, with the Chevrolet Volt specified to take approximately 8 h to charge at Level I. Although Level I charging may be a sufficient solution for many PHEV owners, the lengthy charge times that are necessitated by the much larger EV batteries will likely convince most consumers to opt for higher-power Level II charging. Level II charging is specified at between 208 and 240 V (the voltage used in many homes by electric clothes driers, electric ovens, or well pumps). With the higher power used in Level II charging, EVSEs will have to be permanently mounted. Although Level II EVSEs are specified for charging at between 12 and 80 A, in practice, few vehicles will be able to charge at the maximum amperage rating; most vehicles are being designed to accept a Level II charge at no more than 30 A. Automakers are presumably making a tradeoff between customer preferences on charging times and the cost and weight associated with larger-capacity chargers. For example, the Nissan LEAF, with a 24-kW·h battery pack, is expected to take 4–8 h to charge with a 240-V supply. It should be noted that Level I and Level II charging utilize the same connector interface to the vehicle. The plug that actually plugs into the car is unchanged between the levels. What is different is how those plugs are connected to the grid.

SAE International has defined direct current (DC) fast charging, commonly referred to as Level III charging, as well. Designed for commercial applications, these chargers range from 30 kW to 250 kW, with the goal of a complete charge in less than 10 min. Level III chargers will be significantly more expensive than Level I or II chargers and are expected to be available at commercial charging establishments. For this reason, Level III chargers will be off-board, while Level I and II chargers will reside on board the vehicle. As an example, a Level III charger operating at 50 kW can fully charge a 24-kW·h battery in approximately 25 min and could cost between $25,000 and $50,000. This happens to be about the same as the cost of a typical gas station pump. Fast charging rates will likely not be limited by the details of the standard, but rather by grid infrastructure capability and the tolerance of the battery chemistry. Further details on both AC and DC charging, as defined by SAE, can be found in Table 10.5.

TABLE 10.5 SAE International Charging Configurations and Ratings Terminology [10.32]

Level	Description	Level	Description
AC Level 1 (SAE J1772™)	PEV includes on-board charger 120V, 1.4 kW @ 12 amp 120V, 1.9 kW @ 16 amp Est. charge time: PHEV: 7hrs (SOC*0% to full) BEV: 17hrs (SOC—20% to full)	***DC Level 1**	EVSE includes an off-board charger 200–450 V DC, up to 20 kW (80 A) Est. charge time (20 kW off-board charger): PHEV: 22 min. (SOC*—0% to 80%) BEV: 1.2 hrs. (SOC—20% to 100%)
AC Level 2 (SAE J1772™)	PEV includes on-board charger (see below for different types) 240 V, up to 19.2 kW (80 A) Est. charge time for 3.3 kW on-board charger PEV: 3 hrs (SOC*—0% to full) BEV: 7 hrs (SOC—20% to full) Est. charge time for 7 kW on-board charger PEV: 1.5 hrs (SOC*—0% to full) BEV: 3.5 hrs (SOC—20% to full) Est. charge time for 20 kW on-board charger PEV: 22 min. (SOC*—0% to full) BEV: 1.2 hrs (SOC—20% to full)	***DC Level 2** ***DC Level 3** (TBD)	EVSE includes an off-board charger 200–450 V DC, up to 80 kW (200 A) Est. charge time (45 kW off-board charger): PHEV: 10 min. (SOC*—0% to 80%) BEV: 20 min. (SOC—20% to 80%) EVSE includes an off-board charger 200–600V DC (proposed) up to 200 kW (400 A) Est. charge time (45 kW off-board charger): BEV (only): <10 min. (SOC*—0% to 80%)
***AC Level 3** (TBD)	> 20 kW, single phase and 3 phase		

*Not finalized

Voltages are nominal configuration voltages, not coupler ratings

Rated Power is at nominal configuration operating voltage and coupler rated current

Ideal charge times assume 90% efficient chargers, 150W to 12V loads and no balancing of Traction Battery Pack

Notes:

1) BEV (25 kWh usable pack size) charging always starts at 20% SOC, faster than a 1C rate (total capacity charged in one hour) will also stop at 80% SOC instead of 100%
2) PHEV can start from 0% SOC since the hybrid mode is available.
ver. 080110

10.5.1 Smart Grid

Around the world, modernization of the electric power grid plays a central role in efforts to increase energy efficiency, transition to renewable sources of energy, and reduce greenhouse gas emissions. Billions of dollars are being spent globally to define and develop the elements of what ultimately will be "smart" electric power grids.

Definitions and terminology vary somewhat, but all concepts of an advanced power grid for the 21st Century center around adding many varieties of digital computing and communications technologies, and services with the power-delivery infrastructure. Bidirectional flows of energy and two-way communication and control capabilities will enable an array of new functionalities and applications that go well beyond "smart" meters for homes and businesses. The Energy Independence and Security Act (EISA) of 2007 states that support for creation of a Smart Grid is the national policy. Some of the key characteristics of the Smart Grid cited in the act include [10.33]:

- Increased use of digital information and controls technology to improve reliability, security, and efficiency of the electric grid
- Dynamic optimization of grid operations and resources, with full cyber security
- Deployment and integration of distributed resources and generation, including renewable resources
- Development and incorporation of demand response, demand-side resources, and energy-efficiency resources
- Deployment of "smart" technologies for metering, communications concerning grid operations and status, and distribution automation
- Integration of "smart" appliances and consumer devices
- Deployment and integration of advanced electricity storage and peak-shaving technologies, including plug-in electric and hybrid-electric vehicles, and thermal-storage air conditioning
- Provision to consumers of timely information and control options
- Development of standards for communication and interoperability of appliances and equipment connected to the electric grid, including the infrastructure serving the grid
- Identification and lowering of unreasonable or unnecessary barriers to adoption of Smart Grid technologies, practices, and services

Over the long term, the integration of the power grid with the nation's transportation system has the potential to yield huge energy savings and other important benefits. Estimates of associated potential benefits include:

- Displacement of about half of our nation's net oil imports
- Reduction in U.S. carbon dioxide emissions by about 25%
- Reductions in emissions of urban air pollutants of 40 to 90%

While the transition to the Smart Grid may unfold over many years, incremental progress along the way can yield significant benefits. Table 10.6 provides a comparison of today's grid vs. the envisioned future smart grid for a number of characteristics.

TABLE 10.6 Characteristics and Benefits of the Smart Grid

Today's Grid	Smart Grid
Consumers are informed and non-participative with power systems	Informed, involved, and active consumers—demand response and distributed energy resources
Dominated by central generation—many obstacles exist for distributed energy resources interconnection	Many distributed energy resources with plug-and-play convenience focus on renewables
Limited wholesale markets, not well integrated—limited opportunities for consumers	Mature, well integrated wholesale markets, growth of new electricity markets for consumers
Focus on outages—slow response to power quality issues	Power quality is a priority with a variety of quality/price options—rapid resolution of issues
Little integration of operational data with asset management—business process silos	Greatly expanded data acquisition of grid parameters—focus on prevention, minimizing impact on consumers
Responds to prevent further damage—focus in on protecting assets following fault	Automatically detects and responds to problems—focus on prevention, minimizing impact on consumers
Vulnerable to malicious acts of terror and natural disasters	Resilient to attack and natural disasters with rapid restoration capabilities

In the United States, electric-power generation accounts for about 40% of human-caused emissions of carbon dioxide, the primary greenhouse gas [10.35]. The Electric Power Research Institute has estimated that by 2030, Smart Grid-enabled (or facilitated) applications—from distribution voltage control to broader integration of intermittent renewable resources to electric transportation vehicles—could reduce the nation's carbon-dioxide emissions by 60–211 million metric tons annually [10.36]. The opportunities are many, and the returns can be sizable. If the current power grid were just 5% more efficient, the resultant energy savings would be equivalent to permanently eliminating the fuel consumption and greenhouse gas emissions from 53 million cars [10.34].

10.5.2 Standards Development

On September 24, 2009, U.S. Commerce Secretary Gary Locke unveiled an accelerated plan for developing standards to transform the U.S. power distribution system into a secure, more efficient and environmentally friendly Smart Grid. More than 80 initial standards have been identified [10.37] that will support the interoperability of all the various pieces of the system of systems that the grid is. These systems range from large utility companies and distribution networks to individual homes and electronic devices. A number of key standards-developing organizations were identified in the NIST Framework and Roadmap for Smart Grid Interoperability [10.34]:

- SAE International—Vehicle to Grid (V2G) communication, physical plug
- Zigbee Alliance—Home communications
- American National Standards Institute (ANSI)—Metering
- Institute of Electrical and Electronic Engineers (IEEE)—Electric Vehicle Infrastructure
- National Electrical Manufacturers Association (NEMA) and Underwriters Laboratories (UL)—Building and product

10.5.2.1 *Vehicle-to-Grid Standards*

SAE standards committees have published a number of recommended practices and standards related to V2G operability and continue to develop additional standards. A brief summary of the various published and pending standards follows:

SAE J2293/1: Energy Transfer Systems for Electric Vehicles—Part 1: Functional Requirements and System Architectures.

This document describes the total EV-ETS (Energy Transfer System) and allocates requirements to the EV or EVSE for the various system architectures. It requires an SAE J1850-compliant network for communicating data and control information between an EV and EVSE. This document was published July 2008.

SAE J2293/2: Energy Transfer Systems for Electric Vehicles—Part 2: Functional Requirements and System Architectures.

This document describes the SAE J1850-compliant communication network between the EV and EVSE for this application (ETS Network). It treats the network as a system with the EV and EVSE from Part 1 as external elements using the network. This document was published July 2008.

SAE J2758: Determination of the Maximum Available Power from a Rechargeable Energy Storage System on a Hybrid Electric Vehicle

This document describes a test procedure for rating peak power of the Rechargeable Energy Storage System (RESS) used in a combustion engine Hybrid Electric Vehicle (HEV). This document was published April 2007.

SAE J2344: Guidelines for Electric Vehicle Safety

Technical guidelines relating to safety for Electric Vehicles (EVs) during normal operation and charging that should be considered when designing electric vehicles for use on public roadways. This document covers electric vehicles having a gross vehicle weight rating of 4536 kg (10,000 lb) or less that are designed for use on public roads. This document was published March 2010.

SAE J1711: Recommended Practice for Measuring the Exhaust Emissions and Fuel Economy of Hybrid-Electric Vehicles

Sets recommended practices for measuring the Exhaust Emissions and Fuel Economy of Hybrid-Electric Vehicles, Including Plug-in Hybrid Vehicles. It provides a foundation to assist government regulatory agencies in developing emissions and fuel economy certification and compliance tests for HEVs. This document was published June 2010.

SAE J2711: Recommended Practice for Measuring Fuel Economy and Emissions of Hybrid-Electric and Conventional Heavy-Duty Vehicles

Sets recommended practices for measuring the Exhaust Emissions and Fuel Economy of Hybrid-Electric Heavy Duty Vehicles, Including Plug-in Hybrid Heavy Duty Vehicles. It provides a foundation to assist government regulatory agencies in developing emissions and fuel economy certification and compliance tests for HEVs. This document was originally published September 2001, but is currently being reworked to include plug-in hybrid vehicles.

SAE J2931: Electric Vehicle Supply Equipment (EVSE) Communication Model

This SAE Recommended Practice establishes the digital communication requirements for the Electric Vehicle Supply Equipment (EVSE) as it interfaces with a Home Area Network (HAN), Energy Management System (EMS), or the Utility grid systems. This Recommended Practice provides a knowledge base addressing the communication medium functional performance and characteristics, and interoperability to other EVSEs, Plug-In Vehicles (PEVs) and is intended to complement J1772™ but address the digital communication requirements associated with smart grid interoperability. This document is a work in progress.

SAE J2931/2: Inband Signaling Communication for Plug-in Electric Vehicles

This SAE Recommended Practice J2931/2 establishes the requirements for physical layer communications using Inband Signaling between Plug-In Vehicles (PEV) and the EVSE. This also enables the onward communications via an EVSE bridging device to the utility smart meter or Home Area Network (HAN). This is known as Frequency Shift Keying (FSK) and is similar to Power Line Carrier (PLC) but utilizes the J1772 Control Pilot circuit. This document is a work in progress.

SAE J2841: Definition of the Utility Factor for Plug-in Hybrid Electric Vehicles Using National Household Travel Survey (NHTS) Data

The total fuel and energy consumption rates of a Plug-In Hybrid Electric Vehicle (PHEV) vary depending upon the distance driven. Total distance between charge events determines how much of the driving is performed in each of the two fundamental modes. An equation describing the portion of driving in each mode is defined. This document is a work in progress.

SAE J2836/1: Use Cases for Communication between Plug-in Vehicles and the Utility Grid

Identifies the equipment (system elements) and interactions to support grid-optimized AC or DC energy transfer for plug-in vehicles, as described in SAE J2847/1. Key system elements include the vehicle's rechargeable energy storage system (RESS), power conversion equipment (charger and/or inverter), utility meter, etc. This document was published April 2010.

SAE J2836/2: Use Cases for Communication between Plug-in Vehicles and the Supply Equipment (EVSE)

J2836/2 establishes use cases for communication between plug-in electric vehicles and the off-board charger, for energy transfer and other applications. J2836/2 use cases must be supported by SAE J2847/2. This document is a work in progress.

SAE J2836/3: Use Cases for Communication between Plug-in Vehicles and the Utility Grid for Reverse Power Flow

Establishes use cases for communication between plug-in electric vehicles and public electric power grid, a home branch circuit or a isolated micro grid, for reverse energy transfer and other applications. This document is a work in progress.

SAE J2836/4: Use Cases for Diagnostic Communication for Plug-in Vehicles

Provides information required for diagnostics and J2847/4 will include the detail messages to provide accurate information to the customer and/or service personnel to identify the source of the issue and assist in resolution. This document is a work in progress.

SAE J2836/5: Use Cases for Communication between Plug-in Vehicles and their Customers

Describes how the customer will be able to interact with the PEV as it charges/discharges. Identifies information and control for each session, including status, updates and potential changes. This document is a work in progress.

SAE J2847/1: Communication between Plug-in Vehicles and the Utility Grid

J2847/1 identifies the communication medium and criteria for the Plug-in Electric Vehicle (PEV) to connect to the utility for Level 1 & 2 AC energy transfer. This document was published June 2010.

SAE J2847/2: Communication between Plug-in Vehicles and off-board DC Chargers

Identifies additional messages for DC energy transfer to the PEV. The specification supports DC energy transfer via Forward Power Flow (FPF) from grid-to-vehicle. This document is a work in progress.

SAE J2847/3: Communication between Plug-in Vehicles and the Utility Grid for reverse power flow

Identifies additional messages for DC energy transfer to the PEV. The specification supports DC energy transfer via DC Reverse Power Flow (RPF) from vehicle-to-grid. This document is a work in progress.

SAE J2847/4: Diagnostic Communication for Plug-in Vehicles

Establishes the communication requirements for diagnostics between plug-in electric vehicles and the EV Supply Equipment (EVSE) for charge or discharge sessions. This document is a work in progress.

SAE J2847/5: Communication between Plug-in Vehicles and their Customers

Establishes communication requirements between Plug-in Vehicles and the internet. This document is a work in progress.

SAE J1772: SAE Electric Vehicle Conductive Charge Coupler

General requirements for the electric vehicle conductive charge system and coupler for use in North America. Defines a common electric vehicle conductive charging system architecture, including operational requirements and the functional and dimensional requirements for the vehicle inlet and mating connector. This document was published January 2010.

10.5.2.2 *Home Communications*

ZigBee Smart Energy is the world's leading standard for interoperable products that monitor, control, inform, and automate the delivery and use of energy and water. It helps create greener homes by giving consumers the information and automation needed to easily reduce their consumption and save money, too. This standard supports the diverse needs of a global ecosystem of utilities, product manufacturers, and government groups as they plan to meet future energy and water needs.

All ZigBee Smart Energy products are certified to perform regardless of manufacturer, allowing utilities and consumers to purchase with confidence. ZigBee certified products make it easy for utilities and governments to deploy smart grid solutions that are secure, easy to install, and user-friendly. Some of the world's leading utilities, energy service providers, product manufacturers, and technology companies are supporting the development of ZigBee Smart Energy.

ZigBee Smart Energy version 2.0 is currently under development in cooperation with a number of other standards development groups. It will offer IP-based control for advanced metering infrastructure and home area networks. This version will not replace ZigBee Smart Energy version 1, rather it will offer utilities and energy service providers another choice when creating their advanced metering infrastructure and home area networks (HANs).

ZigBee Smart Energy version 2.0 will feature control of plug-in electric vehicle (PEV) charging, installation, configuration, and firmware download for HAN devices, prepay services, user information and messaging, load control, demand response, and common information and application profile interfaces for wired and wireless HANs.

10.5.2.3 *Other Standards Under Development*

A number of other organizations such as NIST, ANSI, IEEE, ISO, IEC, NEMA, UL, and so on, are in aggregate developing many dozens of standards that will enable interoperability of the Smart Grid, spanning the entire system of systems from power generation to the individual electrical devices in homes and commercial venues. Other charger standards are listed in Table 10.7.

10.5.3 Charging Infrastructure

Electric vehicle supply equipment will be needed to charge the battery in plug-in electric vehicles once depleted. Most commercial vehicles that are good candidates for early adoption of plug-in hybrid technology are operated out of central fleet facilities and have relatively well defined routes. Charging these centrally operated vehicles is relatively straightforward because there is little

TABLE 10.7 List of Other Charger Standards

Region	Standards
US	**UL** • UL2594—AC Charging System • UL2202—DC Charging System • UL2231—Personnel Protection • UL1998—Software Protection Systems **NFPA 70 NEC Article 625** **FCC Part 15—EMI** **NIST SGIP** • PAP11 • V2G PEVWG **NEMA 05EV EVSE Section** **IEEE** • 1547—Interconnection of DER • P1901—Broadband PLC • P2030.1—Electric Transportation Infrastructure
Europe	**IEC** • IEC 61850-1,2-1,2-2,2-3—Charging System • IEC 62196—Charging Connectors • IEC TC-69—EV Charging Task Group • IEC SC-23H—EV Charging Couplers **ISO** • ISO/IEC JWG SC3—Utility Communication
Asian Pacific	**China** • GB/T 18487.1-2001 (eqv IEC 61851-1)—Charging System • GB/T 11918-2001—Charging Connectors (AC & DC) • Q/CSG 11516—Charging System and DC Charge Control • GB/T XXXX-201X—Charging System, Connectors, and DC Charge Control • CISPR—EMC/Wireless **Japan** • JARI±AC Charging System and Connector (identical to SAE J1772™) • CHAdeMO—DC Charging System, Connector, and communication Protocols

immediate need to develop an entire charging infrastructure to support these vehicles. However, there are examples, as well, of early deployment of plug-in vehicles that will not be charged at a central facility. One example is a large Department of Energy funded program to develop and deploy a demonstration fleet of 250 plug-in hybrid-electric Ford F550 based utility trouble trucks. These aerial or "bucket" trucks typically go home with the operator to allow for rapid response in case of downed power lines, power outages, or other emergencies.

While a substantial portion of charging can be done overnight at a centralized fleet facility or at home, public charging options will provide drivers with added flexibility. With limited exceptions, public charging infrastructure does not exist today.

The widespread adoption of electric vehicles beyond those operated from central facilities will depend to a great extent on deployment of a public charging infrastructure. Based on existing battery technology, both PHEVs and EVs will require relatively frequent recharging to benefit from electric propulsion. Because PHEVs will maintain the use of an internal combustion engine, the recharging issue is less of a constraint to mobility. Pure electric vehicles, on the other hand, will most certainly require reliable access to charging units while drivers are carrying out daily commutes and other trips that extend beyond the base range of the battery. In many instances charging will take hours instead of minutes. It is very likely that most vehicles will be equipped with SAE J1772 compliant plugs that will allow them to use every charging facility available in the nation. It is not yet clear who will own and operate the charging facilities, who will provide and be paid for the electrical power, or on what terms and at what rates it will be sold. Level III chargers will provide fast and convenient charging as well as charging for vehicles that are traveling beyond the charge-depleting range of their batteries without time to stop for slower charges.

10.5.3.1 *Level I and II Charging*

Most vehicle charging equipment will fall in the Level I and II category. Level I as a 120-Vac, 15-A platform will likely take from 8 to 20 h for a full charge, depending on battery capacity. Level II, designed as a 208-Vac to 240-Vac, up to 80-A system, could cut charge times to 3 to 6 h (Fig. 10.8 (a)). The battery chargers themselves are carried on board the vehicle and will rectify and convert AC voltage from the grid to a DC voltage level the vehicle battery can accept. The Level I and II public or private Electric Vehicle Service Equipment (EVSE) is simply an interface device between the vehicle and electric grid. In the middle of 2010, Coulomb Technologies announced its Chargepoint America program. With substantial funding support from the U.S. Department of Energy, through the American Reinvestment and Recovery Act, Coulomb will provide nearly 5000 Level I/II charging stations to program participants in Texas, Michigan, California, New York, Florida, Washington State, and Washington DC. Coulombs Chargepoint network also forms the basis for EV rollout plans of Ford and General Motors. Many other companies are offering charging solutions as well: General Electric, Schneider Electric, Eaton, Nissan Motor, AeroVironment, Siemens, Panasonic, Samsung, Sanyo, Leviton, and so on. Residential properties will form the largest market for Level I/II chargers, while public charging facilities will likely be Level II or Level III compliant.

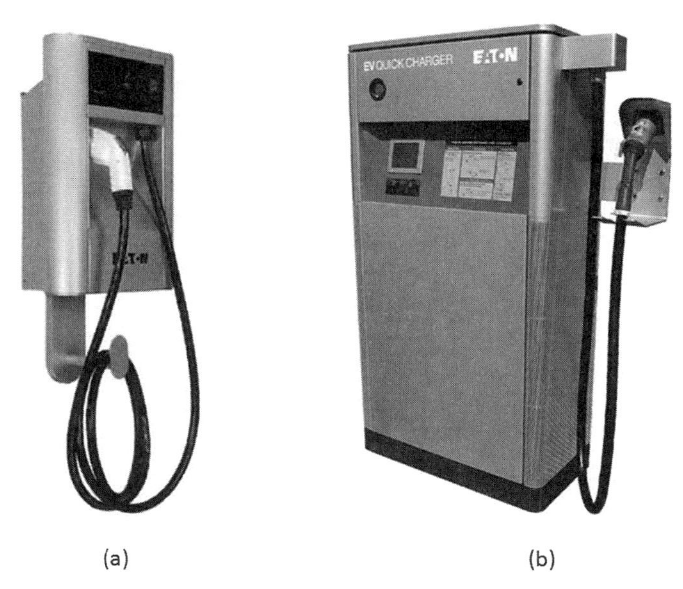

(a) (b)

FIG. 10.8 (a) Level 2 EV Charger; (b) Level 3 DC charger. (Courtesy of Eaton Corporation)

10.5.3.2 Level III Charging

While Level I and II battery chargers are carried on board the vehicle, Level III DC quick chargers (Fig. 10.8 (b)) are located off the vehicle, for both size and cost considerations. Tokyo Electric Power Company (TEPCO) has developed patented technology and a specification for high-voltage (up to 500 Vdc) high-current (125 A) automotive fast charging via a standardized Level III DC fast charge connector. This work formed the basis for the CHAdeMO DC fast charging protocol. CHAdeMO is a partnership that was formed by The Tokyo Electric Power Company, Nissan, Mitsubishi, and Fuji Heavy Industries. Toyota later joined as its fifth executive member. In addition to a safety interlock to avoid energizing the connector before it is safe (similar to SAE J1772), the car transmits battery parameters to the charging station, including voltage at which to stop charging, target voltage, and total battery capacity, and while charging the station has to vary its output current according to signaling from the car.

In the USA, Aker Wade Power Technologies has entered into a licensing agreement with TEPCO to manufacture and market Level III DC fast chargers for electric vehicles. Eaton Corporation has demonstrated a CHAdeMO-compatible DC Quick Charger recharging Mitsubishi iMiEV cars. The Eaton DC charger specification is shown in Table 10.8. Other companies that have developed DC fast charging stations compatible with CHAdeMO protocols are AeroVironment, ABB, and Ecotality.

TABLE 10.8 Eaton DC Charger Specifications

Input voltage	208VAC
Input current	125A 3 phase 3 wire 50/60 Hz
Output voltage	400VDC
Output power	50kW Max
Output current	10A–125A
Connector/cable	CHAdeMO compliant
Cable length	9 foot
Dimensions(H × W × D) in inches (mm)	66.00″ × 44.00″ × 17.75″ (1675 × 1120 × 450)
Charge time	20–30 minutes
Charging station weight	Approximate weight 772lbs (350kg)
Operation	Touch screen interface, start and stop buttons, emergency stop button
Operating environment	Ambient temperature: −10 to 40°C (14 to 104°F) Ambient humidity: 5 to 80% Altitude: 1,000 m (3,281 ft) or lower Atmosphere: Containing no corrosive gas
Enclosure	NEMA Type 3R
Efficiency	90% or greater Secure connector locking system Connector insulation verification system Integrated overcurrent protection

10.6 References

10.1 van Amburg B., "Hybrids hit the Street," Electric Perspectives, ABI/INFORM Global, pp. 18–28, Nov/Dec 2005.

10.2 HTUF fleet fuel economy reference, 2011.

10.3 Environmental Protection Agency Technical Bulletin EPA420-F-04-009, Washington, DC, February 2004.

10.4 Gaines, L., Vyas, A., Anderson, J. L., "Estimation of Fuel Use by Idling Commercial Trucks," 85th Annual Meeting of the Transportation Research Board, Washington, DC, January 22–26, 2006.

10.5 American Transportation Research Institute, Compendium of Idling Regulations, Alexandria, VA, June 2009.

10.6 Lim, H., "Study of Exhaust Emissions from Idling Heavy-Duty Trucks and Commercially Available Idle-Reducing Devices," EPA420-R-02-025, Washington, DC, October 2002.

10.7 California Air Resources Board, California Code of Regulations, Title 13, Division 3, Article 1, Chapter 10, Section 2485, "Airborne Toxic Control Measure to Limit Diesel-Fueled Commercial Motor Vehicle Idling," Sacramento, CA.

10.8 Environmental Protection Agency Verified Idle Reduction Technologies. http://www.epa.gov/SmartwayLogistics/transport/what-smartway/verified-technologies.htm#idle.

10.9 American Transportation Research Institute, Idle Reduction Technology—Fleet Preferences Survey, Prepared for the New York State Energy Research and Development Authority, Alexandria, VA, February 2006.

10.10 Brodrick, C. J., Farshchi, M., Dwyer, H. A., Gouse III, S. W., Mayenburg, M., Martin, J., "Demonstration of a proton exchange membrane fuel cell as an auxiliary power source for heavy trucks," SAE Technical Paper Series: 2000-01-3488, SAE International, Warrendale, PA, 2000.

10.11 Venturi, M., Martin, A., "Liquid fuelled APU fuel cell system for truck application," SAE Technical Paper Series: 2001-01-2716, SAE International, Warrendale, PA, 2001.

10.12 Montemayor, A., "Phased introduction of fuel cells into a Class 8 tractor," Southwest Research Institute, San Antonio, TX, 2002.

10.13 Lutsey, N., Brodrick, C. J., Sperling, D., Dwyer, H. A., "Markets for fuel cell auxiliary power units in vehicles: a preliminary assessment," Transportation Research Board: Energy, Air Quality, and Fuels 2003, 1842:118–26, Washington, DC, 2003.

10.14 Lutsey, N., Brodrick, C. J., Lipman, T., "Analysis of Potential Fuel Consumption and Emissions Reductions from Fuel Cell Auxiliary Power Units in Long-Haul Trucks," *Journal of Energy*, Vol. 32:2428–2438, 2007.

10.15 Liu, D. J., Krumpelt, M., Chien, H. T., Sheen, S. H., "Catalyst and Fuel Mixing for Diesel Reformer in Fuel Cell Auxiliary Power Unit," Electrochemical Technology Program, Argonne National Laboratory, Argonne, Illinois, 2004.

10.16 Lutsey, N., "Fuel Cells for Auxilliary Power in Trucks: Requirements, Benefits, and Marketability," Institute of Transportation Studies, University of California, Davis, 2003.

10.17 Dobbs, H. H., Krause, T., Kumar, R., Krumpelt, M., "Diesel-Fueled Solid Oxide Fuel Cell Auxiliary Power Units for Heavy-Duty Vehicles," Argonne National Laboratory, Argonne, Illinois, 2000.

10.18 Filipi, Z., Louca, L., Stefanopolou, A., Pukrishpan, J., Kittirungsi, B., Peng, H., "Fuel Cell APU for Silent Watch and Mild Electrification of a Medium Tactical Truck," SAE 2004-01-1477, SAE World Congress, Detroit, Michigan, 2004.

10.19 Surampudi, B., Redfield, J., Ray, G., Montemayor, A., Walls, M., McKee, H., Edwards, T., Lasecki, M. "Electrification and Integration of Accessories on a Class-8 Tractor," SAE 2005-01-0016, SAE World Congress, Detroit, Michigan, 2005.

10.20 The California Energy Commission—Distributed Energy Resource Guide: http://www.energy.ca.gov/distgen/equipment/microturbines/microturbines.html.

10.21 Scott, Walter G., "Micro-turbine generators for distribution systems," *Industry Applications Magazine*, pp. 57–62, IEEE, 1998.

10.22 Hamilton, S. L., "Micro turbine generator program," Proceedings of the 33rd Annual Hawaii International Conference on System Sciences, 2000.
10.23 McDonald, C. F., Rodgers, C., "Small Recuperated Ceramic Microturbine Demonstrator Concept," *Applied Thermal Engineering*, Vol. 28, pp. 60–74, 2008.
10.24 California Energy Commission Publication No. CEC-600-2006-001-FS, Sacramento, California, June 2006.
10.25 Perrot, T. and Panich, M., "Truck Stop Electrification (TSE) Market Study and Preliminary Report," Submitted to New York State Energy Research and Development Authority (NYSERDA), 2001.
10.26 Perrot, T. L., Constantino, M. S., Kim, J. C., Tario, J. D., Hutton, D. B., and Hagan, C., "Truck Stop Electrification as a Long-Haul Tractor Idling Alternative," Transportation Research Board 83rd Annual Meeting, Washington, DC, 2004.
10.27 Zietsman, J., Linhua, L., Bochner, B. S., Villa, J. C., Bubbosh, P., "National Deployment Strategy for Truck Stop Electrification," Transportation Research Board 86th Annual Meeting, Washington, DC, 2007.
10.28 Zietsman, J., "National Deployment Strategy for Truck Stop Electrification," http://tse.tamu.edu/, assessed, January 2012.
10.29 "Truck Stop Electrification Sites," http://www1.eere.energy.gov/vehiclesandfuels/facts/2010_fotw628.html, assessed, January 2012.
10.30 Algrain, M. C., Lane, W. H., Orr, D. C., "A Case Study in the Electrification of Class-8 Trucks," Electric Machines and Drives Conference Proceedings, IEEE, 2003.
10.31 Hendricks, T., O'Keefe, M., "Heavy Vehicle Auxiliary Load Electrification for Essential Power System Program: Benefits, Tradeoffs, and Remaining Challenges," SAE Technical Paper Series: 2002-01-3135, 2002.
10.32 Kissel, G. J., "SAE Charging Configurations and Rating Terminology," Plug-in 2010, July 28, 2010.
10.33 Energy Independence and Security Act of 2007 [Public Law No: 110-140] Title XIII, Sec. 1301.
10.34 U.S. Department of Energy, The Smart Grid: an Introduction, 2008.
10.35 Energy Information Administration, U.S. Department of Energy, "U.S. Carbon Dioxide Emissions from Energy Sources, 2008 *Flash* Estimate," May 2009.
10.36 Electric Power Research Institute, *The Green Grid: Energy Savings and Carbon Emissions Reductions Enabled by a Smart Grid,* 1016905 Technical Update, June 2008.
10.37 NIST Framework and Roadmap for Smart Grid Interoperability Standards, Release 1.0, NIST Special Publication 1108, National Institute of Standards and Technology, January 2010.

Chapter 11

Hybrid Powertrain System Modeling, Simulation, Validation, and Certification

11.1 Model-based Control System Development

Model-based design and analysis is a process that enables faster, more cost-effective development of dynamic systems, such as hybrid powertrains. The model can provide a good interface among the relevant engineers who have different technical fields, and the required knowledge can be shared through the model. Model-based control is characterized by a plant model that describes the dynamics of the system to be controlled and allows closed-loop simulation. The number of the model-based control applications, which often can be derived heuristically, has been increasing.

In model-based design, a system model is at the center of the development process, from requirements development, through design, implementation, and testing. The model is an executable specification that is continually refined throughout the development process. After model development, simulation shows whether the model works correctly. The overall goal is to ensure first-pass success when building the physical prototype.

An overview of a model-based development process for hybrid powertrain controls is shown in Fig. 11.1. The process begins at the top left-hand side of the V-shaped diagram with requirement and target setting and proceeds down the left side of the V to design and implementation. Then the process goes up the right side of the V to testing, closed-loop MIL, HIL, and finally to Vehicle Integration and Calibration. MIL and HIL refers to model-in-the-loop and hardware-in the- loop, respectively. In each case, the model, software, and hardware identifiers refer to the particular implementation of the controls that is integrated closed-loop with a plant model representing the balance of the vehicle model.

One of the widely used tools for model-based design and analysis is the Simulink® software [11.1, 11.3]. Simulink® is an environment for multi-domain simulation and model-based design for dynamic and embedded systems. It

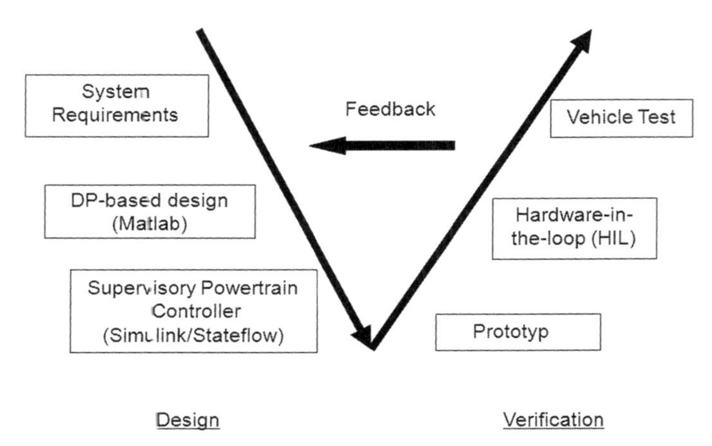

FIG. 11.1 Model-based development process of HEV controls.

provides an interactive graphical environment and a customizable set of block libraries that let you design, simulate, implement, and test a variety of time-varying systems, including communications, controls, signal processing, video processing, and image processing [11.4]. The most commonly used application software tools for model-based design and analysis are explained as follows:

11.1.1 PSAT

The Powertrain Analysis Toolkit (PSAT), developed by Argonne National Laboratory, has been used to perform vehicle simulations [11.5]. PSAT is a forward-looking simulation package (also called driver-driven). A driver model follows any standard or custom driving cycle, sending a power demand to the vehicle controller, which, in turn, sends a demand to the propulsion components. Component models react to the demand and feed their status back to the vehicle controller, and the process iterates to achieve the desired result. Each component model is a Simulink/Stateflow box, which uses the Bond graph formalism. The components boxes are then "assembled" according to the powertrain configuration chosen by the user in the Graphical User Interface (GUI). The user-friendly GUI (Fig. 11.2) allows the user to build a vehicle model within a few minutes. First, the user selects the drivetrain configuration. Each configuration is built according to user input so that vehicle architectures can be compared and the most appropriate one selected. More than 300 preselected configurations are available. Second, the user selects a model for each powertrain component, its initialization file, and tunes the initial parameters. Similarly, the controller strategy is chosen and tuned. The user then selects the combination of cycles to be simulated, and finally launches the simulation. Once the simulation is completed, the GUI provides

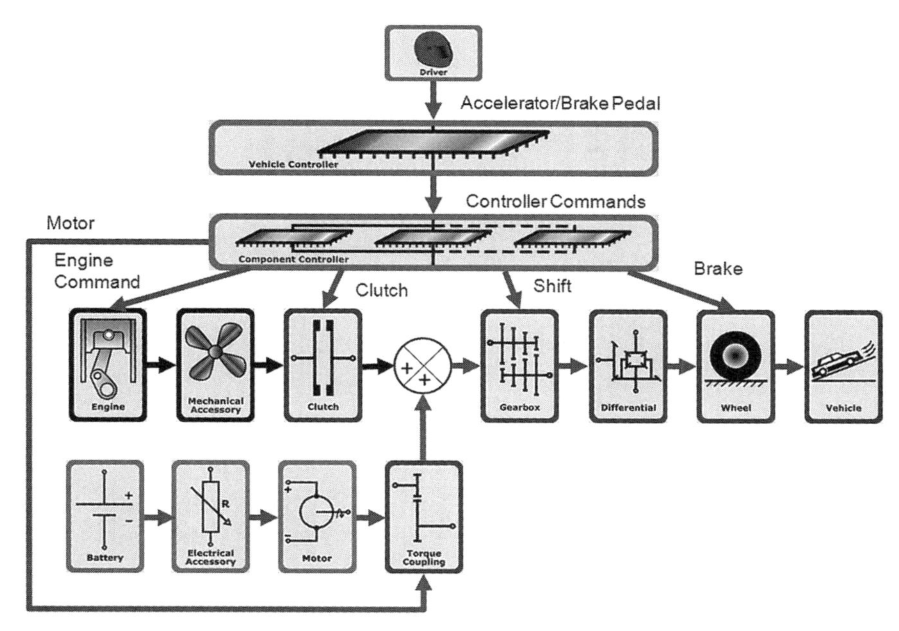

FIG. 11.2 Example of vehicle model in PSAT [11.5, 11.6, 11.7].

the user with a wide range of plots and calculations, particularly useful for analysis.

The plant models used in the study were based on steady-state lookup tables to represent the losses. The shifting algorithm and vehicle-level control logics have been developed to be adapted to any combination of components and validated for several applications with OEMs. All the results are provided for hot operating conditions.

11.1.2 Dymola

Dymola (Dynamic Modeling Laboratory) is a complete tool for modeling and simulation of integrated and complex systems for use within automotive, aerospace, and other applications. Dymola is a commercial modeling and simulation environment based on the open Modelica modeling language [11.2, 11.9].

The architecture of the Dymola program is shown in Fig. 11.3. Dymola is based on the use of Modelica models stored on files and has a powerful graphic editor for composing models. Dymola can also import other data and graphics files and contains a symbolic translator for Modelica equations generating C-code for simulation. The C-code can be exported to Simulink and hardware-in-the-loop platforms.

Dymola has powerful experimentation, plotting, and animation features. Scripts can be used to manage experiments and to perform calculations. Automatic documentation generator is provided.

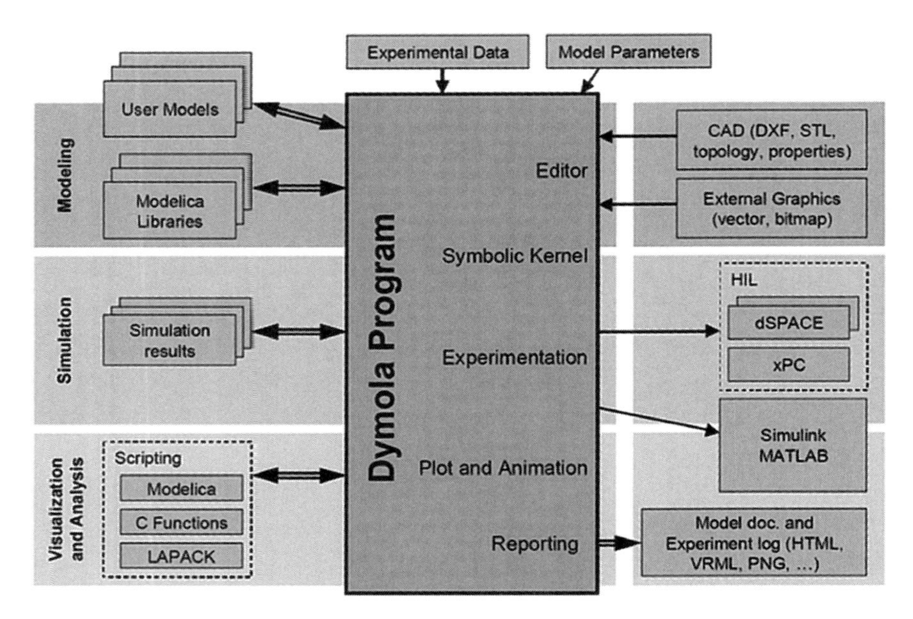

FIG. 11.3 Dymola program features.

11.1.3 ADVISOR

ADVISOR (ADvanced VehIcle SimulatOR) was created by the U.S. Department of Energy's National Renewable Energy Laboratory's (NREL) Center for Transportation Technologies and Systems [11.10]. It's a flexible modeling tool that rapidly assesses the performance and fuel economy of conventional, electric, hybrid, and fuel cell vehicles. The user changes component and vehicle specifications—such as electric motors, batteries, engines, and fuel cells—and ADVISOR simulates the vehicle's response under different driving conditions. This cutting-edge tool runs in the MATLAB/Simulink software environment.

ADVISOR is a "backward" Simulation program with limited forward-looking simulator. It is an empirical model that uses drivetrain component performance to estimate fuel economy and emissions on the given cycle as well as other performance-related parameters such as the acceleration and gradeability. The fuel economy can be assessed on any of the 50 available drive cycles or definitive test procedures under various test conditions.

11.1.4 Hardware-In-the-Loop (HIL) and Model-In-the-Loop (MIL)

Hardware-In-the-Loop is a form of real-time simulation. Hardware-In-the-Loop differs from pure real-time simulation by the addition of a real component in the loop.

The purpose of HIL simulation is to provide an effective platform for developing and testing real-time embedded systems. HIL simulation provides an effective platform by adding the complexity of the plant under control to the test platform. The complexity of the plant under control is included in test and development by adding a mathematical representation of all related dynamic systems.

The typical benefits of a Hardware-In-the-Loop system include: (a) allows for closed–loop dynamic testing; (b) automated test and verification; (c) test safety without damaging equipment or endangering lives.

Model-In-the-loop (MIL) testing refers to the scenario in which the test specimen is part real and part virtual; i.e., a physical subsystem is linked to a real-time computer simulation with closed loops. The MIL simulation was primarily used to verify functionality. It was not used for detailed fuel economy prediction or detailed calibration work. The MIL was more of a development environment rather than a formal verification/validation environment.

The use of MIL and HIL models is important for hybrid powertrain controls and system design because it allows for closed-loop controls development work early in a program when hardware is unavailable, and even later for prototype vehicle testing. These models are also especially useful in cases for which specific testing on hardware is difficult, for example when specific timing of fault injection is required or a high degree of repeatability is necessary. For the hybrid controller development, MIL and HIL models can be used for open-loop or closed-loop design verification (DV) testing to ensure that system-level requirements are met, and for desktop feature and calibration developments.

MIL and HIL models allow DV testing against a large number of requirements with much less time, effort, and cost compared with doing such testing on a vehicle. Moreover, DV testing using MIL and HIL can be automated and rerun as needed for strategy and other updates as the program evolves.

11.1.5 Dynamic Modeling

Dynamic modeling is a mathematical formalization for any fixed "rule" that describes the time dependence of a point's position in its ambient space. The concept unifies very different types of such "rules" in mathematics: the different choices made for how time is measured and the special properties of the ambient space may give an idea of the vastness of the class of objects described by this concept.

Deterministic Modeling: A deterministic system is a system in which no randomness is involved in the development of future states of the system. A deterministic model will thus always produce the same output from a given starting condition or initial state.

Stochastic Modeling: A stochastic system is the counterpart to a deterministic system. Instead of dealing with only one possible reality of how the process might evolve under time, in a stochastic process there are many possibilities the process might go to, but some paths may be more probable and others less so.

Fuzzy Logic Modeling: Fuzzy logic is essentially a logical system that generalizes classical two-valued logic for reasoning under uncertainty [11.8]. A Fuzzy Logic Controller has three main sections, or processes: an input stage, a processing stage, and an output stage. The input stage maps sensor or other inputs, such as driver inputs, to the appropriate membership functions and truth values. The processing stage invokes each appropriate rule and generates a result for each, then combines the results of the rules. Finally, the output stage converts the combined result back into a specific control output value. Fuzzy logic has been successfully applied to many different engineering applications, and is used in the automotive industry by many companies for powertrain, engine, and transmission control.

11.2 Models for Hybrid-electric Powertrains of Commercial Vehicles

The hybrid powertrain is an integrated system that consists of many subsystems such as the IC engine or fuel cell, clutch and transmission, motor, generator, energy storage devices such as battery or supercapacitor, etc. Each subsystem has its own functionality, characteristics, and desired performance. All subsystems must be integrated in an optimal manner to achieve the system objectives, e.g., fuel economy, emissions reduction, drivability, etc. With the increasing complexity of powertrain systems and the need for achieving multiple objectives, an integrated vehicle-level controller is required to accomplish the task [11.11, 11.12].

The HEV system models were built upon the Vehicle Model Architecture (VMA) framework. The Matlab/Simulink® implementation of the Vehicle Model Architecture is shown in Fig. 11.2. The VMA provides a well-defined model structure that facilitates model reuse and sharing and reduces model development time and cost. It defines a high-level modular structure for dynamic vehicle modeling with key vehicle subsystems represented as distinct elements.

The architecture also provides a basis for mapping component model requirements and specifications across application areas and for managing the many configurations of models that are required for program work. The VMA is linked to collections of component and subsystem models that are managed through model libraries. The component and subsystem models are defined and developed according to the requirements of each stage of the hybrid system development process. These models can be integrated into the VMA in a plug-and-play fashion to create vehicle system models appropriate to the stage of the hybrid system development process.

11.2.1 Supervisory Powertrain Controller

Figure 11.4 shows a two-level hierarchical control architecture for controlling the hybrid powertrain. The top level represents the supervisory powertrain control system (SPCS) which coordinates the overall powertrain to satisfy certain vehicle-level performance targets such as fuel economy and emissions reduction. The low-level control systems include the engine electronic control unit (ECU), motor ECU, transmission controller, energy storage system, etc. Based on the driver's demand (e.g., accelerator and brake pedal signals) and the current state of the subsystems (e.g., engine speed, motor speed, SOC, etc.), the SPCS determines the desired outputs to be generated by the subsystems (e.g., engine torque, motor torque, fuel cell power, requested gear, etc.). These desired output signals are sent to the corresponding subsystems and become the commands for the lower-level control system of each subsystem [11.12].

The main objective of the supervisory powertrain control system is to optimize power management to improve efficiency and performance. Due to the nature of multiple power sources, the hybrid vehicle can be operated on different modes depending on the power flow of each subsystem. As shown in Fig. 11.5, five possible operating modes exist in terms of the power flow in a parallel HEV [11.16]:

1. Motor-only mode: The vehicle is solely propelled by the traction motor.
2. Engine-only mode: The vehicle is solely propelled by the combustion engine.
3. Power-assist mode: Both the motor and the engine simultaneously produce propulsion power.
4. Recharging mode: The engine is the prime mover to drive the vehicle while providing additional power to recharge the battery.

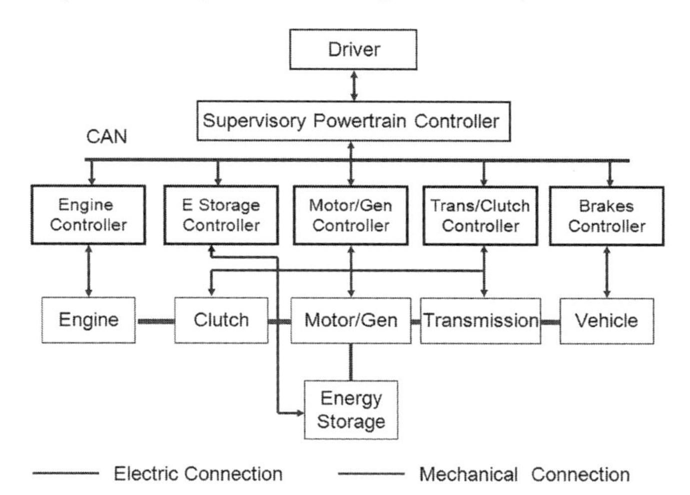

FIG. 11.4 Supervisory control architecture of hybrid powertrain [11.15].

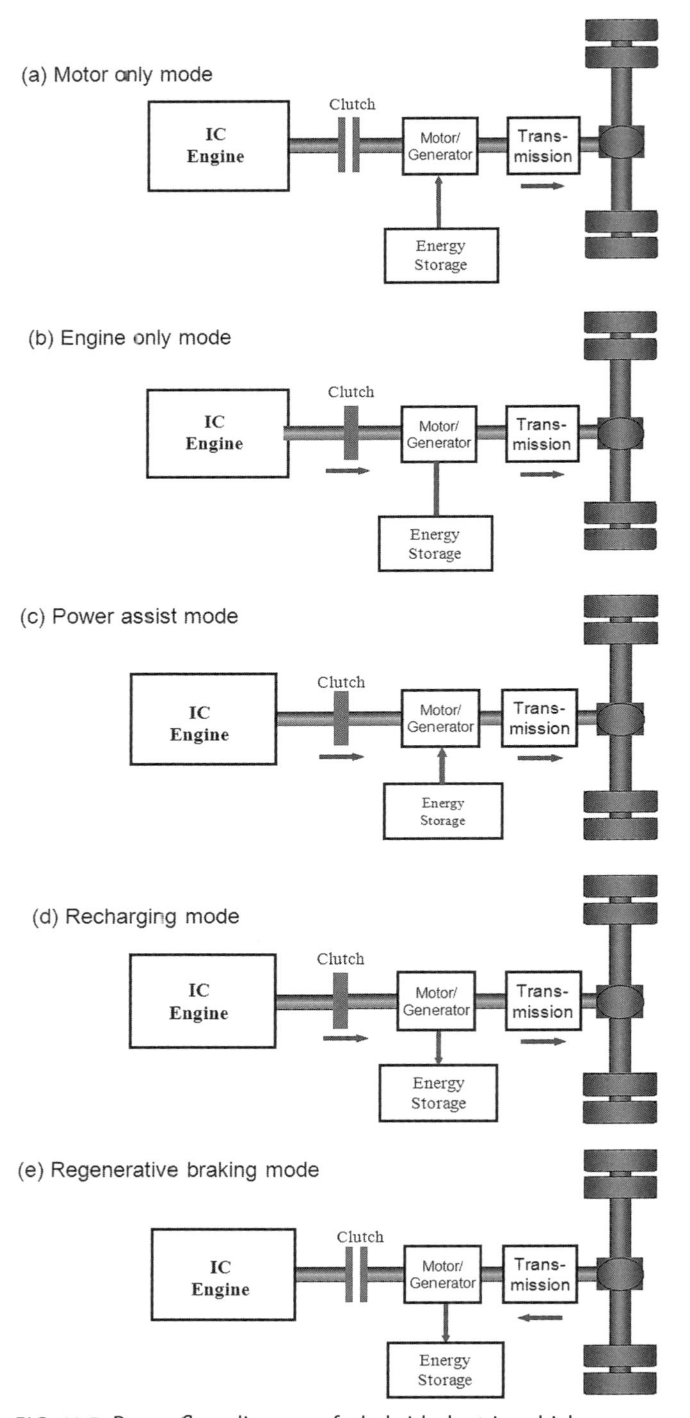

FIG. 11.5 Power flow diagram of a hybrid-electric vehicle.
(a) Motor-only mode; (b) Engine-only mode; (c) Power-assist mode; (d) Recharging mode; (e) Regenerative braking mode

5. Regenerative braking mode: The dissipation of kinetic energy during braking is recovered by operating the motor as a generator to recharge the battery.

With regard to the five possible operating modes, the supervisory power controller of a hybrid vehicle will be able to manage the power flow of each subsystem so that fuel consumption and emissions can be minimized according to the vehicle operating environment.

11.2.2 Driver Model

The driver model is used to model the accelerator or brake pedal position by estimating the torque at the wheels. There are two components in this model. The first is a calculator, which is used to roughly calculate the torque demand at the wheel needed to achieve the desired vehicle speed. It also estimates torque losses due to the friction of braking, rolling resistance, aerodynamic drag, and grade force. The second component is a proportional and integral (PI) controller that requests more or less torque to the vehicle and converts the error between the desired vehicle speed and the current vehicle speed into a demand for torque at the wheels. These two losses combine to produce the torque demand at the wheels, which is shown in Eq. (11.1).

$$T_{dmd} = T_{veh\_loss} + T_{PI} \tag{11.1}$$

where T_{veh_loss} is the torque loss at the wheel and can be calculated as shown in Eq. (11.2); T_{PI} is the torque converted from the error between the desired vehicle speed and the current vehicle speed, as shown in Eq. (11.3).

$$T_{veh\_loss} = T_{inertia} + T_{rolling} + T_{aero} + T_{grade} \tag{11.2}$$

$$T_{PI} = k_p \Delta v + k_i \int \Delta v dt \tag{11.3}$$

where $T_{inertia}$ is the torque of the inertia of the vehicle, $T_{rolling}$ is the torque of the rolling resistance, T_{aero} is the torque of the aerodynamic drag, and T_{grade} is the torque loss due to grade; k_p is the proportional gain, k_i is the integral gain, and Δv is the error between the desired vehicle speed and the current vehicle speed [11.12].

11.2.3 Powertrain Subsystem Models

The key components of a parallel hybrid powertrain, as shown in Fig. 11.5, are internal combustion engine, electric motor and generator, energy storage system, clutch, and transmission. Key aspects of the subsystem models are discussed in the following sections.

11.2.3.1 Internal Combustion Engine

The engine module takes the driver command and external loads as inputs and calculates engine speed response and fuel consumption. Argonne National Laboratory (ANL) has developed a comprehensive internal combustion engine model based on experimental data from more than 350 in-use vehicles, ranging from early model year small cars to recent model year SUVs, as well as some medium-sized diesel trucks [11.7]. The engine model, shown in Fig. 11.6, is divided into four blocks: engine torque calculation, engine thermal calculation, engine fuel rate calculation, and engine emissions calculation.

The engine fuel rate can be estimated by using a lookup table, which is a matrix described in terms of engine torque and engine speed. The engine torque can be calculated by using lookup tables or interpolating between the maximum and minimum torque curves (the curve when the accelerator pedal position is zero), as described in Eq. (11.4):

$$T_{out} = \begin{cases} (1 - T_{cmd}) \cdot T_{e\_\min} + T_{cmd} \cdot T_{e\_\max} & T_{cmd} \geq 0 \;\&\; \omega_e > 0 \\ 0 & \end{cases} \tag{11.4}$$

where T_{out} is the output of engine torque, T_{cmd} is engine command with value range (0, 1), $T_{e\_\min}$ is closed-throttle torque curve, $T_{e\_\max}$ is wide-open-throttle torque curve, and ω_e is engine speed.

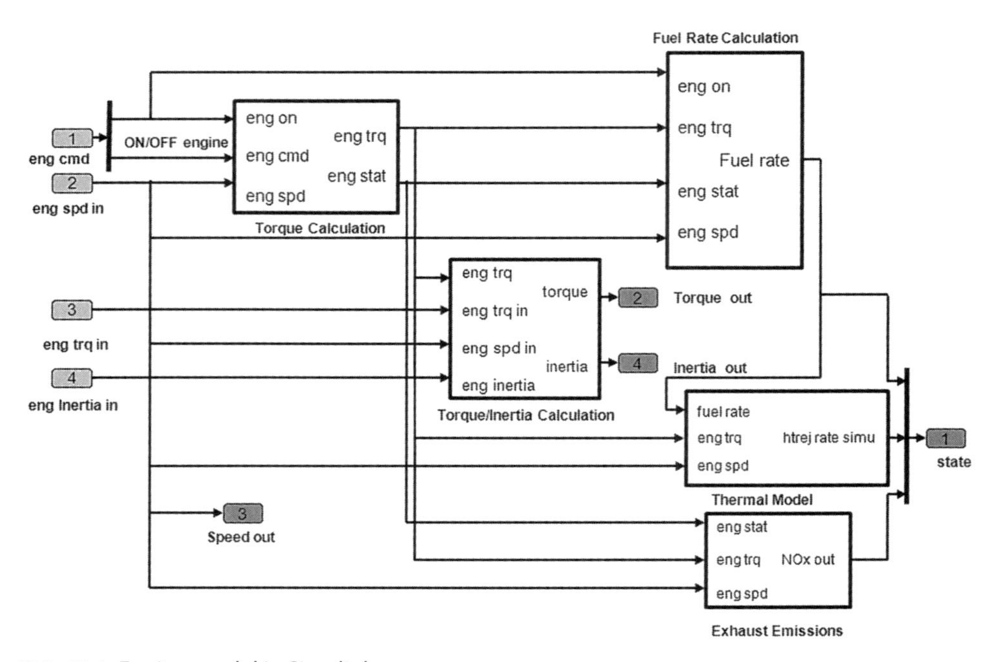

FIG. 11.6 Engine model in Simulink.

The amount of heat rejection, which is the difference between the total fuel energy consumed and the engine power generated, can be calculated in Eq. (11.5) as

$$Q_e = m_{fuel} \cdot q_{lhv} - \omega_e \cdot T_{out} \tag{11.5}$$

where Q_e is the engine heat rejection, m_{fuel} is the fuel mass consumed, and q_{lhv} is total energy released from the fuel by combustion per unit mass of fuel.

The NO_x emissions model can be estimated according to lookup tables or as a function of engine power, speed, torque, and their derivatives [11.11]. The predictive emissions model for NO_x is provided in Eq. (11.6),

$$NO_x = a \cdot P + b \cdot \frac{dp}{dt} + c \cdot T + d \cdot \frac{dT}{dt} + e \cdot \omega + f \cdot \frac{d\omega}{dt} + g \tag{11.6}$$

where P is the engine power, and T and ω are engine torque and engine speed, respectively. The measured NO_x is used to train this linear model. The parameters a, b, c, d, e, f, and g are constant coefficients, which are estimated by the least square method.

11.2.3.2 Fuel Cell System

The prime power of a hybrid vehicle can be provided by a fuel cell system instead of an internal combustion engine. An example of a fuel cell powertrain system, including auxiliary subsystems, is shown in Fig. 11.7. These sub-systems include the hydrogen supply system, air supply system, cooling system, and water management system, which have also been discussed in Chapter 9. The fuel cell system provides electrical power through a DC/DC converter to a system bus, where the battery serves as another power source. The load to the system bus is the power of the electric motor and the power required for the vehicle auxiliary devices. Mechanical propulsion power is provided by the electric motor to drive the vehicle.

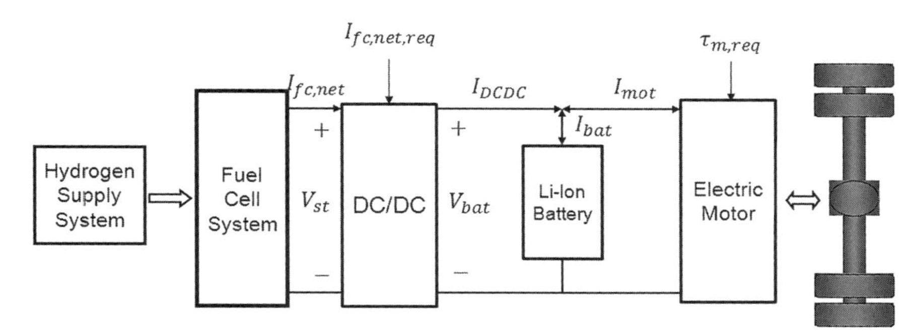

FIG. 11.7 Schematic of the hybrid fuel cell propulsion system.

The fuel cell stack voltage is known to be a function of the fuel cell current, stack temperature, reactant partial pressure, and membrane humidity, which can be modeled as

$$V_{fc} = V_{oc} - V_{act} - V_{ohm} - V_{conv} \tag{11.7}$$

where V_{oc} is the open circuit voltage, V_{act} is the activation overvoltage, V_{ohm} is the ohmic overvoltage, and V_{conv} is the concentration overvoltage [11.18].

Since the fuel cells are stacked in series to form the fuel cell stack, the stack voltage is calculated as the product of the number of cells and the individual cell voltage: $V_{st} = n_{st}V_{fc}$. The stack current I_{st} is equal to the cell current, which can be expressed by $I_{st} = i_{fc}A_{fc}$, where i_{fc} is the current density and A_{fc} is the fuel cell active area. The electrical power generated by the fuel cell stack is $P_{st} = I_{st}V_{st}$.

The net power produced by the fuel cell system is the difference between the fuel cell stack power and the auxiliary power: $P_{fc,net} = P_{st} - P_{aux}$. The fuel cell stack efficiency η_{st} is defined as the ratio of the power produced by the stack to the chemical power for the reaction:

$$\eta_{st} = \frac{P_{st}}{W_{H_{2,react}} LHV_{H_2}} \tag{11.8}$$

where $WH_{2,react}$ is the reacted hydrogen mass flow rate, and $LHVH_2$ is the low heating value of hydrogen. When fuel cell auxiliary components are considered, the fuel cell system efficiency η_{sys} is defined as

$$\eta_{sys} = \frac{P_{fc,net}}{W_{H_{2,react}} LHV_{H_2}} \tag{11.9}$$

Both efficiencies are plotted versus the fuel cell stack current in Fig. 11.8. It is seen that although the fuel cell stack has the highest efficiency in the

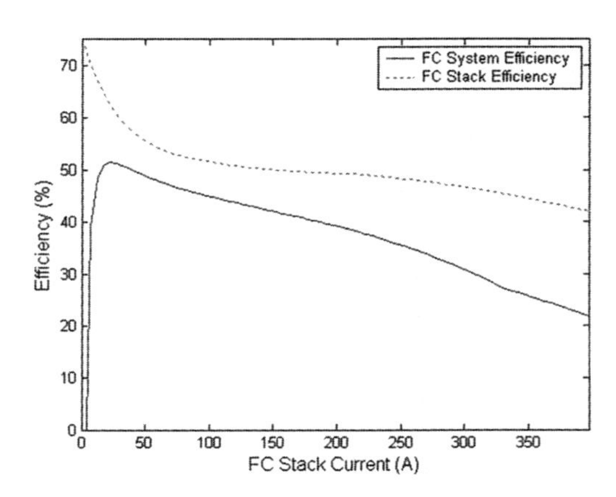

FIG. 11.8 Fuel cell stack efficiency and fuel cell system efficiency.

small current region, the fuel cell system displays very low efficiency due to the parasitic load. For the allowable stack current region, the fuel cell system efficiency is 30 to 50%.

11.2.3.3 Electric Motors and Generators

Motors generally used in HEV systems are DC motors, AC induction motors, or Permanent Magnet Synchronous Motors. Each motor has advantages and disadvantages that determine its suitability for a particular application, as discussed in Chapter 6.

For a permanent magnet DC motor, the efficiency data is a function of motor torque and motor speed, $\eta_m = f(T_m, \omega_m)$, as shown in Fig. 11.9.

The motor dynamics are approximated by a first-order lag. Due to the battery power and motor torque limit, the final motor dynamics have to be obtained from:

Positive Motor Torque:

$$T_m = \min\left(T_{m\_req}, T_{m\_\max}, T_{m\_bat}\right) \cdot \frac{\lambda_m}{s + \lambda_m} \tag{11.10}$$

Negative Motor Torque:

$$T_m = \max\left(T_{m\_req}, T_{m\_\max}, T_{m\_bat}\right) \cdot \frac{\lambda_m}{s + \lambda_m} \tag{11.11}$$

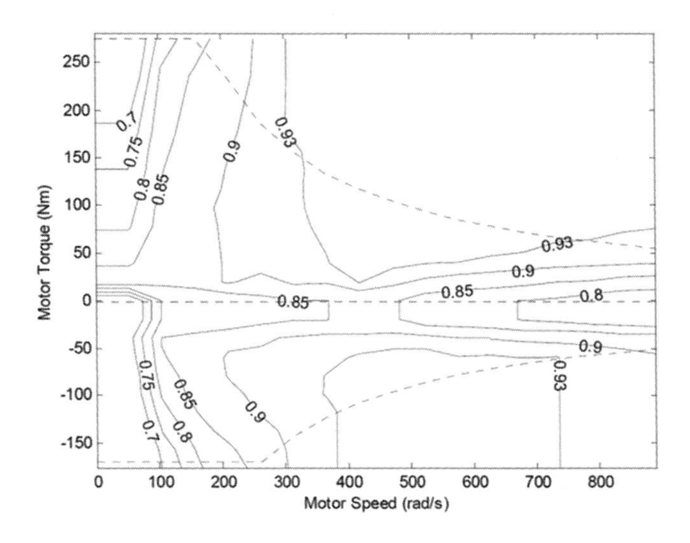

FIG. 11.9 Efficiency map of the DC motor.

where T_{m_req} is the requested motor torque, $T_{m\_\max}$ is the maximum torque the motor can generate under current motor speed, T_{m_bat} is the maximum motor torque due to battery constraint, T_m is the calculated motor torque, and λ_m characterizes the first-order lag of the motor dynamics. The load current for the battery i_{tb} can then be calculated from the following equation:

$$i_{tb} = \begin{cases} \dfrac{\omega_m \cdot T_m \cdot \eta_m}{e_{tb}} & if \quad T_m > 0 \\ \\ \dfrac{\omega_m \cdot T_m}{e_{tb} \cdot \eta_m} & if \quad T_m < 0 \end{cases} \tag{11.12}$$

where e_{tb} is the battery terminal voltage.

11.2.3.4 Electric Energy Storage System

The main considerations in energy storage system design are capacity, discharge characteristics, and safety. Traditionally, a higher capacity is associated with an increase in size and weight. Discharge characteristics determine the dynamic response of vehicle components to extract or supply energy to the energy storage device.

A typical battery model used is an equivalent RC circuit, as shown in Fig. 11.10.

The battery models have been discussed in Chapter 4. The state-dependent internal resistance R_{int} and the open-circuit voltage V_{oc} are lumped representations of complex chemical processes, and are known to be the functions of the battery state of charge (SOC). The terminal ohmic resistance R_t is assumed to be constant. The SOC variables are the only state variables of the battery system.

$$\frac{d\text{SOC}}{dt} = \frac{I_b}{q_m} \tag{11.13}$$

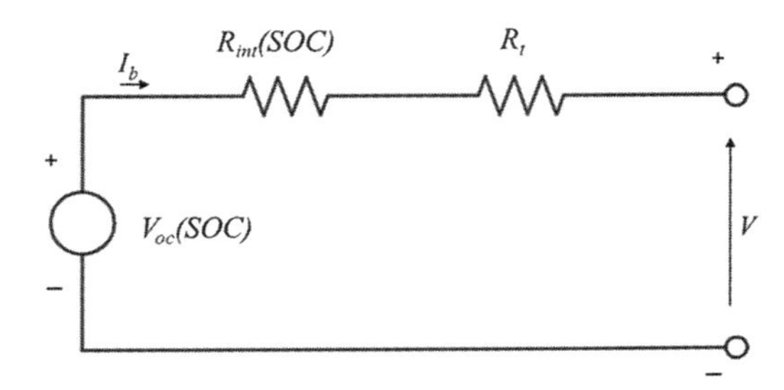

FIG. 11.10 Resistance battery model.

where q_m is the maximum battery charge. The battery current I_b is calculated by the equation

$$I_b = \frac{V_{oc} - \sqrt{V_{oc}^2 - 4(R_{\text{int}} + R_t) \cdot P_b}}{2(R_{\text{int}} + R_t)} \tag{11.14}$$

where P_b is the output power of the battery. The terminal voltage of the battery is given by $V_{bat} = V_{oc} - I_b \cdot (R_{\text{int}} + R_t)$.

11.2.4 Transmission Model

The function of the transmission is to multiply or divide the engine torque/speed by the ratio for the selected gear. A Simulink diagram of a transmission, shown in Fig. 11.11, includes three main blocks: torque calculation, speed calculation, and inertia calculation [11.11]. The speed calculation block calculates the input shaft speed based on the gear ratio, output of shaft speed, and information as to whether the gear is neutral. If the transmission is in gear, the input shaft speed equals the output shaft speed multiplied by the gear ratio selected, as shown in Eq. (11.15). If the transmission is in neutral, the input shaft is free to spin. The spinning shaft produces drag losses that are proportional to its speed:

$$\omega_{in} = \begin{cases} \omega_{out} \cdot i_r & \text{in gear} \\ \int_0^t \frac{T_{in} - T_{drag}}{J_{in}} dt & \text{in neutral} \end{cases} \tag{11.15}$$

where ω_{in} is input shaft speed; ω_{out} is output shaft speed; T_{in} is input torque for the transmission, which is from upstream components; T_{drag} is torque loss when the shaft is free and J_{in} is the input inertia.

The inertia calculation block gives the output inertia of the transmission, which can be calculated by Eq. (11.16). When the transmission is in gear, the output inertia is determined by upstream inertia and the current gear ratio. When the gear is in neutral, the output inertia is zero.

$$J_{out} = \begin{cases} J_{in} \cdot i_r^2 & \text{in gear} \\ 0 & \text{in neutral} \end{cases} \tag{11.16}$$

where J_{out} is the output inertia of the transmission.

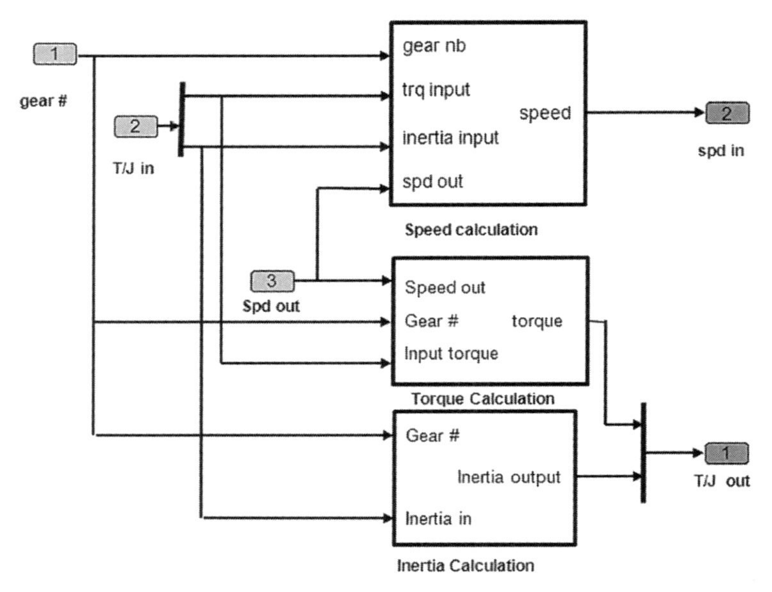

FIG. 11.11 Transmission component Simulink model.

The torque calculation block gives the output torque of the transmission. When the transmission is in gear, the output torque before applying the gear ratio gain equals the input torque minus the torque loss (Eq. (11.17)). When it is in neutral, the output torque is 0.

$$T_{out} = \begin{cases} (T_{in} - T_{loss}) \cdot i_r & \text{in gear} \\ 0 & \text{in neutral} \end{cases} \tag{11.17}$$

where T_{out} is the output torque and T_{loss} is torque loss.

The torque losses are calculated based on a three-dimensional efficiency lookup table, which has the following inputs: gear ratio, input torque, and speed.

11.2.5 Drivetrain and Vehicle Subsystem

The driveline module consists of the clutch, transmission, differential, and driveshafts, as shown in Fig. 11.5. It provides the connection among the engine, the electric motor, and the vehicle dynamics module. The transmission output shaft and the electric motor shaft are connected to the propeller shaft, the differentials, and, via driveshafts, to the driven wheels. The dynamic behavior of the drivetrain is represented by ordinary differential equations that describe the kinematic and dynamic behavior of the real system. These equations are to be integrated with the Simulink model. The vehicle model contains rotational wheel dynamics and linear vehicle dynamics. The inputs to the vehicle model are the shaft torque produced by the electric motor and frictional brake torque. The outputs of the model are the angular wheel speed and vehicle forward speed. The state equation of the wheel speed is written as

$$\frac{d\omega_{wh}}{dt} = \frac{1}{J_r}\left(T_{wh} - T_b - B_{wh}\omega_{wh} - r_{wh}(F_x + F_r)\right) \tag{11.18}$$

where T_{wh} is the driving torque from the motor, T_b is the frictional brake torque, B_{wh} is the viscous damping, r_{wh} is the tire radius, F_x is the tire longitudinal force, F_r is the rolling resistance force, and J_r is the equivalent moment of inertia of rotating components, including the motor, axles, and wheels in the vehicle.

When the vehicle is modeled as a point-mass, the vehicle speed can be calculated as

$$\frac{dv_v}{dt} = \frac{1}{M_v}\left(F_x - \frac{v_v}{|v_v|}\left(F_a(v_v)\right)\right) \tag{11.19}$$

where v_v is the vehicle speed, F_a is the aerodynamic drag force, and M_v is the mass of the vehicle. The aerodynamic drag force is given by $F_a = 0.5C_d\rho_a A_v v_v^2$, where C_d is the aerodynamic drag coefficient, ρ_a is the density of air, and A_v is the frontal area of the vehicle.

11.3 Dynamic Modeling of Hybrid-electric Powertrain System

In the discrete-time format, a model of the hybrid-electric vehicle can be expressed as

$$x_{k+1} = f(x_k, u_k) \tag{11.20}$$

where k is the time index; u_k is the vector of control variables such as desired output torque from the engine, desired output torque from the motor, and gear-shift command to the transmission; and x_k is the state vector of the system [11.15, 11.16]. The sampling time for this main loop control problem is selected to be one second. The optimization goal is to find the control input u_k to minimize a cost function, which consists of the weighted sum of fuel consumption and emissions for a given driving cycle. The cost function to be minimized has the following form:

$$J = \sum_{k=0}^{N-1} L(x_k, u_k) + G(x_N) \tag{11.21}$$

$$J = \sum_{k=0}^{N-1} \left[fuel_k + \mu \cdot \mathrm{NO}_{\mathrm{x}_k} + \nu \cdot PM_k \right] + \alpha \left(SOC_N - SOC_f \right)^2 \tag{11.22}$$

where N is the duration of the driving cycle, and L is the instantaneous cost, including fuel use and engine-out NO_x and PM emissions. SOC_f is the desired SOC at the final time, and α is a positive weighting factor. For a fuel-only problem, the weighting factors are $\mu = \nu = 0$. The case of $\mu > 0$ and $\nu > 0$ represents a simultaneous fuel and emission problem. During the optimization, it is necessary to impose certain inequality constraints to ensure safe/smooth operation of the engine/battery/motor:

$$\begin{aligned} &\omega_{e\_\min} \le \omega_{e,k} \le \omega_{e\_\max} \\ &T_{e\_\min}(\omega_{e,k}) \le T_{e,k} \le T_{e\_\max}(\omega_{e,k}) \\ &T_{m\_\min}(\omega_{m,k}, \mathrm{SOC}_k) \le T_{m,k} \le T_{m\_\max}(\omega_{m,k}, \mathrm{SOC}_k) \\ &\mathrm{SOC}_{\min} \le \mathrm{SOC}_k \le \mathrm{SOC}_{\max} \end{aligned} \tag{11.23}$$

where ω_e is the engine speed, T_e is the engine torque, T_m is the motor torque, and SOC is the battery state of charge. In addition, we impose two equality constraints for the optimization problem, so that the vehicle always meets the speed and load (torque) demands of the driving cycle at each sampling time.

For an HEV system with transmission, a gear-shifting constraint should be added into the cost function to prevent frequent gear shifting. The goal of the control system design is to find a sequence of control actions, including the engine torque, motor torque, and gear selection, to minimize a cost function given by

$$\min J = \min_{\{T_{e,k}\, shift_k\}, k=0,1,\ldots N-1} \left\{ \sum_{k=0}^{N-1} \left[W_{fuel,k} + \mu \cdot NO_{x_k} + \nu \cdot PM_k + L_k \right] + G_N \right\} \tag{11.24}$$

where $W_{fuel,k}$ is the engine fuel flow rate, $G_N(\mathrm{SOC}_N) = \alpha(\mathrm{SOC}_N - \mathrm{SOC}_f)^2$ is the terminal penalty on the SOC of the energy storage system (ESS), and $L_k = \beta \cdot |g_{k+1} - g_k|$ is the dynamic constraint of gear change to prevent frequent gear shifting.

For a driving cycle, the wheel torque $T_{wh,req}$ required to follow the speed profile can be determined for each time step by inversely solving the vehicle dynamics. The required wheel speed $\omega_{wh,req}$ is assumed proportional to vehicle speed through the wheel radius. At every time step, the torque balance equation in the wheel must be satisfied as follows

$$T_{wh,k}(T_{e,k} + T_{m,k}, g_k, \omega_{wh,req,k}) + T_{brake,k} = T_{wh,req,k} \tag{11.25}$$

where $T_{wh,k}$ is the wheel torque produced from the sum of the engine torque, $T_{e,k}$, and motor torque, $T_{m,k}$, through the driveline; g_k is the transmission gear number; and $T_{brake,k}$ is the friction braking torque.

A Simulink model of the parallel hybrid-electric vehicle system is shown in Fig. 11.12. To achieve the desired performance in fuel economy and emissions reduction, subsystem models and control strategies such as power management strategy, regenerative braking control, transmission shift logic, and shifting control must be carefully designed and optimized.

11.3.1 Gear Shifting Schedule

If the transmission is included in the hybrid vehicle design, the gear position of the transmission has a significant effect on fuel economy and emissions because it influences the operating point of the hybrid powertrain. In the supervisory powertrain control system, the requested gear position is a control signal sent to the transmission control system. This requested gear command is determined by a gear-shifting logic based on vehicle status information such as the transmission speed, current gear position, and driver pedal demand. To improve fuel economy, emissions, drivability, and shifting quality simultaneously, the shift logic requires an integrated design approach by considering the overall hybrid powertrain (engine, electric motor, and transmission) together.

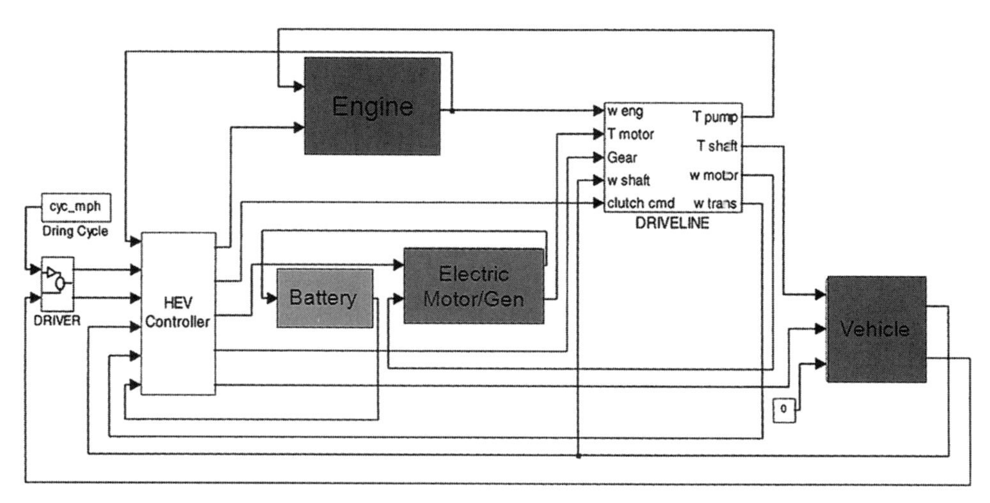

FIG. 11.12 Hybrid-electric vehicle simulation model in Simulink [11.14].

For a six-speed AMT transmission [11.16], the drivability constraint to avoid frequent shifting is:

$$g_{k+1} = \begin{cases} 6, & g_k + shift_k > 6 \\ 1, & g_k + shift_k < 1 \\ g_k + shift_k, & Otherwise \end{cases} \tag{11.26}$$

where g_k is the transmission gear number and $shift_k$ is the gear-shifting command, which is constrained to take on the values of –1, 0, and 1, representing downshift, hold, and up-shift, respectively. At each time step, these constraints are imposed to consider the operating limits of each component. To reduce the possibility of missing synchronization, the restriction of the gear-shifting selection is also taken into account in the optimization:

$$\omega_{wh,req,k} \cdot R_f \cdot R_g(g_{K+1}) > \omega_{in\_min}, \quad \text{if } shift_k = 1 \tag{11.27}$$

$$\omega_{wh,req,k} \cdot R_f \cdot R_g(g_{K+1}) > \omega_{in\_max}, \quad \text{if } shift_k = -1 \tag{11.28}$$

where ω_{in_min} and ω_{in_max} are minimum and maximum allowable input shaft speed, respectively. R_f is the gear ratio of the final drive, and R_g is the gear ratio of the transmission. The optimal gear operational points and upshift/downshift points of a six-speed AMT are plotted on the standard transmission shift-map to identify an optimal shifting schedule, as shown in Fig. 11.13.

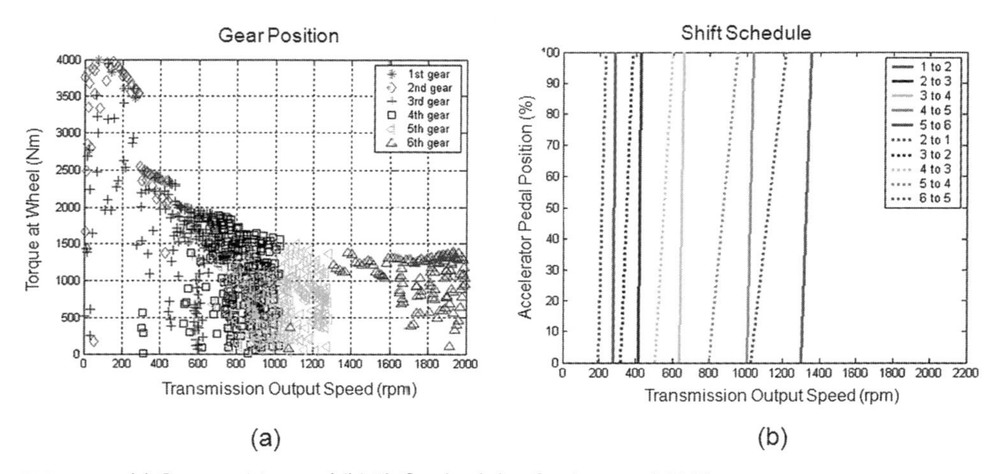

FIG. 11.13 (a) Gear position and (b) Shift schedule of a six-speed AMT.

11.3.2 Power Split Control for Parallel Hybrid System

A power-split-ratio is defined to quantify the positive power flows in the powertrain as

$$PSR = \frac{P_{eng}}{P_{dem}} \tag{11.29}$$

where P_{eng} is the engine power and P_{dem} is the power request from the driver. Four positive-power operating modes are defined: motor-only ($PSR = 0$), engine-only ($PSR = 1$), power-assist ($0 < PSR < 1$), and recharging mode ($PSR > 1$).

Figure 11.14 is the engine efficiency map with an "engine on" power level P_{e_on} and "motor assist" power level P_{m_a}, which are chosen to avoid engine operation in inefficient areas. If P_{dem} is less than P_{e_on}, the electric motor will supply the requested power alone (PSR = 0). Beyond P_{e_on}, the engine becomes the sole power source (PSR = 1). Once P_{dem} exceeds P_{m_a}, engine power is set at P_{m_a} and the motor is activated to make up the difference ($P_{dem} - P_{m\_a}$), therefore, $0 < PSR < 1$. During the deceleration of the vehicle, the engine can operate in recharging mode with $PSR > 1$.

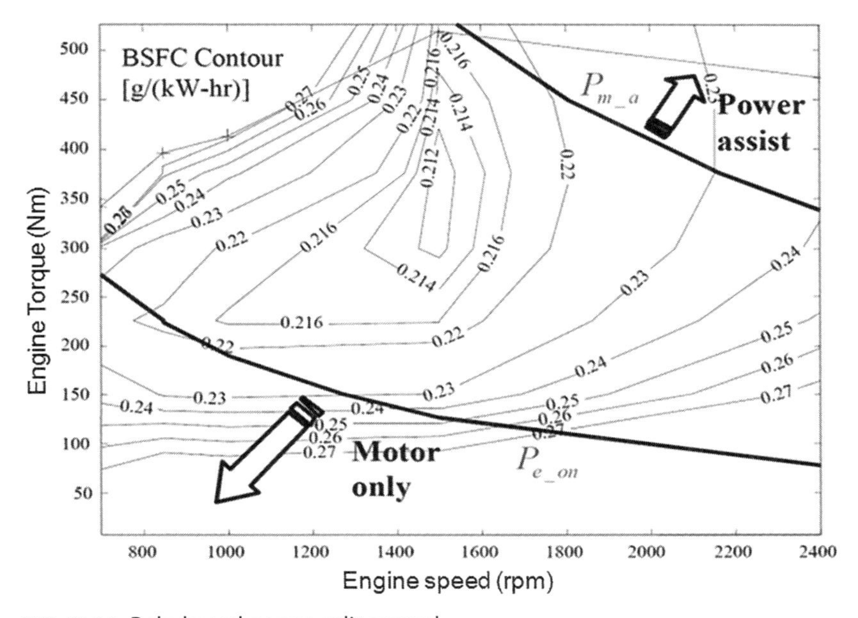

FIG. 11.14 Rule-based power-split control.

11.3.3 Charge Sustaining Strategy

In the recharging control mode, the engine needs to provide additional power to charge the battery in addition to powering the vehicle. Commonly, a preselected recharge power level, P_{ch} , is added to the driver's power demand, which becomes the total requested engine power ($P_e = P_{dem} + P_{ch}$). The motor power command becomes negative ($P_m = -P_{ch}$) in order to recharge the battery. One exception is that when P_{dem} is less than P_{e_on} , the motor alone will propel the vehicle to prevent the engine from operating in the inefficient operation. In addition, when P_{dem} is greater than the maximum engine power, the motor power will become positive to assist the engine.

To maintain SOC of ESS within desired operating range, an additional rule should be developed to prevent the battery from over depleting or overcharging. The strategy for regulating the SOC still needs to be obtained in an approximately optimal manner to satisfy the overall goal: to minimize fuel consumption and emissions. The charge sustaining strategy is closely related to the regenerative braking. With the regenerative energy, the motor can act more aggressively to share the load with the engine, because running the engine at high power is unfavorable for fuel economy and emissions. As a result, knowing the amount of the regenerative braking energy the vehicle will capture in future driving is the key to achieving the best fuel and emissions reduction while maintaining the battery SOC level.

11.3.4 Intelligent Power Management Using GPS Information

Global Positioning Systems (GPS) have become commonplace in vehicle navigation systems, installed in many vehicles around the world. Most systems pinpoint the location of the vehicle overlaid on top of a map of the surrounding area, along with traffic information and topographical data such as elevation, etc. The challenge is to design a structure that utilizes information about speed and elevation in the planned route and to change the instantaneous control strategy to adapt to the supervisory control structure. An instantaneous control strategy is used to determine an optimal ICE torque at the current speed. This is based on the ICE efficiency and emission maps, the battery state of charge, transmission gear ratio, and the total driveline torque required. Another controller, called the navigation controller, varies the instantaneous controller parameters based on implied future states (Fig. 11.15). The inputs to the navigation controller are speed, elevation, and traffic information of sampled points along a predetermined route, from a navigation system [11.27].

The navigation controller consists of four types of information that relate to the vehicle's operating load and energy recovered by regenerative braking. The four types of road information include:

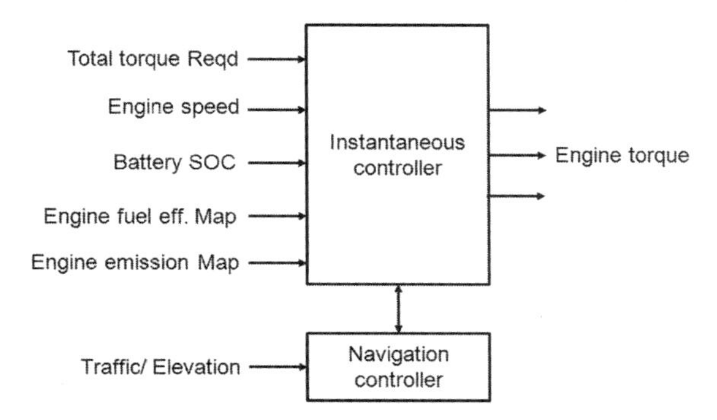

FIG. 11.15 Control strategy structure.

1. Road classification: Six types of roads are considered—inter-city expressways, intra-city expressways, toll roads, national highways, prefectural highways and major roads, and other roads.
2. Congestion level: The four levels of congestion information are congested, crowded, flowing smoothly, and no information.
3. Road grade.
4. Distance of section.

Based on the navigation information, it is possible to predict the SOC at a given engine efficiency value η as

$$SOC_i(k) = SOC(0) + \sum_{l=1}^{k} \delta_l \cdot F(\alpha_l, \beta_l, \gamma_l, \eta_l) \tag{11.30}$$

where k =1,2,...,N is the section number, SOC$_i$(k) is the predicted SOC at the end of section k when η is set to a fixed value η_i, SOC(0) is the present SOC, α_l is the road classification of section l, β_l is the congestion level of section l, γ_l is the average road grade of section l, δ_l is the distance of section l, and F is the lookup table of the predicted battery SOC (%/m) stored in the controller. The series of SOC$_i$(k) corresponds to the predicted time-history of the SOC when the vehicle travels at a fixed value of η_i.

Eq. (11.30) can be used to assess the opportunities of regenerative braking and usage of the engine of the ESS. For example, SOC can be used to a lower level for the upcoming downhill driving condition, and the SOC should be at a higher level when uphill driving is coming. SOC can be used with the uphill and downhill driving conditions. The SOC can be used at a lower level when on local highways or at the congested road condition when driving at cruise speed on interstate highway driving. Therefore, the navigation controller (Fig.

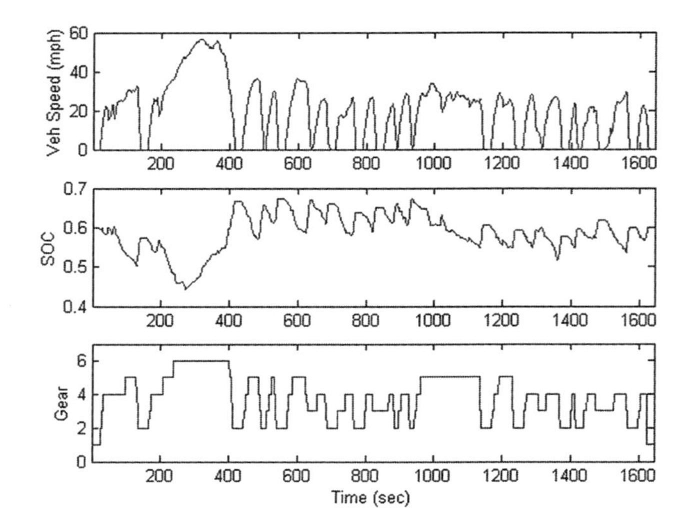

FIG. 11.16 DP simulation results [11.16].

11.15) can bring active SOC management to optimize the engine fuel economy and emissions through the route.

11.3.5 Simulation Results and Performance Evaluation

Dynamic programming (DP) is a powerful tool to solve general dynamic optimization problems for a given driving cycle. For a fuel-only problem, the weighting factors in Eq. (11.22) are $\mu = v = 0$, and the initial and terminal desired SOC were both selected to be 0.6 [11.12, 11.16].

The engine power and motor power trajectories represent the optimal operation between two power movers to achieve the best fuel economy. The initial condition of SOC and gear position in the simulation are 0.6 and first gear, respectively. Because the final desired SOC in Eq. (11.22) was selected to be 0.6, the simulation shows the SOC trajectory returns to 0.6 at the end of the cycle.

11.4 System Control and Optimization for Hydraulic Hybrid Vehicles

The hybrid hydraulic powertrain has been discussed in Chapters 1, 7, and 8. It is characterized by high power density and comparatively high energy conversion efficiency compared to the hybrid-electric powertrain. Particularly, hydraulic accumulators can accept exceptionally high rates of charging and discharging. A combination of high efficiency and high charging/discharging rates enables very effective regeneration and re-use of energy in a heavy vehicle in frequent stop-and-go operating conditions, such as with refuse pickup trucks. However, the energy density of a hydraulic accumulator is relatively low. The system controller has to be carefully designed for specific applications.

Figure 11.17 shows schematics of a series hydraulic hybrid architecture. In the series HHV system, the engine is coupled to the hydraulic pump. The traction pump/motor is connected to the wheels to provide propulsion, and the accumulator allows energy storage. The reservoir is kept at the low pressure and serves primarily to enable transfer of fluid to-and-from the accumulator, with minimal impact on overall energy conversion.

There are two key subsystems that are different from the hybrid-electric system discussed in the previous chapters. The hydraulic hybrid system uses a low-pressure reservoir and a high-pressure accumulator to store the hydraulic energy, and hydro-mechanical motors and pumps instead of electromagnetic motors and generator as in the hybrid-electric system. The hydraulic accumulator has quite different energy storage characteristics from the battery systems of HEVs. Subsystem models of hydraulic accumulators and pumps/motors will be discussed in more detail.

11.4.1 Hydraulic Accumulator

The hydraulic accumulator is in fact the hydro-pneumatic accumulator for energy storage. As the pump transfers the hydraulic fluid into the accumulator, the pressure of the gas sealed inside of it increases, thus storing energy. Since the fluid is transferred through the system during every charging/discharging event, a low-pressure reservoir is included in the system too. The pressure in the reservoir is relatively low, and it does not significantly affect the energy balance, but it is high enough to prevent cavitation in the pump. The pressure difference between accumulator and reservoir determines the maximum torque available from the hydraulic motor. Maximum pressure in the accumulator is determined by its construction and material properties, and it is typically in the range of 35–45 MPa.

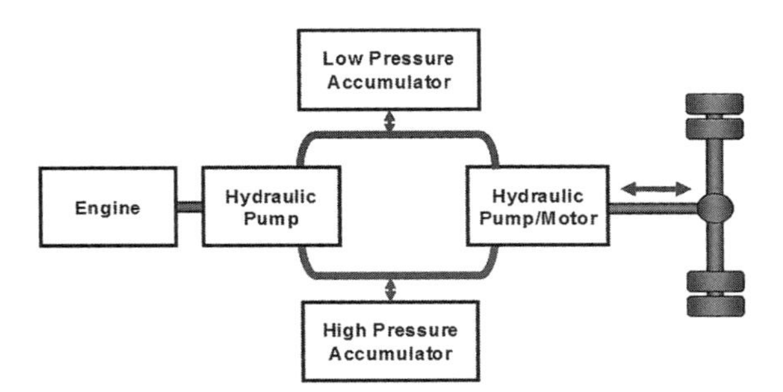

FIG. 11.17 Schematics of a series hydraulic hybrid vehicle.

Modeling of the accumulator and the reservoir is based on the application of the energy conservation equation to the gas charge. A bladder is used to separate the inert gas from the fluid inside the reinforced container; hence, the gas-side can be treated as a closed system. The accumulator bladder, which is filled with nitrogen, is assumed to compress and expand based on the adiabatic gas law:

$$PV^n = P_0 V_0^n = \mathrm{P}_{\mathrm{max}} V_{\mathrm{min}}^n \tag{11.31}$$

where P is the system pressure on the high-pressure side, V is the accumulator gas volume, P_0 is the accumulator precharge pressure, V_0 is the accumulator gas volume at precharge pressure (effective volume), P_{max} is the maximum allowable system pressure, V_{min} is the accumulator gas volume at maximum system pressure, and n is the Adiabatic constant (ratio of specific heat at constant pressure to specific heat at constant volume) for gas in the accumulator.

Further, energy that can be absorbed by the accumulator can be shown as

$$E = \frac{P_{\mathrm{max}}^{1/n} V_0}{n-1}\left[\lambda_p^{1/n} - \lambda_p\right] \tag{11.32}$$

where $\lambda_p = P/P_{\mathrm{max}}$.

For maximum energy capacity of the accumulator, it can also be shown that optimum precharge pressure is

$$P_0 = n^{n/(1-n)} P_{max} \tag{11.33}$$

And maximum accumulator energy is given by

$$E_{max} = P_{max} V_0 n^{n/(1-n)} \tag{11.34}$$

The mass of gas is directly related to the gas pressure range for a given accumulator size. Appropriate accumulator sizing is a tradeoff between several considerations. Larger size adds appreciable mass to the vehicle, but it provides a large buffer to absorb all regenerative braking and supplies additional power during peak accelerations for limited periods. An accumulator that is too small will limit the amount of regenerative energy that can be recovered, and it cannot provide assistance during accelerations for more than very short periods of time. The small accumulator configuration will perform closer to a hydrostatic, continuously variable transmission, in which the engine is coupled somewhat closely to the power demands from the wheel. The larger accumulator configuration allows the engine to be turned off or idled, thus enhancing fuel economy. From a control standpoint, the accumulator is essentially active only during transient intervals of the duty cycle. For prolonged steady-state conditions, it is very desirable for accumulator pressure to reach a steady state. In this condition, the engine-pump system is matching the exact power needs of the drive motor, after taking into account the inefficiencies of the pump, motor, and piping losses.

The SOC is defined as the ratio of instantaneous fluid volume in the accumulator over the maximum fluid capacity. In real vehicle application, the pressure may be used as an indicator of the SOC, as long as the temperature variations are kept low by increasing the thermal capacity of the charge.

11.4.2 Hydraulic Pump/Motors

For an axial-piston, variable-displacement pump/motor, the torque and hydraulic fluid flow is controlled by the displacement factor. The ideal leak-free volumetric flow rate and the ideal frictionless torque of the Pump/Motor (P/M) are

$$Q_i = x\omega D \tag{11.35}$$

$$T_i = x\Delta p D \tag{11.36}$$

where x is the displacement factor, i.e. the ratio of current P/M displacement to the maximum P/M displacement. Its sign is negative for pump operation and positive for motor operation; ω is angular velocity; D is the maximum P/M displacement per radian; and, Δp is the P/M pressure difference.

The ratio of actual (Q_a) and ideal (Q_i) volumetric flow determines the volumetric efficiency as

$$\eta_{v,pump} = \frac{Q_a}{Q_i} = 1 - \frac{C_s}{x\,s} - \frac{\Delta p}{\beta} - \frac{C_{st}}{x\,\sigma} \tag{11.37}$$

$$\eta_{v,motor} = \frac{Q_i}{Q_a} = \frac{1}{1 + \frac{C_s}{x\,s} + \frac{\Delta p}{\beta} + \frac{C_{st}}{x\,\sigma}} \tag{11.38}$$

where $\frac{C_s}{x\,s}$, $\frac{\Delta p}{\beta}$, and $\frac{C_{st}}{x\,\sigma}$ represent laminar leakage loss, compressibility and turbulent leakage loss, respectively; β is the oil bulk modulus of elasticity; $S = \mu\omega/\Delta p$ and $\sigma = \omega D^{1/3}/(2\Delta p/\rho)^{1/2}$ are dimensionless numbers, in which ρ is oil density and μ is oil dynamic viscosity.

Torque efficiency relates actual and ideal torque, and can be defined as

$$\eta_{t,pump} = \frac{T_i}{T_a} = \frac{1}{1 + \frac{C_v S}{x} + \frac{C_f}{x} + C_h x^2 \sigma} \tag{11.39}$$

$$\eta_{t,motor} = \frac{T_a}{T_i} = 1 - \frac{C_s S}{x} - \frac{C_f}{x} - C_h x^2 \sigma \tag{11.40}$$

where $C_v S/x$, C_f/x, and $C_h x^2 \sigma$ represent the viscous loss, mechanical friction, and hydrodynamic loss, respectively.

The coefficients, which are used for calculation of volumetric and torque efficiencies, are calibrated to match the particular P/M for the propulsion purpose.

For the optimization work in this study, the loss coefficients are assumed to be the same when scaling, the maximum P/M displacement.

11.4.3 Power Management Strategy

There are two main elements of an effective series system power management. The top-level strategy determines the engine power demand based on the state-of-charge (SOC) in the accumulator. The strategy needs to meet the driver request under any condition, including hard acceleration or hill climbing, and maintain the SOC levels conducive to effective regeneration. Secondly, for any given power level, the actual engine speed/load point needs to be chosen in a way that ensures minimum specific fuel consumption. In the practical implementation, the engine fueling is determined based on target torque, and its speed is controlled via the adjustment of the P/M displacement factor.

The goal of the hybrid hydraulic control strategy is to find the optimum power split between the two power sources at any point during operation. Methodology such as load leveling, where the internal combustion engine is forced to operate around its most efficient point, has been demonstrated for its potential benefits. The additional level of optimization is possible by minimizing engine transients and further controlling engine operation in specific regions of this line. By properly controlling the pump displacement and engine rack, the engine can be guided to operate on the best efficiency line for most duty cycles except when the pressure is very low on steep gradients. In these situations, the engine may have to operate close to its torque curve above the efficiency line.

The energy recoverable during a braking event is the kinetic energy minus the energy losses due to wind, grade, and rolling resistance:

$$E_{braking} = \frac{1}{2}m_v v^2 - \frac{v^2}{a_b}\left[c_0 + c_1\frac{v}{3} + c_2\frac{v^2}{4}\right] \tag{11.41}$$

When combining Eq. (11.32) and Eq. (11.41) and solving the nonlinear equation, the result is the desirable accumulator pressure P_{des} (green line in Fig. 11.18) for the system, which should be maintained at every motor or vehicle speed.

The power of the engine is selected to allow the accumulator pressure to track the green line, as shown in Fig. 11.18. When the pressure is above the value of the green line, it should only be due to regeneration during braking. The engine should be either off or idling above the green line, to save fuel. Eq. (11.41) and Eq. (11.32) illustrate that the green line is dynamic and can change with changes in wind, grade, and rolling resistance. Staying on the line before a braking event ensures that there is adequate room to recover brake energy (except on downgrades).

The motor displacement is directly proportional to the driver pedals, except on grades. Optimal motor displacements can be computed, based on the grade, to maximize torque output. It is also blended with friction brakes to allow for safe stopping and slow-down characteristics.

11.4.4 Simulation Model and Results

Because the fuel economy of the vehicle depends on the system efficiency, rather than component efficiency, investigation on the effect of threshold power on system behavior is needed to propose optimal strategies. In addition, engine transients and dynamic interactions in the system play a role. A predictive series hydraulic hybrid propulsion system simulation model (Fig. 11.19) enables detailed investigations and can lead to the optimized strategy such as motor gear ratio and transmission shift logic.

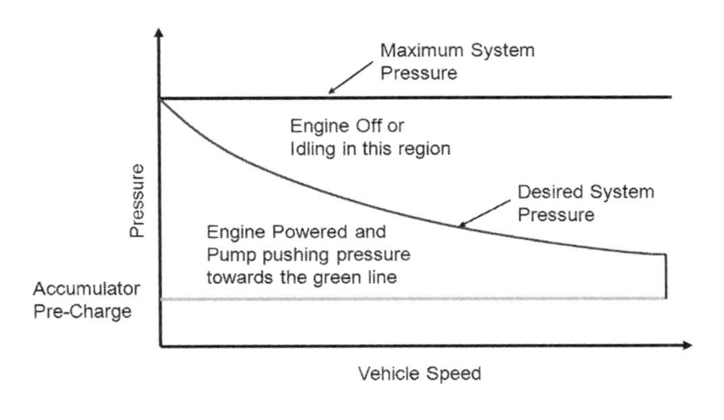

FIG. 11.18 Engine ON and OFF (or idle) is determined by where the actual system pressure lies with reference to the desired system pressure line P_{des}.

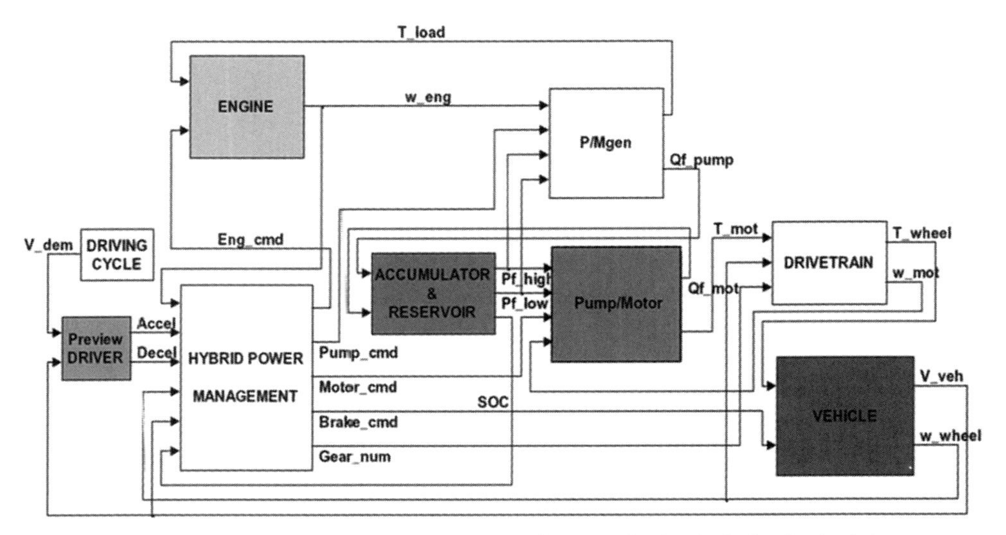

FIG. 11.19 Integrated, forward-looking simulation of a series hydraulic hybrid vehicle in Simulink.

Figure 11.20(a) shows the shift of operating points from relatively inefficient regions typically observed in a conventional vehicle, and Fig. 11.20(b) illustrates the proximity of the optimal brake specific fuel consumption (BSFC) trajectory by decoupling the engine from the wheels. The color scale indicates the amount of fuel consumed in each zone. The results demonstrate impressive improvements of fuel economy over the city driving schedule, e.g., more than 49% with engine idling, and over 68% with engine shut-downs. The improvements predicted for highway driving are not nearly as high, but they are still tangible and offer promise of favorable efficiency under real-life mixed driving conditions. The numbers discussed above suggest that a significant amount of savings in city driving can be attributed to regeneration (and shut-downs, if applicable), and a smaller fraction to optimization of engine operation [11.21].

11.5 Model-based Test and Validation

The hybrid powertrain control system can be tested and redesigned in phases as shown in Fig. 11.1. Multiple validation steps are included for control system software validation, including unit testing, HIL tests, and vehicle testing, as shown on the right-hand side of the V. The validation process was developed to ensure that a variety of issues can be captured during the development process to provide a quality software release.

The supervisory powertrain controller (SPC) developed in the previous sections can be first evaluated by using the simulation model. This simulation

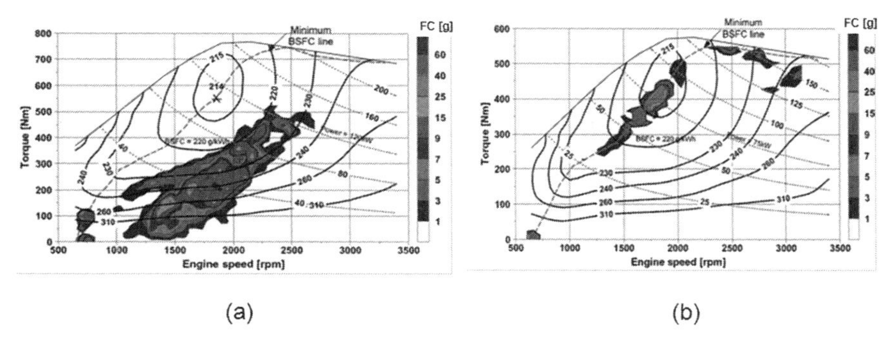

FIG. 11.20 (a) A conventional vehicle BSFC map and operating zone. (b) Engine visitation points on the BSFC map for a series hydraulic hybrid (V6 engine) [11.21, 11.17].

phase allows the algorithms and parameters in the SPC to be examined and tuned before the hardware prototype is available. To reduce the development time and cost, a PC-based rapid control prototyping tool can be used to implement the SPC in the prototype vehicle. The Simulink/Stateflow-based SPC model on the host computer can be built and downloaded to the supervisory control computer on the prototype vehicle via an Ethernet connection. Real-time data can be captured by using a host computer and plotted for later analysis. This rapid prototyping system enables the engineers not only to test and operate the real components in the hardware-in-the-loop phase, but also to test the vehicle on the road in a real driving phase. From the real-time measurement, the engineers could quickly analyze the performance of the SPC, modify the controller model, and build and download the modified code to the prototype vehicle computer in a fast and cost-effective manner.

11.5.1 Dynamometer Test

To measure the emissions and fuel economy of the hybrid powertrain on a chassis dynamometer, a driving cycle has to be selected to represent the real-world operations. The vehicle is mounted on the chassis dynamometer so that it can be driven through the test cycle. A visual display of the desired and actual vehicle speed will be provided to the driver to allow the driver to operate the vehicle on the prescribed cycle. SAE J2711 recommends that the vehicle be tested using three different driving cycles, representative low-, intermediate- and high-speed operations. Various driving cycles are listed in Appendix 11.8 [11.16].

Emission and fuel economy measurements were made on a full-scale chassis dynamometer; the load at the wheel was controlled to provide required road load such as rolling resistance and aerodynamic drag.

For hybrid-electric Class 4–6 urban pickup and delivery vehicles, a Combined International Local and Commuter Cycle (CILCC) is best suited to evaluate vehicle performance. The CILCC cycle, which consists of four International Local cycles and one shortened Commuter cycle, as shown in Fig. 5.2, has been discussed in Chapter 5. The International Local cycle is a city-suburban drive cycle used by International Truck & Engine Corporation to conduct application analysis for their heavy-duty vehicles. The CILCC includes 54% city, 29% suburban, and 17% highway driving, which is close to the driving pattern reported by typical end users. The cycle's number of stops per mile is also similar to that reported by fleet customers. The CILCC provides ease of SOC balancing, requires an appropriate amount of vehicle testing time, and gives the most accurate representation of user driving for this vocation among available candidates. It also can be used as one of the standard drive cycles to evaluate and test other Class 4–6 heavy-duty HEVs [11.25].

11.5.2 Road Test

Road testing provides the final validation of the performance of hybrid vehicles. The SAE International Standard J2711 outlines the test method on heavy-duty hybrid-electric vehicle fuel economy and exhaust emissions. Figure 11.21 shows a setup for onboard measurements for a parallel hybrid city bus [11.26].

Specifications of the test vehicle are listed in Table 11.1. Portable vehicle emissions measurement devices, including the Horiba OBS-2200 and DEKATI ELPI, were used for real-time measurement of all exhaust pollutants (CO, HC, NO, CO_2, PM). FLOWTRONIC 210 fuel consumption meter was used to measure diesel fuel consumption. The detailed description of the measurement instruments are discussed in Ref. [11.26].

A typical urban driving cycle of the public bus in China is shown in Fig. 11.22. It synthesizes characteristics of the urban driving cycles of public buses of metropolitan areas such as Beijing, Shanghai, and Wuhan, and it has become the national standard of performance tests of public buses in China [11.29]. The typical urban driving cycle of public buses is often used in field tests at specified automobile quality test centers to acquire national certification.

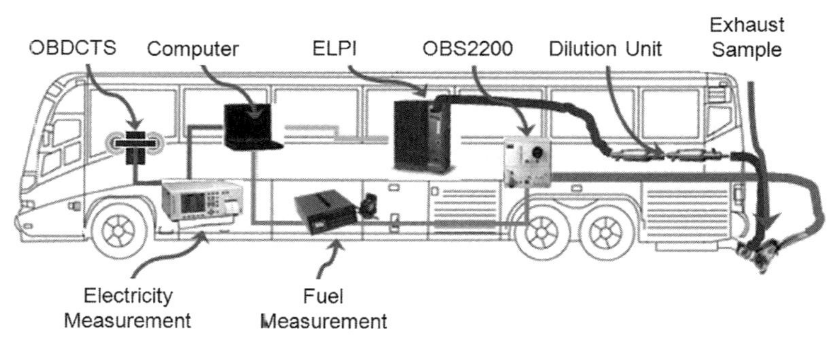

FIG. 11.21 Test vehicle setup.

TABLE 11.1 Test Vehicle Specifications

Vehicle Type	HEV
Length (mm)	11400
Width (mm)	2550
Height (mm)	3150 (with Air)
Axle Distance (mm)	5650
Chassis Type	BJ6113C7M4D
Chassis Maker	Beiqi Foton
Max Weight (kg)	15500
Vehicle Weight (kg)	11500
Max Capacity/Seats	61/23-42
Engine Type	ISBE 220 30
Displacement (L)	5.9
Rated Power (kW/rpm)	162/2400 (220HP/2400rpm)

The testing method follows China's GB/T 19754—2005 "Heavy-duty Hybrid Electric Automotive Energy Consumption Test Methods" [11.31]. For heavy-duty hybrid-electric vehicles, two runs are required to minimize the impact of State of Charge of the energy storage system and other random operating variations.

China's GB/T 19754—2005 "Heavy-duty Hybrid Electric Automotive Energy Consumption Test Methods" references the SAE International J2711 "Heavy-duty Hybrid Electric Vehicle and the Traditional Vehicle Fuel Economy and Exhaust Emission Test Method," in hybrid-electric vehicles on energy consumption. The GB/T 19754—2005 standard uses different test cycles, loading

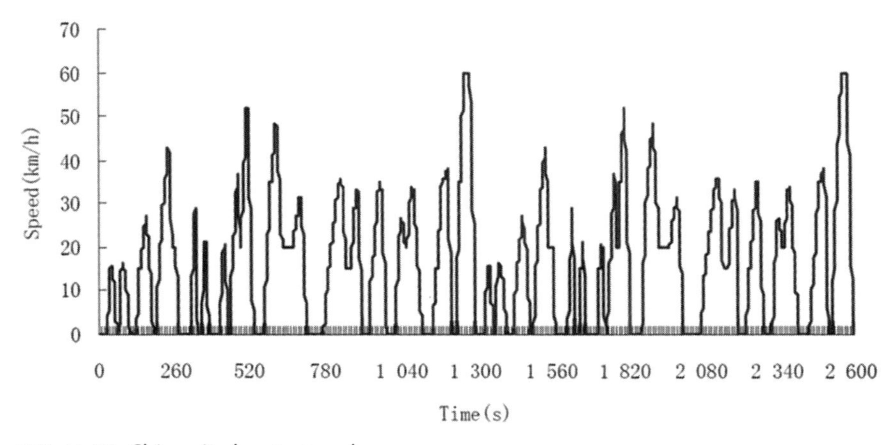

FIG. 11.22 China city bus test cycle.

distribution, and test equipment requirements in accordance with China's national standards. A typical test procedure is:

1. Load the test vehicles with sandbags to reach the required weight. For each vehicle, the tests are carried out with and without turning on the air conditioning system.
2. Check the power status of the test instruments and preheat equipment to reach the required temperature.
3. Drive the bus in accordance with the predetermined test cycle. Test the hybrid vehicle at least three times to reduce or eliminate the variation of the battery SOC. Record speed, exhaust flow, CO, HC, NO_x, CO_2, PM, hybrid charge-discharge data, as well as how the above parameters affect the relevant data (such as ambient temperature, atmospheric pressure, etc.).
4. Complete the data recording, and check instruments and equipment in accordance with the relevant requirements.

Other typical field test cycles representing different vehicle operating conditions are listed in Appendix 11.8.

11.6 Hybrid Vehicle Certification

A significant hurdle for adoption of hybrid technologies lies in the current limitations of engine and full-vehicle certification. The engine-only certification process prevents the medium- and heavy-duty truck manufacturers from certifying a full-vehicle-level platform, unlike their counterparts in the passenger car market. This is further complicated by the wide range of modifications that adapt the basic truck chassis platform to accommodate the individual end-user requirements. The problem with most hybrid technologies is that they affect more than just the engine and after-treatment. Hybrid technology generally reduces inefficiencies elsewhere in the drivetrain, as discussed elsewhere in this book.

Hybrid technologies are not the only ones under development to improve vehicle fuel efficiency and emissions. Truck manufacturers and their suppliers are investigating a range of options, including aerodynamics and rolling resistance improvements. Many of these technologies are highly duty-cycle dependent. For example: aerodynamic improvements and low-rolling-resistance tires have a significant positive effect on Class 8 long-haul truck fuel efficiency, but hybrid technologies tend to have very little impact on vehicle efficiency in this class of vehicles. Quite the opposite is true for many vocational vehicles. Operating conditions and duty cycles are critical differentiators for the success of select technologies in a particular market segment. The challenges for the regulatory agencies quickly become apparent: how does one certify vehicles that may use any combination of technologies to improve vehicle fuel efficiency and emissions? Depending on which segment of industry one talks to, the answer is to continue with established engine dynamometer procedures that have been used for decades, to switch to chassis-based certification, or even go to track testing. Another solution more recently brought forward is to use hardware-in-the-loop testing (essentially a combination of hardware and simulation testing) or certify entirely based on computational modeling. Each of these methods has its advantages and disadvantages; a brief discussion follows.

11.6.1 Engine Dynamometer

Engine dynamometer-based testing has been applied for decades to certify engines and engine families to meet established emissions regulations. Power and torque certification enables manufacturers to assure customers that engines deliver their claimed performance. Manufacturers have the choice of certifying engines to the SAE International standards J1349 (for net power) or SAE J1995 (for gross power) [11.32, 11.33]. The witnessed certification procedure (carried out in an approved ISO9000/9002 facility) is fully described in the new SAE J2723 standard [11.34]. The advantage of continuing with this method is that an established infrastructure exists to perform certification testing. Another benefit to the manufacturers is that there is an established link between certification test data and real-world vehicle emissions. A significant disadvantage is that a limited set of technologies can effectively be tested using an engine dynamometer-based protocol.

11.6.2 Chassis Dynamometer

Chassis dynamometer testing is the protocol used to certify passenger cars and other vehicles below 10,000 lb GVW. This procedure tests a complete vehicle as it drives a predefined drive cycle on a stationary roller dynamometer. An example of chassis dynamometer certification is the US EPA SmartWay certification program [11.35]. Every vehicle the EPA looks at, whether it is a car, an SUV, or a truck, gets two different scores, an air pollution score and a greenhouse gas score. Respectively, these scores rate emission levels and fuel economy values. The air pollution score rates the emissions coming out of a vehicle's tailpipe. The greenhouse gas score is based on a vehicle's fuel economy, or how efficiently an engine burns fuel. Both of these scores range from 0 to 10, where 0 is the worst and 10 is the best. To get the SmartWay certification, both scores must be at least 6, and the scores have to add up to at least 13.

The obvious benefit is that this is a far more representative test to determine vehicle-level emissions and fuel economy. The disadvantage for truck manufacturers lies in the earlier-mentioned fact that the medium- and heavy-duty truck market is highly segmented and diverse in applications and duty cycles. To effectively test for emissions and fuel economy, many different drive cycles would have to be defined. Where today a single engine family can be certified on a single engine dynamometer-based test, dozens or more certification tests would have to be performed on a chassis dynamometer. Additionally, a significant capital investment would be required to establish the testing infrastructure to support this certification protocol for medium- and heavy-duty trucks.

11.6.3 Power Pack Testing

Power pack testing describes a procedure for simulating a chassis test with a post-transmission hybrid system for A to B testing. The hardware that must be included in these tests is the engine, the transmission, the hybrid-electric motor, the power electronics between the hybrid-electric motor and the energy storage system, and the energy storage system. The test setups can be modified to allow testing non-electric hybrid vehicles. CO_2 emissions will be measured while operating the system over specific test cycles.

Although the Power Pack testing does not represent all vehicle operating conditions on the road, it provides a method to assess the benefits of the hybrid powertrain for given operating conditions.

11.6.4 Track Testing

This proposed test protocol goes one step beyond chassis dynamometer-based testing in testing the entire vehicle on the road [11.36]. The added benefit, compared to chassis dynamometer-based testing, is that aerodynamic effects can be properly accounted for. While this type of certification protocol comes closest to representing real-life operating conditions, cost and complexity of this type of testing are prohibitive.

11.6.5 Simulation-based Certification

Certifying vehicles or individual technologies based on computational simulations is an intriguing concept, as this could, in theory, provide a cost-effective means of certifying technologies and vehicles for emissions and fuel economy over a wide variety of operating conditions.

EPA has developed a vehicle simulation model, GEM, to serve as the primary tool to certify Class 7/8 combination tractors and Classes 2b–8 vocational vehicles in meeting the EPA's and NHTSA's proposed vehicle greenhouse gas (GHG) emission levels and fuel efficiency requirements.

The GEM model has six systems (Driver, Ambient, Electric, Engine, Transmission, and Vehicle) used to describe different vehicle categories. An input profile of the GEM model is shown in Fig. 11.23 [11.37]. The EPA has validated GEM based on the chassis test results from "SmartWay"-certified tractors tested at the Southwest Research Institute. Because many aspects of one tractor configuration (such as the engine, transmission, axle configuration, tire sizes, and control systems) are similar to those used on a manufacturer's sister models, the validation work conducted on these vehicles is representative of the other Class 8 tractors.

11.6.6 Certifications of Commercial Vehicles in the United States

The U.S. EPA and NHTSA, on behalf of the Department of Transportation, have established a comprehensive Heavy-Duty National Program that will reduce greenhouse gas emissions and fuel consumption for on-road heavy-duty vehicles to take coordinated steps to produce a new generation of clean vehicles [11.38]. NHTSA's final fuel consumption standards and EPA's final carbon dioxide (CO_2) emissions standards are tailored to each of three regulatory categories of heavy-duty vehicles: Combination Tractors; Heavy-duty Pickup Trucks and Vans; and Vocational Vehicles. The rules include separate standards for the engines that power combination tractors and vocational vehicles. Certain rules are exclusive to the EPA program, such as EPA's final nitrous oxide (N_2O) and methane (CH_4) emissions standards that apply to all heavy-duty engines, pickup trucks, and vans.

EPA's final greenhouse gas emission standards under the Clean Air Act will begin with model year 2014. NHTSA's final fuel consumption standards

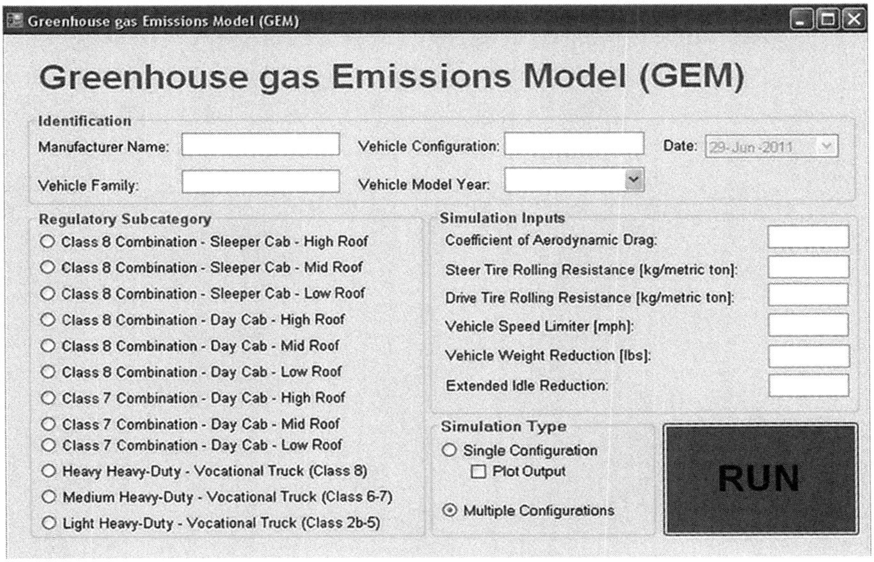

FIG. 11.23 Input of greenhouse gas emission model (GEM).

under the Energy Independence and Security Act of 2007 will be voluntary in model years 2014 and 2015, becoming mandatory with model year 2016 for most regulatory categories. Commercial trailers are not regulated in this phase of the Heavy-Duty National Program.

Figure 11.24 shows the certification processes of both engines and vehicles. Dynamometers were used to test the greenhouse gas emissions of an engine of each engine family for certification. The selected engine configuration must represent at least one percent of total actual U.S. directed production volume for the engine family. The test engine must meet the GHG standards, also called Family Certification Levels (FCLs) of this tested configuration and N_2O standards with engineering analysis in 2014 MY and earlier. The engine family is considered in compliance with the standards if all emission-data engines representing the tested configurations have official test results and deteriorated emission levels at or below the standard levels of FCLs.

The engine emissions credited or de-credited will be assessed based on the total amount of CO_2 emissions [11.39].

Manufacturers of Class 7/8 combination tractors and Classes 2b–8 vocational vehicles must calculate a projected production-weighted average CO_2 performance for each family based on the modeled (GEM) results from at least ten subfamilies to certify the vehicle for sales in the United States. As shown in Fig. 11.25, the certification process includes four steps: 1) precertification preparation; 2) certification application; 3) end of year compliances; and 4) post model year [11.40].

The precertification preparation includes: a) define vehicle families, subfamilies, and configurations; b) conduct aerodynamic evaluation, collect tire data, other testing; c) determine other GEM inputs, such as advanced vehicle technologies, weight reduction, etc.; and d) conduct GEM modeling of the vehicle subfamilies under a specific driving cycle which would reasonably

FIG. 11.24 Certification process for both engine and vehicle OEMs.

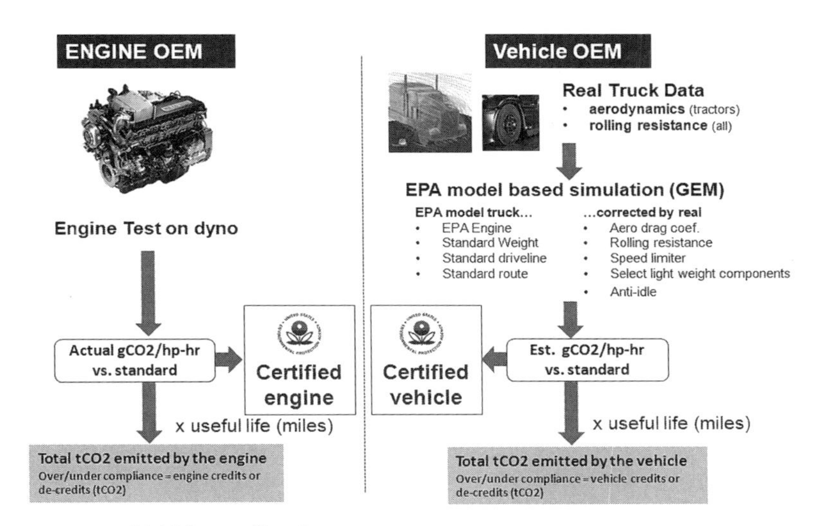

FIG. 11.25 Vehicle certification process.

represent the primary actual use. After completing the GEM simulation, an application template will be completed and submitted to the U.S. EPA for certification approval. The vehicle family is considered in compliance with the emission standards if all vehicle configurations in that family have modeled CO_2 emission rates at or below the applicable standard levels. If vehicle emissions controls are likely to deteriorate during the useful life, a deterioration factor must be applied to calculate the CO_2 emissions during the useful life of the vehicles. For example, a deterioration factor may be applied to address deterioration of battery performance for a hybrid-electric vehicle. If all regulatory requirements are satisfied, the U.S. EPA and NHTSA will issue a certificate of conformity, allowing the vehicle family to be introduced into U.S. commerce.

The certificate is valid from the effective date until the end of the model year for which it is issued. A model year must include January 1 of the calendar year for which the model year is named and may not begin before January 2 of the previous calendar year; it must end by December 31 of the named calendar year, and the certificate must be renewed annually for vehicles that continue to be produced.

11.6.7 Certification of Hybrid Powertrains for Commercial Vehicles

The U.S. EPA awards advanced technology emission credits for hybrid vehicles and recommends the certifications process that follows [11.41]:

(1) Measure the effectiveness of the advanced system by chassis testing or power pack testing a vehicle equipped with the advanced system and an

equivalent conventional vehicle. A conventional vehicle or a test prototype must be created for comparison purposes. A vehicle is considered to be equivalent if it has the same footprint, vehicle service class, aerodynamic drag, and other relevant factors not directly related to the hybrid powertrain. To quantify the benefits of a hybrid system for power takeoff (PTO) operation, the conventional vehicle must have the same number of PTO circuits and have equivalent PTO power [11.42]. The conventional vehicle is considered Vehicle A, and the advanced vehicle is considered Vehicle B. An improvement factor and g/ton-mile benefit can be calculated using the following equations and parameters:

(i) Improvement Factor = [(Emission Rate A) – (Emission Rate B)]/(Emission Rate A)

(ii) g/ton-mile benefit = Improvement Factor × (GEM Result B)

(iii) Emission Rates A and B are the g/ton-mile CO_2 emission rates of the conventional and advanced vehicles, respectively, as measured under the test procedures specified in this section.

The emission benefits of the hybrid vehicle can be converted to advanced technology credits. The U.S. EPA allows Advanced Technology Credits to apply to either engines or vehicles of the applicable subcategories shown in Fig. 11.26.

In summary, the certification procedures of medium- and heavy-duty on-road commercial vehicles for both emissions and vehicle fuel economy in the United States have been finalized by the regulatory agencies for MY 2014 and beyond. Certification procedures for hybrid, fuel cell, and other advanced powertrain technologies are also recommended. These certification procedures are significant steps forward to accelerate the harmonization

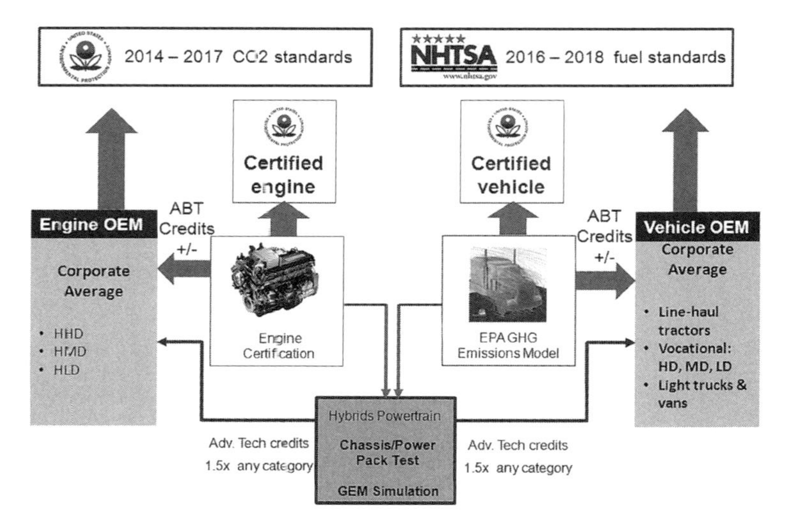

FIG. 11.26 Hybrid commercial vehicle certification procedure.

of global certification procedures of vehicles with hybrid and other advanced powertrain technologies.

11.7 References

11.1 Karbowski, D., Delorme, A., and Rousseau, A., "Modeling the Hybridization of a Class 8 Line-Haul Truck," SAE Paper No. 2010-01-1931, SAE International, Warrendale, PA, 2010.

11.2 Rousseau, A., Sharer, P., and Besnier, F., "Feasibility of Reusable Vehicle Modeling: Application to Hybrid Vehicle," SAE Paper No. 2004-01-1618, SAE International, Warrendale, PA, 2004.

11.3 Surampudi, B., Redfield, J., Ray, G., Montemayor, A. et al., "Electrification and Integration of Accessories on a Class-8 Tractor," SAE Paper No. 2005-01-0016, SAE International, Warrendale, PA, 2005.

11.4 Fang, W. F. et al., "Modeling of Li-ion Battery Performance in Hybrid Electric Vehicles," SAE Paper No. 2009-01-1388, SAE International, Warrendale, PA, 2009.

11.5 Karbowski, D. et al., " 'Fair' Comparison of Powertrain Configurations for Plug-In Hybrid Operation Using Global Optimization," SAE Paper No. 2009-01-1334, SAE International, Warrendale, PA, 2009.

11.6 Delorme, A., Karbowski, D., Sharer, P., "Evaluation of Fuel Consumption Potential of Medium and Heavy Duty Vehicles Through Modeling and Simulation," October 2009, www.ipd.anl.gov/anlpubs/2010/05/66884.pdf, assessed, April 2011.

11.7 An, F. et al., "Integration of a Modal Energy and Emissions Model into a PNGV Vehicle Simulation Model, PSAT," SAE Paper No. 2001-01-0954, SAE International, Warrendale, PA, 2001.

11.8 Matheson, P. et al., "Modeling and Simulation of a Fuzzy Logic Controller for a Hydraulic-Hybrid Powertrain for Use in Heavy Commercial Vehicles," SAE Paper No. 2003-01-3275, SAE International, Warrendale, PA, 2003.

11.9 Argonne National Laboratory, PSAT (Powertrain Systems Analysis Toolkit), www.transportation.anl.gov/ modeling_simulation/PSAT, assessed April, 2011.

11.10 http://www1.eere.energy.gov/vehiclesandfuels/pdfs/success/advisor_simulation_tool.pdf.

11.11 Wang, L., Clark, N, and Chen, P., "Modeling and Validation of an Over-the-Road Truck," SAE Paper No. 2010-01-2001, SAE International, Warrendale, PA, 2010.

11.12 Lin, J. et al., "Development Control System Development for an Advanced-Technology Medium-Duty Hybrid Electric Truck," SAE Paper No. 2003-01-3369, SAE International, Warrendale, PA, 2003.

11.13 Lin, J. and Peng, H., "Modeling and Control of a Power-Split Hybrid Vehicle," IEEE Transactions on Control Systems Technology, VOL. 16, NO. 6, November 2008.

11.14 Lin, C. C. et al., "A Stochastic Control Strategy for Hybrid Electric Vehicles," Proc. Of the 2004 American Control Conference, Boston, MA, June 30–July 2, 2004.
11.15 Lin, C. C., Feng, H., Grizzle, J. W., and Kang, J., "Power Management Strategy For A Parallel Hybrid Electric Truck," IEEE Trans. Control Syst. Technol., vol. 11, no. 6, Nov. 2003, pp. 839–849.
11.16 Lin, C. C., "Modeling and Control Strategy Development for Hybrid Vehicles," Doctor Dissertation, Department of Mechanical Engineering, University of Michigan, 2004.
11.17 Surampudi, B. et al., "Design and Control Considerations for a Series Heavy Duty Hybrid Hydraulic Vehicle," SAE Paper No. 2009-01-2717, SAE International, Warrendale, PA, 2009.
11.18 Suh, Kyung-Won et al., "Effects of Control Strategy and Calibration on Hybridization Level and Fuel Economy in Fuel Cell Hybrid Electric Vehicle," SAE Paper No. 2006-01-0038, SAE International, Warrendale, PA, 2006.
11.19 Rajagopalan, A. and Washington, G., "Intelligent Control of Hybrid Electric Vehicles Using GPS Information," SAE Paper No. 2002-01-1936, SAE International, Warrendale, PA, 2002.
11.20 Schyr, C. et al., "Model-based Development and Calibration of Hybrid Powertrains," SAE Paper No. 2007-01-0285, SAE International, Warrendale, PA, 2007.
11.21 Kim, Y. J. et al., "Simulation Study of a Series Hydraulic Hybrid Propulsion System for a Light Truck," SAE Paper No. 2007-01-4151, SAE International, Warrendale, PA, 2007.
11.22 Deguchi, Y. et al., "HEV Charge/Discharge Control System Based on Navigation Information," SAE Paper No. 2004-21-0028, SAE International, Warrendale, PA, 2004.
11.23 Wu, B. et al., "Optimal Power Management for a Hydraulic Hybrid Delivery Truck," Vehicle System Dynamics, 2004, Vol. 42, Nos. 1–2, pp. 23–40.
11.24 Dawood, V. and Emadi, A., "Performance and Fuel Economy Comparative Analysis of Conventional, Hybrid, and Fuel Cell Heavy-Duty Transit Buses," Vehicular Technology Conference, IEEE, 2003, Vol.5, pp. 3310–3315.
11.25 Zou, Z. et al., "A New Composite Drive Cycle for Heavy-Duty Hybrid Electric Class 4–6 Vehicles," SAE Paper No. 2004-01-1052, SAE International, Warrendale, PA, 2004.
11.26 Zeng, X. et al., "Analysis and Simulation of Conventional Transit Bus Energy Loss and Hybrid Transit Bus Energy Saving," SAE Paper No. 2005-01-1173, SAE International, Warrendale, PA, 2005.
11.27 O'Keefe, M. P. et al., "Duty Cycle Characterization and Evaluation Towards Heavy Hybrid Vehicle Applications," SAE Paper No. 2007-01-0302, SAE International, Warrendale, PA, 2007.
11.28 Qin, K. et al., "On-Road Test and Evaluation of Emissions and Fuel Economy of the Hybrid Electric Bus," SAE Paper No. 2009-01-1866, SAE International, Warrendale, PA, 2009.

11.29 Hu, H. et al., "On-board Measurements of City Buses with Hybrid Electric Powertrain, Conventional Diesel and LPG Engines," SAE Paper No. 2009-01-2719, SAE International, Warrendale, PA, 2009.
11.30 SAE Standard J1666 "Hybrid Electric Vehicle Acceleration, Gradeability, and Deceleration Test Procedure," SAE International, Warrendale, PA, 2004.
11.31 GB/T19754-2005 (2005) "Heavy-Duty Hybrid Electric Automotive Energy Consumption Test Methods," National Standard of People's Republic of China.
11.32 SAE J1349: "Engine Power Test Code—Spark Ignition and Compression Ignition—As Installed Net Power Rating," SAE International, September 2011.
11.33 SAE J1995, "Engine Power Test Code—Spark Ignition and Compression Ignition—Gross Power Rating," AE International, September 1995.
11.34 SAE J2713: "Engine Power Test Code—Engine Power and Torque Certification," SAE International, August 2007.
11.35 U.S. EPA: "SmartWay Certified Vehicles," http://www.epa.gov/smartway/vehicles/fuel-options.htm, assessed January 2012.
11.36 Kies, A. et al., "Option for a European Certification Procedure for CO2 Reduction of Heavy Duty Vehicles," SAE Paper No. 2011-01-2192, SAE International, Warrendale, PA, 2011.
11.37 "Greenhouse Gas Emissions Model (GEM) User Guide," http://www.epa.gov/otaq/climate/documents/420b11019.pdf, assessed, Feb. 19, 2012.
11.38 "Greenhouse Gas Emissions Standards and Fuel Efficiency Standards for Medium- and Heavy-Duty Engines and Vehicles," Federal Register /Vol. 76, No. 179 /Thursday, September 15, 2011.
11.39 "Greenhouse Gas (GHG) Emission Requirements Heavy-Duty Engines," Industry/EPA/NHTSA Workshop, Washtenaw Community College, November 3, 2011.
11.40 "Greenhouse Gas (GHG) Emission Requirements Combination Tractors and Vocational Vehicles," Industry/EPA/NHTSA Workshop, Washtenaw Community College, November 3, 2011.
11.41 "Special procedures for testing post-transmission hybrid systems," Electronic Code of Federal Regulations, Title 40, Part 1037, subpart F, § 1037.550.
11.42 "Special procedures for testing hybrid vehicles with power take-off," Electronic Code of Federal Regulations, Title 40, Part 1037, subpart F, § 1037.525.

11.8 Appendix: Drive Cycles

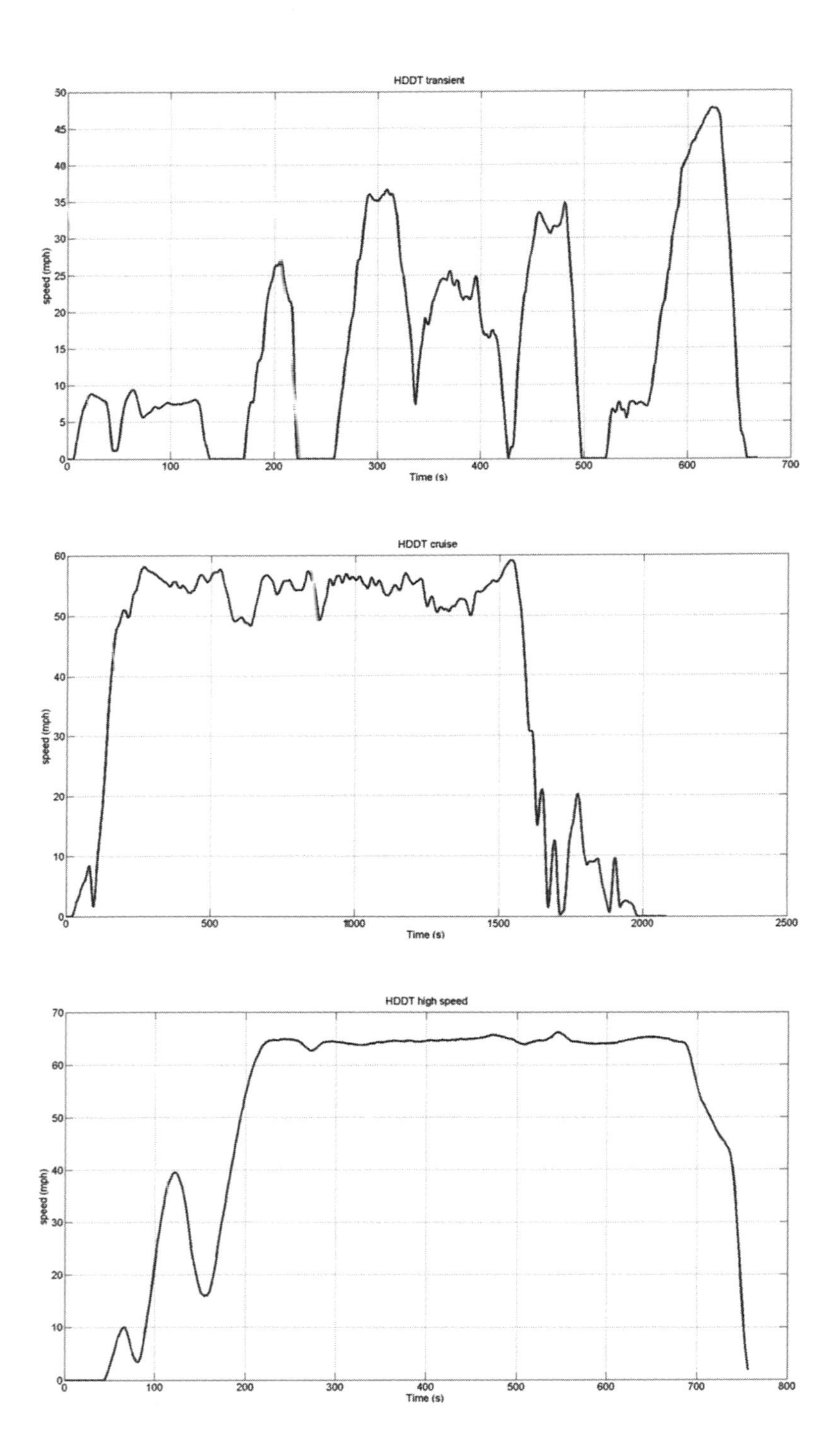

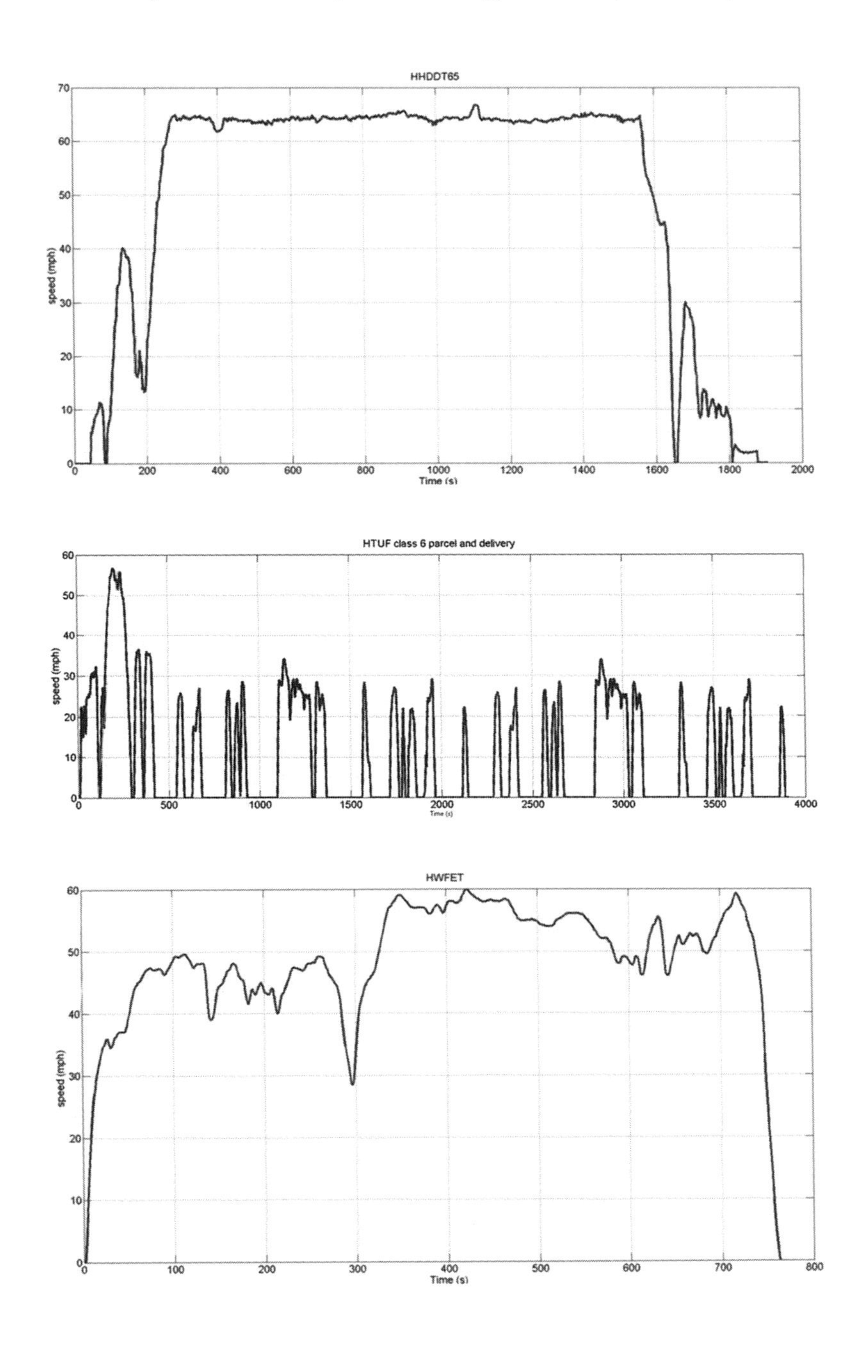
HHDDT65
speed (mph)
Time (s)
HTUF class 6 parcel and delivery
speed (mph)
Time (s)
HWFET
speed (mph)
Time (s)

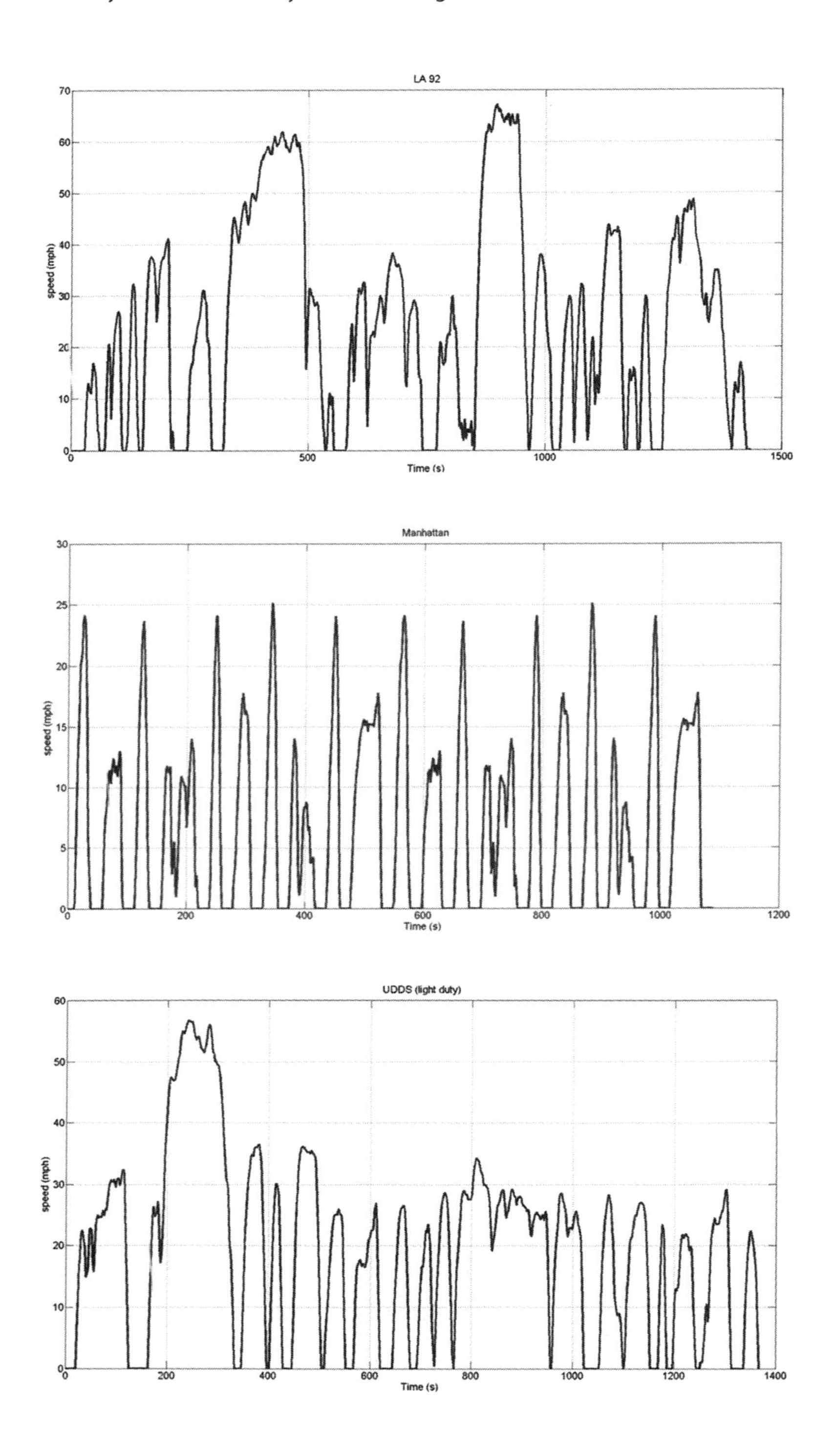
LA 92
speed (mph)
Time (s)
Manhattan
speed (mph)
Time (s)
UDDS (light duty)
speed (mph)
Time (s)

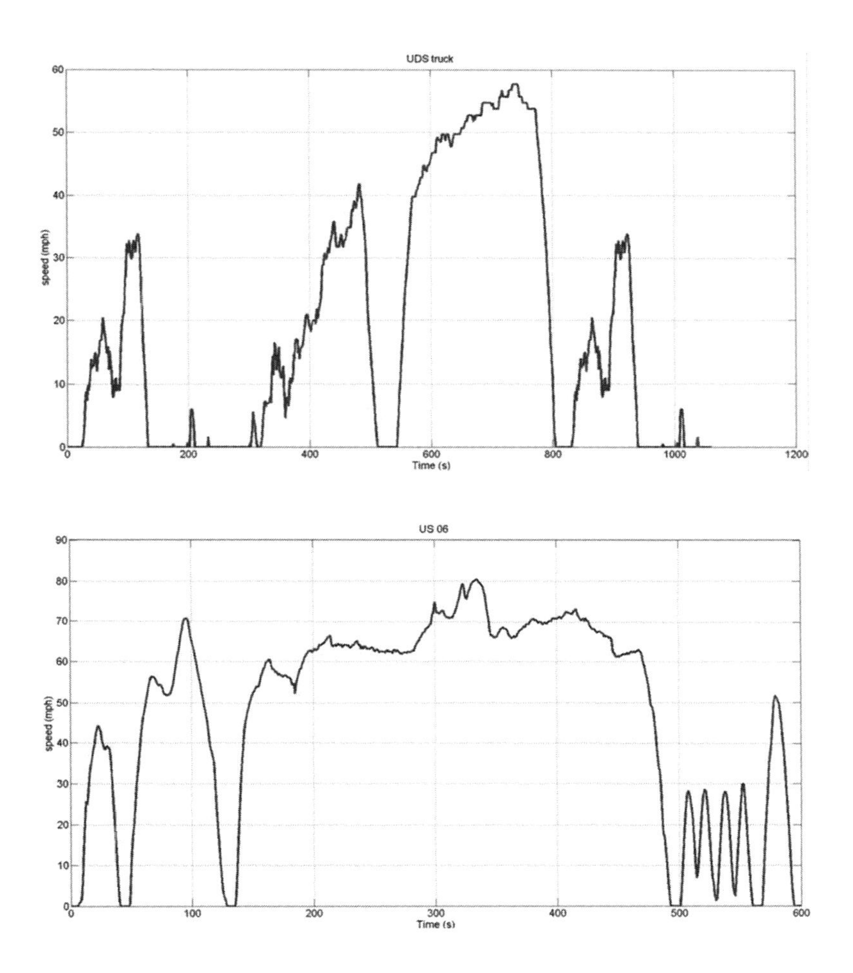
UDS truck
speed (mph)
Time (s)
US 06
speed (mph)
Time (s)

Chapter 12

Sustainable Transportation and Future Powertrain Technologies

12.1 Introduction

Powertrains for commercial vehicles over the last 100 years have been developed following the late 19th Century invention of the internal combustion engine and the realization of the transport fuels potential of light petroleum products (such as gasoline and diesel fuels) produced by the distillation of crude oil. The recent development of hybrid powertrain technologies, as discussed in previous chapters, can reduce CO_2 and fuel consumption by up to 50% compared to the conventional IC engine powertrain. However, the growing demand of energy consumption outpacing energy production (Fig. 12.1), worsening pollution, and traffic congestion in major cities indicate that sustainable transportation solutions must be explored.

What is sustainable transportation? There are many definitions of sustainable transportation and sustainable mobility. One such definition, from the European Union Council of Ministers of Transport [12.1], defines a sustainable transportation system as one that:

- Allows the basic access and development needs of individuals, companies, and society to be met safely and in a manner consistent with human and ecosystem health, and promotes equity within and between successive generations.
- Is affordable, operates fairly and efficiently, offers a choice of transport mode, and supports a competitive economy as well as balanced regional development.
- Limits emissions and waste within the planet's ability to absorb them, uses renewable resources at or below their rates of generation, and uses non-renewable resources at or below the rates of development of

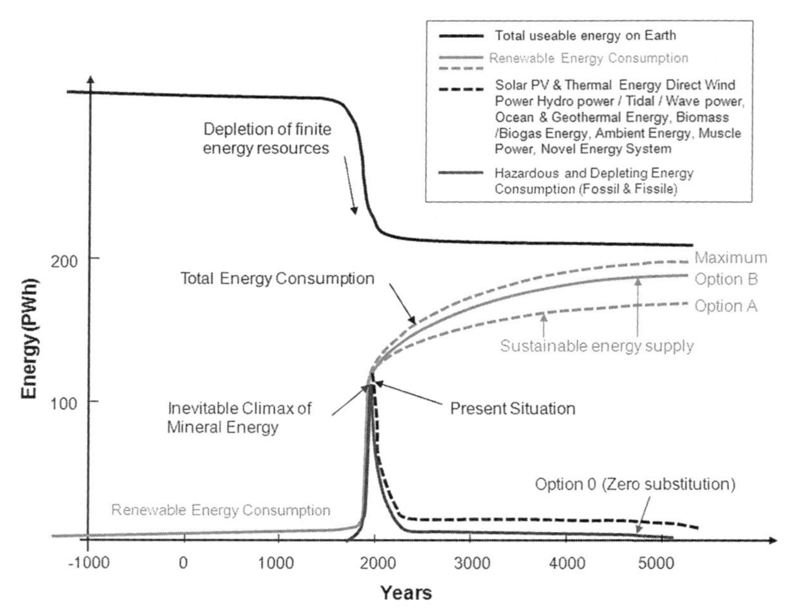

FIG. 12.1 Energy history and forecast [12.2].

renewable substitutes, while minimizing the impact on the use of land and the generation of noise.

The process of achieving sustainable transportation may depend on three factors: a) increase energy supply; b) reduce energy demand; c) improve energy conversion efficiency of the transportation process [12.3].

The production of crude oil has peaked in many countries (Fig. 12.1). With rapid industrialization and population growth in developing countries, the energy consumption is far outpacing the energy generation [12.2]. The finite reserve of crude oil will be depleted in the future. Neither biofuels nor hydrogen can ever fully replace fossil fuels. Hence, a possible way to meet the future energy demands of transportation is electricity from various sources of renewable and nuclear energy.

Reducing energy demand is also an important factor of sustainable transportation solutions. The reduction of energy demand seems contradictory to increasing industry activities and the energy needs of growing populations. However, it can be done by government policy and incentives, advanced transportation planning, and public awareness and participation. For example, public transportation, car pooling, and the development of multimodal transportation can significantly reduce energy consumption and improve industry activities. The experience of the Eco-city in Tianjin, China [12.7] shows that it is feasible to achieve economic growth and reduce energy consumption simultaneously.

The third factor is to improve transportation energy conversion efficiency. In addition to the internal combustion engine hybrid powertrain, which has been discussed in the previous chapters, the future powertrain technologies will be capable of using various fuel sources with optimized energy conversion efficiency. The future powertrain technologies for sustainable transportation will include fuel cell and fuel cell hybrid, electrohydraulic powertrain; multimodal transportation with intelligent transportation technologies; plug-in hybrid, range extender electric vehicles, and pure electric vehicles; smart grid and intelligent electric charging infrastructure, etc. [12.5, 12.6]. In this chapter, sustainable transportation and future powertrain technologies will be explored in the following four areas:

1. Plug-in Hybrid, Range Extender, and Electric Powertrain
    - PHEV, Range Extender, and Electric Powertrain
    - Charging Infrastructure
2. ITS, Smart Grid, and Inter-Modal Transportation
    - ITS
    - Smart Grid
    - Inter-Modal
3. Government Regulations and Incentives
4. Future Powertrain Technologies and System Solutions

12.2 PHEV and Battery Charger

A plug-in hybrid-electric vehicle is a hybrid-electric vehicle with a larger battery that is charged both by the vehicle's engine and by an electric grid from plugging into an electrical outlet. As with the HEVs, PHEVs use both battery-powered motors and the engine to propel the vehicle, and use regenerative braking to recuperate the kinetic energy during the braking. The large battery capacity makes PHEVs possible to drive using only electricity for some distance, commonly referred to as the "all-electric range" of the vehicle. For commercial vehicle applications, the energy stored onboard can be harnessed to do work normally done by the drive engine. Using the battery to drive an electric motor and power the power takeoff (PTO) or other onboard systems while eliminating an idling engine is efficient and practical, especially when the vehicle is parked and working on-station.

During urban driving, most of the PHEV's power comes from stored electricity. For example, a light-duty PHEV driver might drive to and from work on all-electric power, charge the vehicle at night, and be ready for another all-electric commute the next day. For longer trips or periods of higher acceleration, the internal combustion engine is used. Heavy-duty PHEVs sometimes work just the opposite, using the internal combustion engine while a worker drives to and from a job site and using electricity to power the vehicle's equipment or to keep the vehicle's cab at a comfortable temperature at the job site. Like the HEVs, the configuration of PHEVs can be parallel, series, or a combination of both.

12.2.1 Plug-in Hybrid Utility Vehicles

Most electric utilities use a large number of light-, medium-, and heavy-duty vehicles in many configurations for both onroad and offroad applications and to accomplish many specific missions. The bulk of the working fleet is made up of medium-duty service trucks, material handling trucks, and heavy-duty bucket and digger derrick-type trucks. These trucks are used to construct and maintain the vast network of overhead and underground electric distribution and transmission lines that make up the utility electrical system. The overhead line work requires specialized trucks equipped with aerial devices and lifts that are operated using hydraulics and allow linemen safe and easy access to the poles, wires, and other equipment that are part of the distribution system, as shown in Fig. 12.2.

These vehicles typically power the hydraulic system through operation of the diesel engine that runs a PTO device that is part of the transmission. This PTO-driven hydraulic system operates the whole time the vehicle is on-station and the aerial device is being used.

Based on the service territory for any specific line crew, the vehicles can be on station working anywhere from 5–7 hours per day, which translates to as much as eight hours of engine run time, if not more, for multi-shift operations. Many times these vehicles must also operate at night and on weekends for emergency response or for operational needs when work cannot be done during normal work hours due to noise restrictions and other customer satisfaction issues.

FIG. 12.2 Utility hybrid bucket truck.

Figure 12.3 shows a schematic of the plug-in hybrid powertrain for utility vehicle application. The battery energy capacity is much larger compared with the HEV system, and it can be charged by the hybrid drive system in a typical charge-sustaining hybrid or off the grid in a plug-in hybrid [12.26].

The large energy storage system can provide an all-electric driving range and supports the various work-related electrical needs such as lights, hand tools, and other specialized equipment, as shown in Fig. 12.4 (a). When the battery is depleted during operation, the diesel engine will turn on automatically to provide power for operation and to charge the battery when external charging from grid power is not available (Fig. 12.4 (b)).

This battery energy can be used both directly and indirectly. The direct usage is to convert the stored energy in the battery with an inverter into 120 Vac to power lights, charge portable power tools, and run higher-power specialty tools. The indirect usage is to use the battery power to run the electric drive motor that would normally propel the truck to power the PTO without operating the diesel engine.

The benefits of plug-in hybrids in medium- and heavy-duty applications are many. Some of the key ones are listed below [12.26]:

- Approx. 20–30% improvement in on-road fuel economy
- Approx. 30–70% reduction in fuel usage (includes idle reduction)
- Approx. 50–90% reduced emissions and greenhouse gases (GHGs)
- Reduced maintenance costs

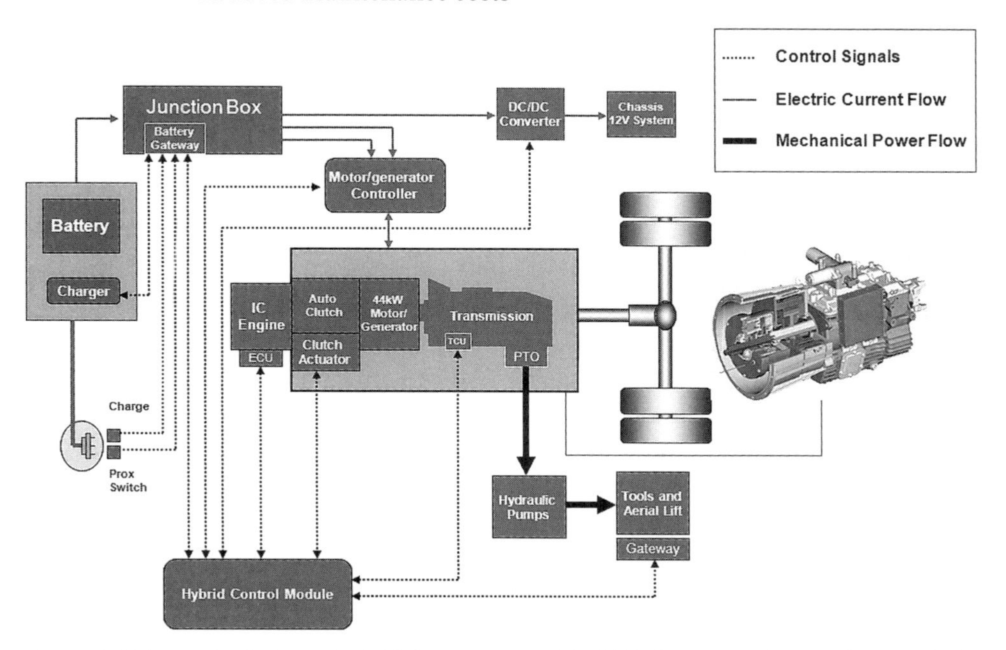

FIG. 12.3 Schematic of plug-in hybrid drive system with ePTO. (Courtesy of Eaton Corporation)

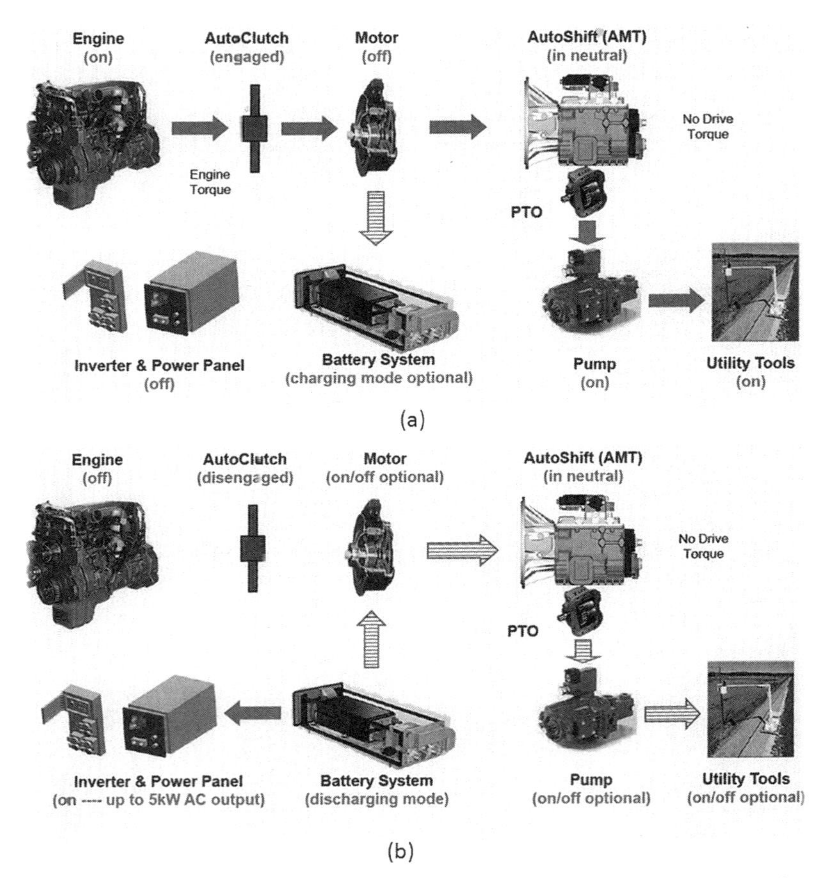

FIG. 12.4 (a) Battery power operation (engine off). (b) Engine on, mechanical PTO mode. (Courtesy of Eaton Corporation)

- Improved low-end performance
- Lower noise levels in operation
- Standby power capability
- No idle/engine off "on-station" operation
- Stealth "EV" mode capable and dedicated EV range for PHEV
- Operational flexibility due to quiet operation
- Green and environmentally friendly corporate image
- Healthier work environment for employees and customers
- Potential vehicle-to-grid (V2G) capability with onboard energy storage

The benefits of plug-in hybrids in commercial vehicle applications are not limited to the utility sector. Any business that operates a vehicle with a PTO device or has a need to idle for long periods of time to power auxiliary equipment can benefit from this technology. Other applications of PHEV include dump trucks, tow trucks, delivery vehicles, and city transit buses [12.27].

12.2.2 Onboard and Off-board Chargers

Onboard chargers take AC power from an electric grid and convert the AC power into DC power to charge a DC battery pack. Off-board units make this same conversion and deliver DC power to the vehicle. Communication between the battery management system and the charger must occur to ensure energy is delivered efficiently and safely. Power-quality standards for chargers are being developed with the goal of minimizing detrimental impacts to grid operation.

Chargers and associated cords are categorized by voltage and power levels, as shown in Table 12.1. The SAE standards specify as: Level I is 120 Vac up to 20 A (2.4 kW), Level II is 240 Vac up to 80 A (19.2 kW), and Level III (which is yet to be defined fully) will likely be 240 Vac and at power levels of 20–250 kW. It is expected that similar definitions will be created to categorize charging with DC power delivery by other countries such as Japan and China. The value of each charge power level is tied directly to the size of the onboard battery pack and the time available for recharging. Industry standards related to the

TABLE 12.1 PHEV/EV Chargers

Level	Original definition	Technical Specification	Standards
Level 1/ Mode 1	AC energy to the vehicle's on-board charger; from the most common U.S. grounded household receptacle, commonly referred to as a 120 volt outlet.	120 V AC; 16 A 208–240 V AC; Max 80 Amps	SAE J1772 (16.8 kW) IEC 61851-1 (3kW)
Level 2/ Mode 2 and 3	AC energy to the vehicle's on-board charger; 208–240 volt, single phase or 380 volt, three phase. The maximum current specified is 32 amps (continuous) with a branch circuit breaker rated at 63 amps.	12 A to 80 A 220V Single phase AC; 32A; 380V Three-phase; 380V AC; 32A to 63A Max 500VDC, 125 Amps	SAE J1772 (16.8 kW) IEC 62196 (44 kW)
Level 3/ Mode 4	DC energy from an off-board charger; there is no minimun energy requirement but the maximum current specified is 400 amps and 240 kW continuous power supplied.	Max 600VDC, 300 Amps	China GBStandards CHAdeMO (62.5 kW)

- SAE J1772 is a North American standard for electrical connectors for electric vehicles maintained by SAE International.
- CHAdeMO is an abbreviation of "CHArge de Move," equivalent to "charge for moving." CHAdeMO was formed by The Tokyo Electric Power Company, Nissan, Mitsubishi and Fuji Heavy Industries (the manufacturer of Subaru vehicles). Toyota later joined as its fifth executive member.
- GB standards are the Chinese national standards issued by the Standardization Administration of China.
- The International Electrotechnical Commission (IEC) is a non-profit, non-governmental international standards organization that prepares and publishes International Standards for all electrical, electronic, and related technologies.

PHEV/EV charger in the United States, Europe, and Asia Pacific have been discussed in Chapter 10.

12.2.3 Battery Swapping Stations

Due to large battery capacities and charging timing restrictions of commercial vehicles such as transit buses, the batteries can be modularly designed for quick installation and removal for off-board charging. Fig. 12.5 shows the transit EV buses in Beijing with modular battery design. A typical 12-m transit EV bus has ten battery packs of total energy capacity from 100 to 140 kW·h. The battery can be removed and installed in a few minutes with a robotic arm, as shown in Fig. 12.6. The modular designed battery is in the shelves for charging. It will take about three hours to fully charge the battery with a C rate of 0.3. The fully charged battery can be installed in the vehicle in less than 5 min.

While plug-in hybrid-electric vehicles (PHEVs) are a promising vehicle technology, there are many broad energy and environmental considerations that must be examined before they become widely available. For example, while a PHEV might be less costly for the consumer to drive than a gasoline-powered vehicle, it would draw power from the electrical grid while charging. In many parts of the United States and worldwide, much of that electricity would likely be generated by coal-burning power-generation plants. The costs to extract and transport the coal, as well as the environmental considerations associated with burning the coal, are all part of the overall cost of using plug-in technology. These issues decrease in importance as the amount of renewable energy in the electricity mix increases. There is also the question of how used batteries will be recycled, and how much that recycling will cost on a per-vehicle basis once all transport, processing, and disposal costs are considered.

FIG. 12.5 Modular battery design for quick battery change.

FIG. 12.6 Battery swap station and battery charger for transit city buses in Beijing.

Significant technical barriers, such as cost, battery size and performance, durability, and safety etc., also must be overcome before plug-in hybrid-electric vehicles are available for broad applications for commercial purposes.

12.3 Intelligent Transportation System, Smart Grid, and Multimodal Transportation

In addition to hybrid powertrain and plug-in hybrid powertrain technologies, other technologies are also integral parts of achieving sustainable transportation. In the following chapters, three key technologies will be discussed: intelligent transportation system (ITS), smart grid, and multimodal transportation.

12.3.1 Intelligent Transportation Systems (ITS)

Intelligent transportation systems (ITS) include telematics and all types of communications in vehicles, between vehicles (e.g., vehicle-to-vehicle), and between vehicles and fixed locations (e.g., vehicle-to-infrastructure) in an effort to optimize transportation times and fuel consumption and to improve their safety, reliability, efficiency, and quality. The deployment of intelligent transportation systems and the provision of corresponding services are not limited to the road transport sector only, but include other domains such as railways, aviation, and maritime [12.38, 12.39, 12.40, 12.41, 12.42].

Intelligent transportation systems vary in technologies applied, from basic management systems such as vehicle navigation; traffic signal control systems; container management systems; variable message signs; automatic number plate recognition or speed cameras to monitor applications, such as security closed-circuit television systems; to more advanced applications that integrate live data and feedback from a number of other sources, such as parking guidance and information systems; weather information; and bridge deicing systems. Additionally, predictive techniques are being developed to allow advanced modeling and comparison with historical baseline data. Some of the technologies typically implemented for commercial vehicle application, such as bus rapid transit, can be described as follows [12.40]:

Vehicle Assist and Automation (VAA) Technologies: This technology includes the uses of permanent magnet, stereo vision and Differential Global Positioning System (DGPS) based lateral guidance technologies, and radar, laser, and sonar for longitudinal control technologies to achieve precision docking at bus stations, vehicle guidance on the running way between stations, automatic platooning of buses at close separations, and fully automated vehicle operation. An Integrated GPS/INS (Inertial Navigation System) navigation system is shown in Fig. 12.7. The proposed system has been demonstrated to achieve 10-cm lane-keeping accuracy at highway speeds and 5-mm precision docking accuracy in 2003 in San Diego under the California Partners for Advanced Transit and Highways (PATH) Program [12.38].

Adaptive Transit Signal Priority (ATSP) Technologies: An ATSP system concept is aimed at reduction of delays for buses at traffic signals, while minimizing the impact on the rest of the traffic and maintaining

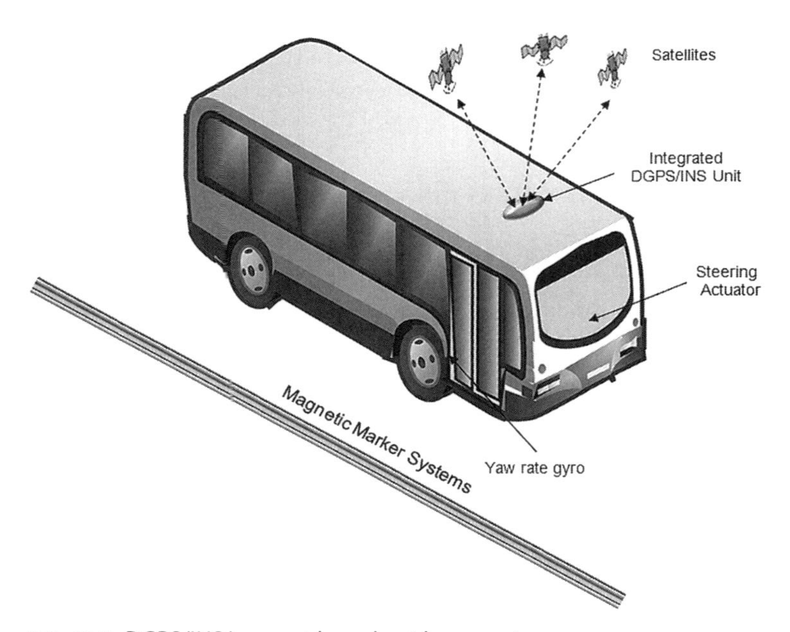

FIG. 12.7 DGPS/INS/magnet-based guidance system.

pedestrian safety. The ATSP concept uses real-time GPS positioning, bus movement information, and historical bus travel behavior data to predict the bus arrival time at the intersection (Fig. 12.8). The predicted bus arrival information provides a much longer lead time to allow the traffic controller to be "adaptive" to the bus arrival as well as to the traffic situation and makes it possible to distribute the impact over several signal control phases prior to the "hit" phase. Many transit buses have already been instrumented with GPS-based advanced communication systems (ACS). The ATSP would allow all buses instrumented with GPS/ACS to become signal-priority capable without additional equipment on the buses. It is therefore an integrated and cost-effective approach for deployment of bus signal priority.

Integrated Collision Warning Systems (ICWS): Transit safety is a major concern for operators of commercial vehicles. Bus crashes have resulted in property damage, service interruptions, and personal injuries. In addition to collision damage, passenger falls resulting from emergency maneuvers contribute to an increased potential for passenger injuries and liability.

A study by California transit agencies indicated that 30–50% of transit bus accidents could have been prevented if an integrated collision warning system (ICWS) had been deployed [12.40].

ICWS is an integration of collision warning systems for vehicle front (FCWS), side (SCWS), and rear (RCWS) [12.40]. The challenges in developing a frontal collision warning system (FCWS) for transit buses reside in the

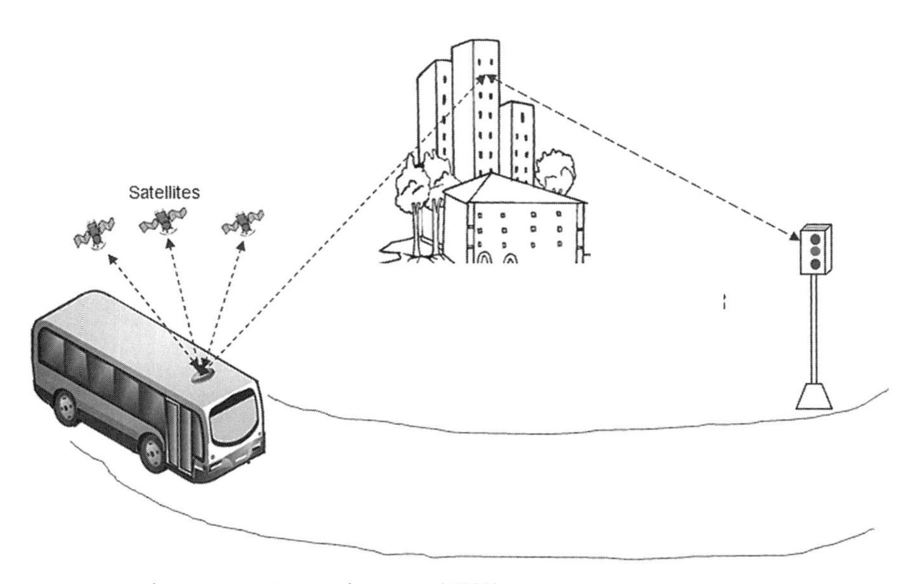

FIG. 12.8 Adaptive transit signal priority (ATSP) system.

complexity of the urban traffic environment and the requirement for accurate real-time detection of true hazards. Too many false warnings to the drivers may cause drivers to ignore the system or disable it. Prototypes of the Integrated Collision Warning System (ICWS) in the San Francisco Bay Area and in Pittsburgh indicate that the introduction of the warning system appeared to lead to more consistency of driving behavior across the sample of operators. More significantly, these data show that drivers with somewhat more aggressive and inconsistent driving behaviors before the introduction of the warning system changed their driving behaviors and became noticeably more consistent and less aggressive after the warning system was activated.

In 1993, the US Department of Transportation (USDOT) initiated a program to define and develop a national Intelligent Transportation Systems (ITS) architecture. The national ITS architecture defines a single framework that may effectively guide the development and implementation of ITS user services over the next 20 years. As defined by the USDOT [12.48], "A system architecture is the framework that describes how system components interact and work together to achieve total system goals. It describes the system operation, what each component of the system does, and what information is exchanged among the components. A system architecture is different from a system design. Within the framework of an architecture, many different designs can be implemented." The benefits of this synergy to ITS system designers and planners appear in two areas [12.48]:

- Efficiencies in functional performance: In the architecture, a given function may appear in multiple packages that in a system design only needs to be performed once. This means that potential redundancies in data collection, processing, and dissemination in an ITS system design can be avoided. These efficiency benefits take the form of cost savings in avoiding functional duplication.
- Cost savings from common technology. When implementing packages with common functions, only one package containing the given function is required. In this way, the second (or third, etc.) market package that uses these common functions can leverage an existing technology (software, hardware, communications, etc.) investment.

The ITS System framework can be applied on any size of network, with benefits to the user in reduced costs and improved efficiencies. There is much to be gained by integrating systems in an enclosed environment, such as a multimodal interchange, PHEV/EV charging infrastructure and smart grid, as well as for city-wide implementations in which network management becomes much simpler due to the unified interface and operations, as shown in Fig. 12.9.

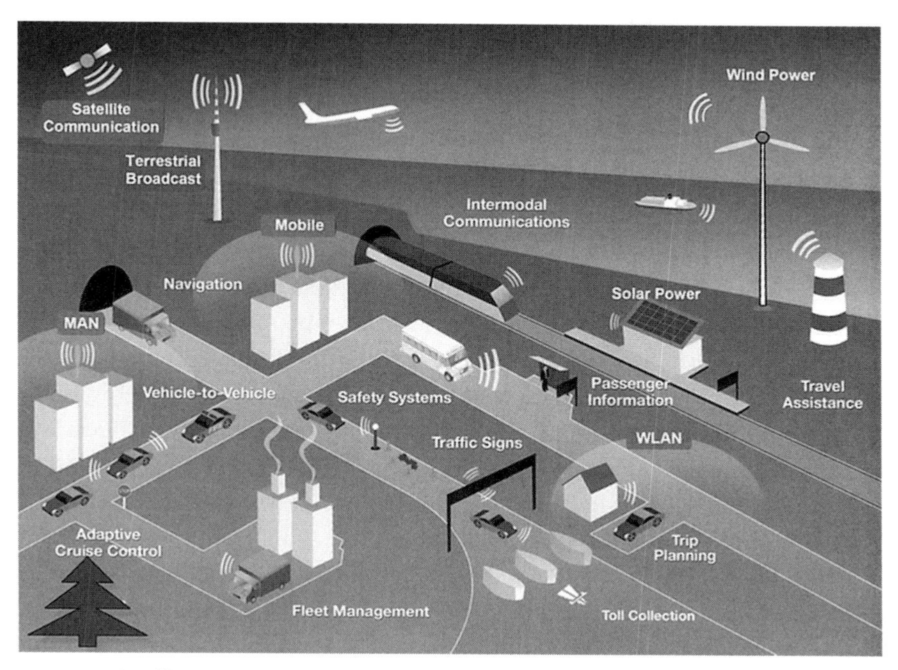

FIG. 12.9 Intelligent transportation systems [12.41].

12.3.2 Multimodal Transportation

Multimodal transportation is the shipment of cargo and the movement of people involving more than one mode of transportation during a single, seamless journey to reduce energy consumption and air pollution and to promote sustainable economic development [12.52, 12.53].

For cargo transportation, trucking lags far behind rail and marine in efficiency on a ton/mile basis. Table 12.2 shows estimates of the energy efficiency of alternative transport modes by US DOT Maritime demonstration. From the fuel efficiency viewpoint, freight transport by rail and marine will be the preferred methods to achieve sustainable transportation. However, transport modes are, and will remain, very commodity- and distance-sensitive, and the on-highway delivery will be eventually needed for the "last-leg" distribution [12.56].

TABLE 12.2 US DOT-Maritime Administration Estimates of the Energy Efficiency of Alternative Transport Modes [12.54]

	Truck	Rail	Inland Barge
Number of Miles One Ton can be Carried per one Gallon Fuel	59	202	514

In general, trucking is the preferred method of shipping for transport routes of less than 1000 km, in addition to small-volume and time-sensitive deliveries. These routes encompass the final product delivery. The inherent flexibility of trucking, ability to handle smaller loads, reach all destinations, and efficiently make multiple deliveries/pickups ensures that trucking will maintain its dominance in this segment.

Continued efficiencies in marine, rail, and truck transport will occur to enhance/maintain their viability in their current core markets, but even these markets are regionally different. A rail network that is somewhat less conducive, due to the shorter distances involved, to freight movements than that of the United States makes the European and Japanese relatively more dependent on trucking, although rail freight improvements are actively under consideration in those regions due their greater traffic congestion problems.

There are challenges in deployment of multimodal transportations. In Europe, a high percentage of freight (70–90%) is moved over the highways. The alternative of rail, however, is difficult to implement. In addition to the fact that trucks will still be needed at each end of the journey, and the journey beginning- and end-points are usually where traffic congestion is at its worst, there is also the issue of the compatibility of passenger and freight rail service.

However, studies of multimodal transportation have been conducted in many cities. Halifax Regional Municipality [12.57], Canada, and Sino-Singapore Tianjin eco-city project [12.7, 12.8] summarize their design and deployment of multimodal transportation in the following five items.

Transportation Hubs: Transit hubs and terminals are locations where community, commerce, and the transportation system interact. They should be permanent structures and form an integral part of the community they serve. Stations are protected from the weather, provide amenities, and are comfortable and vibrant meeting places. They are well connected to the surrounding community by multi-use active transportation routes and public open-space networks.

Dedicated Rights-of-Way: A dedicated right-of-way for transit, high-occupancy, and active transportation ensures that alternatives to single-occupancy automobiles are a fast, convenient, and affordable choice. A fixed route also communicates the sense of permanence essential to attract new investment and development over the long term. The actual mode of transportation and exact route selected for these corridors would be appropriate to the situation, e.g., bus, train, light rail, or ferry.

Efficient Mode Switching: Implementing advanced ITS facilitates efficient mode switching, with faster loading and unloading of cargo and people from one mode to another. Telematics, tracking trains with GPS instead of the hard-wired signaling systems, and other technology advances enable efficient mode switching.

Land-Use: Multimodal transportation planning must be fully integrated with the planning of sustainable transportation systems.

Local Feeder Routes: Not all residents will live within walking distance of a transit hub, so a feeder system is required to serve the main transportation corridors. Appropriate modes of travel would be selected based on the local situation, e.g., shared taxis, van-pools, ferries, or local buses.

12.3.3 Smart Grid

A smart grid refers to an electricity distribution system that uses bi-directional digital technology to control and monitor devices located at homes, and commercial and industrial sites, such as the PHEV charging stations. The goals of a smart grid are to save energy, reduce cost, and increase reliability and transparency. A smart grid incorporates the capability of integrating alternative sources of electricity such as solar and wind [12.59, 12.60].

Smart grid technologies open a new door for system optimization. The smart grid allows utilities to better understand their needs and resources and optimize system use. Various levels of implementation likely will exist, from data monitoring and remote controls throughout the entire network to basic automation of meter reading [12.64].

The smart grid updates the existing power grid to employ real-time, two-way communication between power suppliers and customers. Power suppliers provide their customers with pricing information that is based on electricity demand for a given period during the day. Customers use that information to guide power usage. For example, PHEV/EV owners may decide to recharge their cars late at night, when electricity rates are much lower compared to the early afternoon. As energy use is shifted to off-peak periods, fewer power plants will need to be built, and electricity needs will be met by cleaner and more efficient sources.

The smart grid will facilitate the use of large-scale energy storage systems that will become more important as renewable energy technologies spread. Large-scale energy storage will mitigate disruptions in electricity flow across the grid as variable sources like solar and wind power ramp up and down. With this capability, the smart grid would not only reserve power, but would route it even more efficiently.

Electric-based vehicles can become part of the grid in the future with flexible energy storage to manage power flow and help deploy energy sources (Fig. 12.10), as it:

- Communicates vehicle location and charge status to the utility operator, who transmits energy mix, real-time pricing, and availability information to the vehicle
- Allows the utility operator to wirelessly advise the vehicle to maximize the charge rate when a surplus of clean energy is available, and to minimize charge rate when it is not

Smart grid allows V2G vehicles to provide power to help balance loads by "valley filling" (charging at night when demand is low) and "peak shaving" (sending power back to the grid when demand is high). It can give utilities new ways to provide regulation services (keeping voltage and frequency stable) and provide spinning reserves (meet sudden demands for power). In future development, it has been proposed that such use of plug-in electric vehicles could buffer renewable power sources such as wind power, for example, by storing excess energy produced during windy periods and providing it back to the grid during high load periods, thus effectively stabilizing the intermittency of wind power.

There are regulatory and technological barriers that limit the implementation of smart grid technology. The main barrier limiting the deployment of this technology is the lack of consistent standards and protocols. There currently are no standards for these technologies in the United States. This limits the interoperability of smart grid technologies and limits future choices for companies that choose to install any particular type of technology. The US

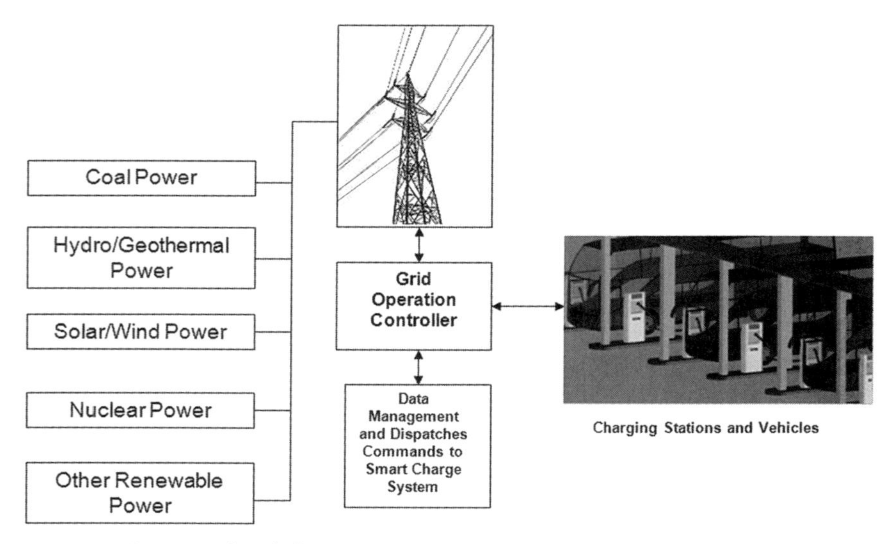

FIG. 12.10 Smart grid and charge system [12.61].

Department of Energy lists five fundamental technologies that will drive the smart grid [12.59]:

- Integrated communications, connecting components to open architecture for real-time information and control, allowing every part of the grid to both "talk" and "listen"
- Sensing and measurement technologies, to support faster and more accurate response such as remote monitoring, time-of-use pricing, and demand-side management
- Advanced components, to apply the latest research in superconductivity, storage, power electronics, and diagnostics
- Advanced control methods, to monitor essential components, enabling rapid diagnosis and precise solutions appropriate to any event
- Improved interfaces and decision support, to amplify human decision-making, transforming grid operators and managers quite literally into visionaries when it comes to seeing into their systems

In summary, sustainable transportation requires advanced ITS technologies, efficient intermodal switching, and smart grids for renewable energy development and implementation. Figure 12.11 illustrates the future of a sustainable transportation system.

FIG. 12.11 Sustainable transportation system with advanced ITS, efficient intermodal switching, and smart grid with renewable power supply. (Courtesy of Eaton Corporation)

12.4 Government Regulations and Incentives

Regulatory actions and incentives taken by governments worldwide are now clearly pushing the transportation industry toward much more aggressive adoption of hybrid vehicles and vehicle electrification. Many of these initiatives can be traced back to rising concerns about greenhouse gas concentrations and the Kyoto Protocol of 1997 (note that CO_2 and fuel economy regulations are essentially the same, because each gallon of gasoline/diesel burned will always produce 19.4/22.2 pounds of CO_2) [12.68].

12.4.1 United States and Canada

In the United States, incentive grants for the purchase of fuel-efficient heavy vehicles are available through several sources.

The EPA administers loans through its SmartWay program, which has a US$30m fund, and there are Clean Diesel Grant Programs at federal and state levels with funds of US$156m and US$88m, respectively. Federal incentives available for heavy-duty PEVs and HEVs range from US$7500 to US$15,000, with, for example, a Kenworth T270 hybrid attracting US$6000 and a T370 hybrid, US$12,000. California incentives range from US$10,000 to US$35,000, with a further US$5000 available with certification. In total, grants can be accumulated for up to 25% of the purchase price of the vehicle.

For fuel cell buses, the U.S. National Fuel Cell Bus Technology Program supports research and development programs, and WestStart-CALSTART, a consortium of 155 transportation technology firms and organizations worldwide, has launched five fuel cell bus programs in collaboration with the Federal Transit Administration (FTA). The program has US$12m in funding from the FTA and US$12m from other participants.

In October 2010, the U.S. EPA and NHTSA announced a first-ever program to reduce GHG emissions and improve fuel efficiency of medium- and heavy-duty vehicles, such as the largest pickup trucks and vans, semi trucks, and all types and sizes of work trucks and buses in between. The proposed regulation starts to phase in during model year 2017. In May 2009, the U.S. essentially adopted California's vehicle emissions regulatory policy for light-duty vehicles.

To achieve the government mandates on fuel efficiency improvement, HEV/EVs and improvements to internal combustion engines are required.

In 2009, the Obama Administration allocated $2 billion to a Plug-in Vehicle Tax Credit in its 2009 stimulus bill. The credit acts as a subsidy on vehicles that are propelled by a battery of 4 kW·h or more ($2500 for any 4kW·h vehicle, plus $417 for each additional 1 kW·h up to a maximum credit of $7500 for a vehicle with a battery of 16 kW·h or more). Each automaker will get 100%

credit for their first 200,000 eligible vehicles sold, 50% credit for the next two quarters, and 25% credit for the final two quarters.

In terms of HEVs, credits of up to $3400 have been available since 2005. But the credits phase out over a one-year period for a given manufacturer once it has sold over 60,000 eligible vehicles. This tax credit expired on December 31, 2010. There are no known initiatives to extend or expand the program.

State and local governments are also studying major consumer-based incentive systems to provide further assistance to the automakers in terms of selling advanced vehicles, as well as providing a pricing offset to the additional content required to meet the standards.

In Canada, Ontario has launched an incentive program for hybrid and alternative-fuel commercial vehicles.

12.4.2 European Union

To achieve its central policy objective of reducing GHG emissions by 20% by 2020 against 1990 levels, the EU has put together an energy-policy package encompassing all of these dimensions and affecting all areas of the economy. Transport, accounting for about 20% of European CO_2 emissions, is obviously one of the targeted areas for improvement, with passenger cars (12% of total) presenting the biggest contributor. In its effort to become the leading low-carbon society, the EU has put a tough regulatory framework in place, requiring Europe to take the global lead in fuel economy improvements.

Besides the regulatory push, we note that there is active government support for vehicle electrification in nearly all major European countries.

As can be seen in Table 12.3, many European countries already offer substantial consumer sales incentives for EVs—among the most aggressive comes from Denmark, which exempts EVs from vehicle taxes (ICEs pay a tax of 105% on the first 76,500 DKK ($14,000), and 180% for each additional krona). On a volume-adjusted basis, we estimate that European governments

TABLE 12.3 PHEV/EV Incentives of Selected EU Countries

France	15% of fleet zero carbon within 2–3 years, 5mm charge spots with 3 years	€5,000 bonus for <60/g/km vehicles up to 20% of purchase price. Free registration / free parking in select region	€650mm in government loans for low emisssion vehicle development
Germany		EV's exempt from annual circulation tax for 5 years	€750mm in various research, electric mobility, smart grid programs
UK	1mm xEV's by 2014	Gov't credits on EV / PHEV between £2,000–6000	Approx £75mm in charging infrastructure/EV development funding

offer on average €3,000 per EV already, and this excludes substantial support in Germany, which we see as likely soon.

In the UK, low-emissions vehicles are eligible for road fund fee exemptions, and the government is researching the possibility of incentive schemes for low-carbon buses.

12.4.3 China, Japan, and South Korea

China: China has announced an extensive incentive program to encourage the use of EVs and HEVs, and has set a target of having 60,000 alternative-energy vehicles, including heavy commercial vehicles, operating in 13 cities by 2012. The State Council, or China's cabinet, adopted a development plan for energy-saving and new-energy vehicles (2012-20) on April 18, 2012. According to the plan, China's accumulative output of pure electric and plug-in hybrid electric vehicles will rise to 500,000 units by 2015 and 5 million units by 2020.

To support the long-term development of Chinese new-energy vehicles, China's Ministry of Finance (MOF) and the country's Ministry of Science and Technology (MOST) jointly launched subsidy policies for energy-saving and new-energy vehicles in 13 pilot cities, e.g., Beijing, Shanghai, and Chongqing. The sole focus is on public transportation vehicles such as buses (up to 1000 buses for each of 13 cities will receive a $70,000 subsidy) and taxis (up to $8500 subsidy). The government incentives of city buses are listed in Table 12.4. In addition, China's State Grid Corporation has recently announced that it will

TABLE 12.4 China: Government Incentive for City Buses [12.69]

TYPE	HEV/PHEV Gas Saving	Electricity Saving as % of Overall Power	Government Incentive In Thousand RMB
HEV/PHEV	5–10%	BSG	4
	10–20%	BSG	4
		10–20%	28
		20–30%	32
	20–30%	10–20%	32
		20–30%	36
		30–100%	42
	30–40%	20–30%	42
		30–100%	45
	>40%	30–100%	50
BEV		30–100%	60
Fuel Cell		30–100%	250

speed up the construction of electric vehicle charging stations in Shanghai, Beijing, Tianjin, and other large cities as a first step.

Japan: The Japanese auto industry has made significant efforts to improve fuel economy over the last ten years, and the government has already set a 15% average fuel efficiency improvement target for 2015 vs. 2007 (16.8 km/l under a new fuel economy measurement methodology, which compares with 14.6 km/l calculated for 1997 under the same methodology). In addition, on June 09, 2009, the former Japan Prime Minister also announced a new mid-term CO_2 reduction plan for Japan that calls for a 15% CO_2 emission reduction target by 2020. As part of this policy, the government also projected increased penetration of next-generation vehicles—pure hybrids, PHEVs, and EVs—to 40% of new vehicle sales by 2020, up from 10% in 2009.

The Japanese government has rolled out tax policies to stimulate demand for more fuel efficient vehicles through an Eco Tax program (Table 12.5). Under this program, next-generation vehicles such as EVs, PHEVs, HEVs, clean diesels, and natural gas vehicles are exempt from acquisition and tonnage taxes. For example, a Prius buyer would save approximately ¥140k ($1550).

Japan has also launched programs that subsidize the cost of electrifying a standard car up to a maximum value of the cost of the base car. It uses a different formula to calculate the subsidy on a mini-car based EV (MMC, Subaru) and a non-mini based EV (not yet launched Nissan Leaf). The subsidy offsets 50% of the cost of electrification for a mini car, and 25% of the cost of electrification for a non-mini.

South Korea: In July 2009, the Korean government announced a long-term plan to inject W107tr ($90 bn) into green growth over the next five years. In conjunction with this plan, the government announced a 2015 corporate average fuel economy target of 17 km/l (CO_2 emission levels 140 g/km), representing a 16.5% increase from current levels. Non-compliance would result in OEMs paying fines starting in 2013.

12.5 Future Powertrain Technologies and System Solutions for Sustainable Transportation

12.5.1 Future Powertrain Technologies

Today's motorized transportation system has evolved from the petroleum-based fuels powertrain over the last 100 years. With finite petroleum reserves and increasing concerns of global warming exacerbated by the GHG emissions, alternative fuels and power technologies are now being explored (Fig. 12.12).

TABLE 12.5 Japan: Tax Incentives of Three Major Vehicle Taxes for Electric Drive Vehicles [12.69]

		2003	2004	2005	2006	2007	2008	2009	2010	2011
HYBRID	One time acquisition tax	−2.20%	−2.20%	−2.20%	−2.20%	−2.00%	−1.80%	exempted	exempted	exempted
	Annual tonnage tax							exempted	exempted	exempted
EV & FC	One time acquisition tax		−2.70%	−2.70%	−2.70%	−2.70%	−2.70%	exempted	exempted	exempted
	Annual tonnage tax									
	Annual automobile tax							−50%	−50%	−50%
PHEV	One time acquisition tax							exempted	exempted	exempted
	Annual tonnage tax							exempted	exempted	exempted

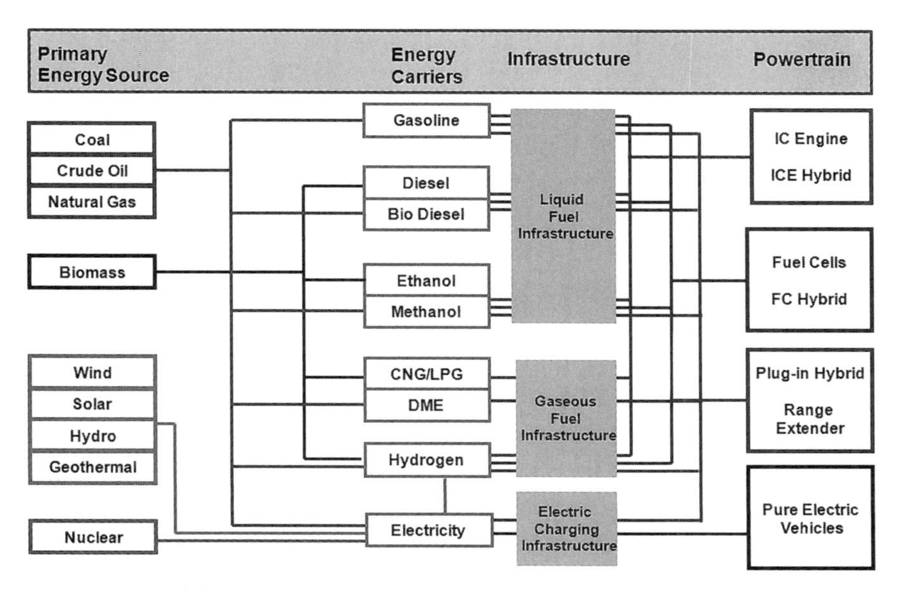

FIG. 12.12 Possible powertrain energy pathway [12.1].

In Fig. 12.12, the first column identifies various sources of primary energy available to propel transport vehicles. The second column shows energy carriers in use at present or proposed for use in the future as transport fuels. The lines connecting the first and second columns show some of the many possible ways that different primary energy sources can be transformed into energy carriers.

For an energy carrier to be used widely as a transport fuel, there must be an infrastructure capable of distributing it. The third column identifies three major categories of transport energy distribution systems: 1) liquid fuels, 2) gaseous fuels, and 3) electricity. The lines connecting the second and third columns show which energy carriers are capable of being distributed by each category of energy infrastructure. The fourth column of Fig. 12.12 shows the four major categories of propulsion systems either presently being used or likely to be used in road, rail, and waterborne vehicles. These are ICEs (including ICE hybrids), fuel cells (including fuel cell hybrids), plug-in hybrid and range extenders; and pure electric vehicles. As with the diversity of the energy sources and applications, the four types of powertrains for commercial application will coexist in the foreseeable future.

Over the next 30 years, internal combustion engine technologies will continue to improve, given the availability of suitable and appropriate, cleaner enabling fuels. For the spark ignition (SI) engine technology, downsized SI engines are expected to take a much greater share of the SI engine market in the near future. Gasoline direct injection (DI) engines are likely to be more important than conventional port fuel injection engines by 2020. Spark ignition engines with variable valve trains and other reduced friction technologies, displacement-on-demand, turbocharging, controlled auto ignition (CAI), and multispeed transmissions will enable improved energy utilization.

Diesel engines will still be the preferred powertrains for long-haul transport in the foreseeable future. The dominant diesel engine technology will be direct injection (up to 2500 bar) with high turbocharging, EGR, and intercooling. Active emission control technologies, such as low-temperature combustion technology, will become commercially available after 2015.

Fuel cell systems have been discussed in Chapter 9. A fuel cell vehicle (FCV) using hydrogen derived from carbon-neutral sources would offer the highest overall propulsion system energy efficiency (more than 40%) and the lowest GHG and conventional emissions. The proton exchange membrane (PEM) fuel cell has demonstrated its technical maturity for commercialization. However, its impact will be limited because of the availability of hydrogen infrastructure and onboard hydrogen storage, and the level of high-cost precious metals required for the fuel cell stacks. The solid oxide fuel cell (SOFC) based powertrain is still in the development stage; its main challenges will be the long warm-up time, limited life due to thermal cycling, and system cost.

Hybrid technology will become increasingly important for commercial transportation. With technology advancement and cost reduction in each area of hybrid components, including electric motor controllers, high-efficiency electric motors, high power/energy density batteries, and optimized system control and integration, the hybrid powertrain will be widely adopted by the commercial vehicle industry by 2017.

Vehicles powered by electricity generated using renewable energy (solar/hydro/wind/geothermal) and nuclear energy has the potential to be truly "zero emissions"—one that produces no emissions of either greenhouse gases or local pollutants. The electric-powered fast-speed train has been available in Japan, China, France, and other countries. Electric city buses and delivery vehicles are rapidly emerging in major cities in China and other cities worldwide. With the advancement in energy storage technologies, electric vehicles, along with range extender electric vehicles, plug-in hybrid vehicles will be the key element of the sustainable transportation solutions in the foreseeable future.

12.5.2 System Solutions and Sustainable Transportation

Future sustainable transportation concerns systems, policies, and technologies. In addition to powertrain technologies, future transportation may include clean vehicle technology and system integration, advanced energy storage technology, renewable energy and management, smart grid and charging infrastructure, telematics and intelligent transportation technologies, and multimodal transportation, etc. Figure 12.13 illustrates a sustainable transportation framework. It shows that an in-depth, cross-border project syndication and deployment capabilities of partnership are required. More detailed discussion of the requirements of sustainable transportation is provided in Fig. 12.13. It shows that gaps in systems planning could lead to a bottleneck in the deployment of sustainable transportation. Many cities, regions, and nations want to adopt new-energy vehicles and multimodal transportation, but have not planned and developed the transportation infrastructure and energy management requirements for sustainable transportation.

The solutions to these problems are not as simple as to deploy the efficient transportation technologies. This is not just a technology issue, it is also a system issue. To address fundamental sustainable transportation systems issues, all of the following measures must be taken, in concert, to drive lasting, meaningful change. In addition, no one company exists to bring these solutions to market. Collaboration is a necessity, not a choice. To address systems issues, at least six foundational partners are needed, but a larger supporting cast is the key to driving large-scale sustainable transportation systems goal attainment.

Many industry alliance organizations have been formed to deploy sustainable transportation. China's State-owned Assets Supervision and Administration Commission (SASAC) of the State Council has formed an alliance of 16 Chinese government-owned businesses, aimed at unifying EV standards and speeding up research and development [12.71]. The alliance includes the country's top three oil companies, top two power grid operators, battery and charging equipment makers, as well as the automakers China FAW Group Corporation, Dongfeng Motor Corporation, and China Changan Automobile Group. Another alliance, The New Energy Sustainable Transportation International Alliance (NESTIA) is the first new industry-driven cross-border economic initiative focused on the development of sustainable transportation solutions in the United States and China. NESTIA

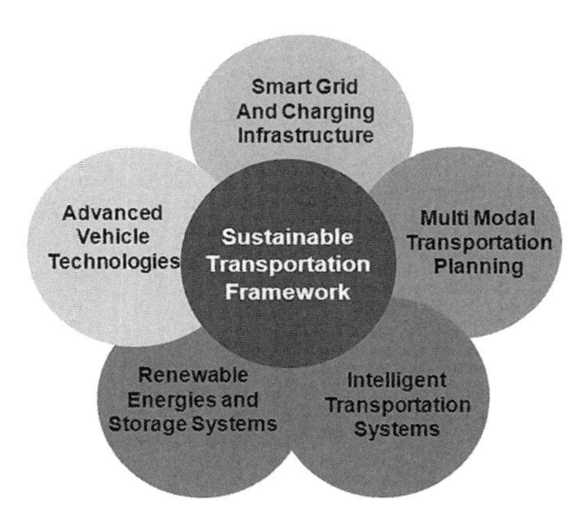

FIG. 12.13 Framework of sustainable transportation.

provides project development and coordination for joint research and development, joint ventures, partnerships, and alliances between network members who are seeking to bring sustainable transportation technologies to market within projects that have the potential to lead to mass commercialization and sustainable economic development in the U.S. and China. With expertise in clean vehicle technology, clean vehicle integration and optimization, advanced battery chemistry and energy storage, renewable energy, smart grid infrastructure and information technology informatics and integration, and sustainable urbanization planning and development, the NESTIA may have the most in-depth, cross-border capabilities of any research and development partnership in the industry. NESTIA also has a great ability to move innovation from the lab to mass commercialization, which will drive sustainable economic development in the United States and China [12.70].

The NESTIA alliance includes three U.S.-based companies and three China-based companies: Eaton, AECOM, IBM, Foton Auto, MGL Battery, and Broad-Ocean Motors; the objective is to advance the adoption of sustainable transportation from plug-in to full-electric across public and private transportation sectors in key markets worldwide. Its approach provides a balanced, holistic approach to the transportation and urban planning process. Accessibility of charging infrastructure and modes of transportation, an indicator of the ability to efficiently reach destinations, is a product of mobility and proximity. As such, destination accessibility can be enhanced by either increasing the speed of getting from one point to another (mobility), by bringing the two points closer together (proximity), or through some combination thereof,

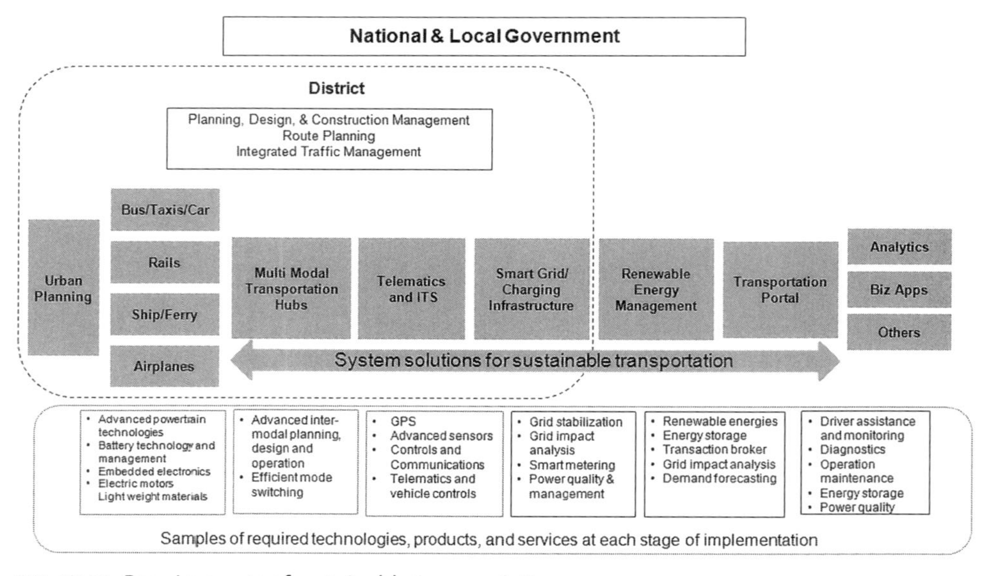

FIG. 12.14 Requirements of sustainable transportation.

and must be supported not only by the mix of new-energy vehicles and other modes of transportation, but also with the inclusion and accessibility of other service facilities, such as the electrical charging infrastructure. This dual focus allows development of transportation solutions that enable people to access the goods, employment, and services they need and want, while eliminating range anxiety and without creating mobility bottlenecks while accessing the charging infrastructure.

The system solutions proposed by NESTIA incorporate the following elements into an overall plan (Fig. 12.14):

- Master planning, ITS, power management, and multimodal transportation hub planning to drive the most efficient deployment of sustainable transportation and infrastructure
- Smart grid and renewable energy systems.
- Advanced vehicle and transportation technologies and deployment
- Service infrastructure of sustainable transportation

The primary objective of NESTIA is to deploy sustainable transportation in cities in China, along with its strategic partners in China. The long-term objective of the NASTIA alliance will lead the way to deployment of sustainable transportation solutions around the world.

As the world moves toward sustainable transportation, more alliances will be formed jointly to develop and deploy sustainable transportation solutions. In 2009, a U.S.-China Energy Cooperation Program was established to leverage private sector resources for project development work to develop and deploy clean energy transportation solutions in the United States and in China. In 2010, the U.S.-China Clean Energy Research Center (CERC) was established to facilitate joint research and development of clean energy technologies by teams of scientists and engineers from the United States and China to contribute to dramatic improvements in technologies with potential to reduce the dependence of vehicles on oil and improve vehicle fuel efficiency [12.72]. The success of the collaborative CERC program between U.S. and China, the World's top energy consumers, energy producers, and greenhouse gas emitters will be a milestone toward sustainable transportation.

12.6 References

12.1 Gilbert, R., "Defining Sustainable Transportation," The Center for Sustainable Transportation, #T8013-4-0203, Transport Canada, March 31, 2005.

12.2 "Mobility 2030," WBCSD (World Business Council for Sustainable Development) report, 2004.

12.3 Richardson, B., "Sustainable Transport: Analysis Framework," *Journal of Transport Geography*, Vol. 13, pp. 29–39, 2005.

12.4 Steg, L. and Gifford, R, "Sustainable Transportation And Quality Of Life," *Journal of Transport Geography*, Vol. 13, pp. 59–69, 2005.

12.5 Yaegashi, T., Abe, S., and Hermance, D., "Future Automotive Powertrain—Does Hybridization Enable ICE Vehicles to Strive Towards Sustainable Development?" SAE Paper No. 2004-21-0082, SAE International, Warrendale, PA, 2004.

12.6 Gott, P., "Is Mobility As We Know It Sustainable?" SAE Paper No. 2009-01-0598, SAE International, Warrendale, PA, 2009.

12.7 "CHINA: GEF Sino-Singapore Tianjin Eco-City," World Bank, 2010.

12.8 "Sino-Singapore Tianjin Eco-City: A Case Study of an Emerging Eco-City in China," World Bank, 2009.

12.9 Olson, W. W., "What Does Sustainability Mean For the Mobility Industries?" SAE Paper No. 2005-01-0535, SAE International, Warrendale, PA, 2005.

12.10 Fedewa, E. A., "Sustainable Mobility: The Business Case for Global Vehicle Electrification," SAE Paper No. 2010-01-2311, SAE International, Warrendale, PA, 2010.

12.11 Parent, M., "Cybercars for Sustainable Urban Mobility—A European Collaborative Approach," SAE Paper No. 2010-01-2345, SAE International, Warrendale, PA, 2010.

12.12 "OECD Proceedings: Towards Sustainable Transportation," The Vancouver Conference, Conference organized by the OECD hosted by the Government of Canada, 1996.

12.13 "China's Pathway Towards a Low Carbon Economy," China Council for International Cooperation Environment and Development, 2009.

12.14 Parkash, M., "Sustainable Development in the People's Republic of China," Asian Development Bank, 2008.

12.15 Pan, J. and Zhu, X., "Energy and Sustainable Development in China," Helio International, 2006.

12.16 "China's 12th Five-Year Plan How it actually works and what's in store for the next five years," APCO Worldwide, 2010.

12.17 Deakin, E., "Sustainable Development and Sustainable Transportation: Strategies for Economic Prosperity, Environmental Quality, and Equity," University of California, Berkeley, 2001.

12.18 Lagan, C. and McKenzie, J., "Sustainable Cities, Sustainable Transportation," the EMBARQ Background Paper on Global Transportation and Motor Vehicle Growth in the Developing World—Implications for the Environment, 2004.

12.19 Pan, H., et al., "Mobility for Development: Shanghai, China," Case Study, World Business Council for Sustainable Development (WBCSD), 2008.

12.20 "Shifting to Sustainable Transportation, A Sustainable Transportation Framework for HRM," HRM Sustainable Environmental Management Office, 2009.

12.21 Litman, T. and Burwell, D., "Issues in sustainable transportation," *Int. J. Global Environmental Issues*, Vol. 6, No. 4, 2006.
12.22 Casalegno, F. and Chiu, D., "White Paper on Sustainable Transportation MIT Mobile Experience Laboratory, 2008.
12.23 Ornelas, E., "The Case For Medium and Heavy Duty Plug-in Hybrid Vehicles—A Utilities Perspective," EVS-23, 2008.
12.24 Elgowainy, A. et al., "Well-to-Wheels Analysis of Energy Use and Greenhouse Gas Emissions of Plug-in Hybrid Electric Vehicles," Argonne National Laboratory, ANL/ESD/10-1, 2010.
12.25 Markel, T., "Plug-in Electric Vehicle Infrastructure: A Foundation for Electrified Transportation," MIT Energy Initiative Transportation Electrification Symposium, NREL/CP-540-47951, 2010.
12.26 Cornils, H., "Medium-Duty Plug-in Hybrid Electric Vehicle for Utility Fleets," SAE Paper No. 2010-01-1933, SAE International, Warrendale, PA, 2010.
12.27 "DaimlerChrysler Ramps Up its Plug-In Sprinter Development Program," http://www.greencarcongress.com/2007/01/daimlerchrysler_1.html, assessed, Sept. 2011.
12.28 Tate, E. and Savagian, P., "The CO_2 Benefits of Electrification E-Revs, PHEVs and Charging Scenarios," SAE Paper No. 2009-01-1311, SAE International, Warrendale, PA, 2009.
12.29 Pesaran, A. et al. "Battery Requirements for Plug-In Hybrid Electric Vehicles: Analysis and Rationale," EVS-23, NREL/CP-540-42240, Dec. 2007.
12.30 Brooker, A. et al., "Technology Improvement Pathways to Cost-Effective Vehicle Electrification," SAE Paper No. 2010-01-0824, SAE International, Warrendale, PA, 2010.
12.31 SAE Standards, "SAE Electric Vehicle and Plug in Hybrid Electric Vehicle Conductive Coupler," SAE J1772, Jan 2010.
12.32 China GB Standards, "Electric vehicle conductive charge coupler," GB/T ××××—201X, 2010.
12.33 CHAdeMO Standards, "Technical Specifications of Quick Charger for the Electric Vehicle" Rev. 0.9, 2010.
12.34 "Standards for PHEV/EV Communications Protocol," http://www1.eere.energy.gov/vehiclesandfuels/pdfs/merit_review_2010/veh_sys_sim/vss024_kintermeyer_2010_o.pdf, 2010.
12.35 Ornelas, E., "PG&E's view: PHEVs, V2G and the Progress so far," PHEV2007 Conference "Where the Grid Meets The Road," 2007.
12.36 "Plug-in Hybrid Electric Vehicles Value Proposition Study; Final Report," ORNL/TM-201-/46, 2010.
12.37 Guezennec, Y. G. et al., "A U.S. Perspective of Plug-in Hybrids and an Example of Sizing Study, Prototype Development and Validation of Hybridized FC-NEV with Bi-directional Grid Inter-connect for Sustainable Local Transportation," SAE Paper No. 2006-01-3001, SAE International, Warrendale, PA, 2006.
12.38 Gregg, R., "Overview of Vehicle Assist and Automation (VAA)Technologies and Applications," 2009 International Bus Roadeo, 2009.

12.39 Alexander, L. and Donath, M., "Differential GPS Based Control of a Heavy Vehicle," Minnesota Department of Transportation, MN/RIC-2000-05, 2000.
12.40 Zhang, W. and Shladover, S. E., "PATH Innovative Research on ITS Technologies and Methodologies for Multimodal Transportation Solutions," 2006 IEEE Intelligent Transportation Systems Conference, pp. 23–29, 2006.
12.41 "Intelligent Transport Systems," http://www.etsi.org/ITS, assessed, April 2011.
12.42 "Intelligent transportation system Transport Systems," http://en.wikipedia.org/wiki/Intelligent_transportation_system, assessed Sept. 2011.
12.43 Houghton, J. et al.," Intelligent transport: How cities can improve mobility," IBM Global Services, 2009.
12.44 "Action Plan for the Deployment of Intelligent Transport Systems in Europe, " 2935th TRA_SPORT, TELECOMMU_ICATIO_S and E_ERGY Council meeting Brussels, 30 March 2009.
12.45 Shimura1, A., "Proposal of an Autonomous Group-Management Model and its Application to Intelligent Transport System," Proceedings of the Sixth International Symposium on Autonomous Decentralized Systems, IEEE, 2003.
12.46 Drilo, B. et al., "The role of telecommunications in development of New-Generation Intelligent Transport Systems," Wireless VITAE'09, IEEE, 2009.
12.47 Najm, W. G. et al.,"Frequency of Target Crashes for IntelliDrive Safety Systems," DTNH22-09-V-00030, National Highway Traffic Safety Administration, 2010.
12.48 Hickman, M. et al., "Assessing the Benefits of a National ITS Architecture," UCB-ITS-PWP-96-10, California PATH Research Report, 1996.
12.49 Dessouky, M. et al., "Benchmarking Best Practices of Demand Responsive Transit Systems," UCB-ITS-PRR-2003-1, California PATH Research Report, 2003.
12.50 Rajagopalan, A. and Washington, G. "Intelligent Control of Hybrid Electric Vehicles Using GPS Information," SAE Paper No. 2002-01-1936, SAE International, Warrendale, PA, 2002.
12.51 Deguchi, Y. et al., "HEV Charge/Discharge Control System Based on Navigation Information," SAE Paper No. 2004-21-0028, SAE International, Warrendale, PA, 2004.
12.52 Litman, Todd, "Sustainable Transportation and TDM," Online TDM Encyclopedia. Victoria Transport Policy Institute, http://www.vtpi.org/tdm/tdm67.htm. Retrieved 2009-04-07. 2009.
12.53 Liu, P. et al., "Health Assessment and Environmental Impacts of Modal Shift in Transportation, Ceans2006–Asia pacific," IEEE, 2006.
12.54 Nealer, R. et al., "Modal Freight Transport Required for US Goods and Services Production," Sustainable Systems and Technology (ISSST), IEEE, 2010.
12.55 Kumar, A. et al., "A Vision System for Monitoring Intermodal Freight Trains," Applications of Computer Vision, WACV '07. IEEE, 2007.

12.56 "Multimodal Operations," http://www.unescap.org/ttdw/CapBuild/Module%20Multimodal%20Transport%20Operations.pdf, assessed Sept. 2011.
12.57 "Shift to Sustainable Transport," http://www.halifax.ca/environment/documents/Shifting_to_Sustainable_Transport_report_June2009.pdf, assessed, Sept. 2011.
12.58 Grob, G. R., "Future Transportation with smart grids & sustainable energy," 2009 6th International Multi-conference on Systems, Signals and Devices, 2009.
12.59 "SMART GRID: an introduction.," http://www.oe.energy.gov/SmartGridIntroduction.htm.
12.60 "The Smart Grid," http://www.anl.gov/Media_Center/News/2010/FactSheet-Smart_Grid2010.pdf.
12.61 "Vehicle-Grid Interface Key to Smart Charging Plug-in Vehicles," http://www.transportation.anl.gov/pdfs/HV/617.PDF.
12.62 Morrow, K. et al., "Plug-in Hybrid Electric Vehicle Charging Infrastructure Review: Final Report," Battelle Energy Alliance, INL/EXT-08-15058, 2008.
12.63 "IEC Smart Grid Standardization Roadmap," SMB Smart Grid Strategic Group (SG3), 2010.
12.64 http://www.strategictelemetry.net/Documents/ImsaArticle.pdf.
12.65 "EPA and NHTSA Propose First-Ever Program to Reduce Greenhouse Gas Emissions and Improve Fuel Efficiency of Medium- and Heavy-Duty Vehicles: Regulatory Announcement," EPA-420-F-10-901, 2010.
12.66 Farla, J. et al., "The Transition toward Sustainable Mobility in the Netherlands: Analysis of Policy Initiatives, Barriers and Risks," Infrastructure Systems and Services: Building Networks for a Brighter Future (INFRA), IEEE, 2008.
12.67 Beijderwellen, F. N. et al., "Toward sustainable transportation: more as just less," Logistics and Industrial Informatics, IEEE, 2009.
12.68 Lache, R. et al., "Electric Cars: Plugged In 2 A mega-theme gains Momentum," Deutsche Bank Securities Inc., http://gm.db.com/IndependentResearch, 2009.
12.69 Lloyd, A. C., "Legislation and International Trends Effecting Development of Hybrids," SAE 2011 Hybrid Vehicle Technologies Symposium + Electric Vehicle Technologies Day, February 2011.
12.70 "Electric Vehicles to Advance in China with Help from New Alliance between Eaton, IBM, AECOM, Beiqi Foton and Other Technology Leaders," http://www.eaton.com/EatonCom/OurCompany/NewsandEvents/NewsList/NewsArticle/PCT_204235, assessed January 2012.
12.71 "China Creates State Owned EV Industry Alliance," http://www.thetruthaboutcars.com/2010/08/china-creates-state-owned-ev-industry-alliance/, assessed, January 2012.
12.72 "US-China Clean Energy Cooperation," http://energy.gov/pi/office-policy-and-international-affairs/office-policy-and-international-affairs/office-policy-a-8, assessed January, 2012.

Index

About the Authors

Dr. Haoran Hu

Dr. Haoran Hu is the Chief Scientist at Eaton Corporation. He has over 20 years of experience in research and development of internal combustion engines, advanced powertrain systems, and emission control technologies. Prior to joining Eaton in 2002, Dr. Hu held engineering leadership positions at Caterpillar Inc., Detroit Diesel Corporation, and Jacobs Vehicle Systems.

Dr. Hu has a Master of Engineering degree from Huazhong University of Science and Technology (华中科技大学), Wuhan, China, a Doctor of Science (Sc.D.) degree in Thermodynamics from Massachusetts Institute of Technology (MIT), and an MBA from The Ohio State University.

Dr. Hu is the recipient of the SAE Arch T. Colwell Merit Award, and the Eaton Corporation "Engineer of the Year" Award. He has been awarded 30 U.S. patents.

Dr. Rudy Smaling

Dr. Rudy Smaling currently serves as executive director of systems engineering at Cummins, responsible for implementation of systems engineering principles and processes across the corporation. Dr. Smaling previously held the position of Chief Engineer with global responsibility for hybrid system architecture and new product development in Eaton Corporation's Hybrid Power Systems Division. Dr. Smaling also holds a position as adjunct Professor in Mechanical and Aeronautical Engineering at Western Michigan University.

Dr. Smaling's academic background includes a degree in Mechanical Engineering from the Technische Universiteit Delft, a Masters Degree in Mechanical Engineering from Michigan Technological University, a dual degree in Engineering and Management from MIT, as well as a PhD in Engineering Systems from MIT.

Dr. Smaling has over 20 years of experience in new product development in the automotive and commercial vehicle industry and has been awarded 11 patents.

Simon J. Baseley

Simon Baseley recently retired from Bosch Rexroth Corporation after over 25 years of working in leadership positions on automotive pumps and hydraulic systems, including vehicle hybrid systems. He now works part time at the University of Michigan as a Visiting Research Investigator. He previously worked for over 20 years in design and development of aircraft engines for Rolls Royce Ltd.

Simon Baseley has a BSc in Mechanical Engineering from the University of Nottingham and an MSc from Cranfield University, England.

He teaches seminars on Hydraulic Hybrid Vehicles for SAE and is named in over 10 patent applications (granted and pending).